Sustainable Design in Building and Urban Environment

Sustainable Design in Building and Urban Environment

Editor

Farshid Aram

MDPI

Basel • Beijing • Wuhan • Barcelona • Belgrade • Novi Sad • Cluj • Manchester

Editor
Farshid Aram
Urmia University
Urmia, Iran

Editorial Office
MDPI
St. Alban-Anlage 66
4052 Basel, Switzerland

This is a reprint of articles from the Special Issue published online in the open access journal *Designs* (ISSN 2411-9660) (available at: https://www.mdpi.com/journal/designs/special_issues/SDBUE).

For citation purposes, cite each article independently as indicated on the article page online and as indicated below:

Lastname, A.A.; Lastname, B.B. Article Title. *Journal Name* **Year**, *Volume Number*, Page Range.

ISBN 978-3-0365-8566-6 (Hbk)
ISBN 978-3-0365-8567-3 (PDF)
doi.org/10.3390/books978-3-0365-8567-3

Cover image courtesy of Farshid Aram

Contents

About the Editor

Farshid Aram

Dr. Farshid Aram obtained his PhD in the Sustainability and Urban Regeneration program at Politecnica Universidad de Madrid. He is a faculty member of the Urban Planning Department at Urmia University. In particular, he has published in four high-ranked journals (in the top 10% of Scopus rank): *Habitat International*, the *Journal of Climate Risk Management*, the *Journal of Urban Studies*, and the *Journal of Engineering Applications of Computational Fluid Mechanics*. Dr. Aram has invented software (Aram Mental Map Analyzer) to measure citizens' mental cognition in urban spaces through mental maps. He has two other patents in the field of architecture and urban design. He is a member of the editorial board of journals indexed in WOS in the field of sustainability and environmental resources, and has been the guest editor of four reputable journals during 2021–2023. Dr. Aram has reviewed articles for high-ranking journals in the field of Urban Studies and Architecture.

designs

Editorial

Sustainable Design in Building and Urban Environment

Farshid Aram

Faculty of Architecture and Urbanism, Urmia University, Urmia 5756151818, Iran; f.aram@urmia.ac.ir

The basic objectives of sustainability are to reduce the consumption of non-renewable resources, minimize waste, and create healthy, productive environments. Sustainable design in construction seeks to reduce adverse impacts on the environment and the health and comfort of people in buildings and urban areas, thereby improving the performance of buildings and urban spaces.

Sustainable design principles include the ability to optimize site potential, minimize non-renewable energy consumption, use environmentally preferable products, protect and conserve water, protect and enhance green resources, enhance the indoor and outdoor environmental quality, and optimize operational and maintenance practices.

Utilizing sustainable design principles encourages decisions at each phase of the design process, aiming to reduce adverse impacts on the environment and people's health without compromising the bottom line. It is the integrated, holistic approach that encourages compromise and tradeoffs, with such an integrated approach positively impacting all phases of an urban environment's life cycle, including design, construction, operation, and decommissioning.

As the editor of this Special Issue "Sustainable Design in Building and Urban Environment", I was pleased to receive several interesting research papers and review articles. This collection consists of a wide range of studies from different parts of the world, from South American to European and Asian cities, all accomplished in the fields of civil engineering, architecture, and urban planning. These studies introduced different methods regarding sustainable design from new techniques using computer simulation and artificial intelligence to studies on the traditional method of historical sites. In this Editorial paper, we present an overview of the main findings and conclusions of the included articles.

da Costa et al. [1] highlighted the issue of the large number of retired containers stacked in ports worldwide and the need for sustainable strategies for their use. Repurposing these containers into permanent structures, such as container houses, has become a popular trend. However, due to the urgency in disaster situations, container houses are often built quickly without considering energy efficiency principles. This can lead to performance issues, including overheating, corrosion, and rust, which further impact the vulnerable populations they serve. The objective of this study was to compare the performance of two thermal insulators applied to a temporary shelter container designed to promptly serve vulnerable populations. The researchers used Building Information Modeling (BIM) software and Building Energy Simulation (BES) software to simulate and analyze the technical and economic viability of the model. The results indicate that thermal insulators, particularly mineral wool, can significantly reduce energy consumption and improve long-term performance [1].

In a similar study regarding containers and tiny houses, Nezzi et al. [2] examined the relationship between the perceived knowledge and sustainability in the evaluation of a sustainable product, specifically a tiny house prototype. The study utilized a questionnaire to assess the quality, creativity, appropriateness, and sustainability of the tiny house. Unlike previous research that focused on direct sustainability-related information, this study provided sustainability information in an indirect and diffuse form. The majority

Citation: Aram, F. Sustainable Design in Building and Urban Environment. *Designs* **2023**, *7*, 99. https://doi.org/10.3390/designs7040099

Received: 3 August 2023
Accepted: 8 August 2023
Published: 9 August 2023

of evaluators were ordinary people, with a limited number of experts in the field, making the sample representative of the general population. The findings indicated that prior knowledge and background did not significantly influence evaluations. However, gender and age had an impact, with women and younger participants rating the tiny house higher in terms of sustainability and other factors. The evaluation criteria were found to be significantly correlated, particularly in terms of perceived sustainability, preference, and creativity [2].

Cabeza-Lainez et al. [3] introduced a system called DianaX for architectural simulations that considers volumetric and three-dimensional properties, as well as energy sources involved in energy exchanges within and around buildings and urban spaces. The system is based on advances in optics theory, building upon assumptions of different studies in this field. The system utilizes complex integral equations to solve for radiated energy and offers advantages in terms of clearer visualization and analysis of building performance. The software can be considered a Design Tool, enabling the assessment of heritage building paradigms and the potential of new projects with unconventional lighting approaches. The main finding of this research was the feasibility and appropriateness of this method for addressing the problems at hand. The authors express their intention to expand the catalog of designs that can benefit from the use of their tool for scientific design in the future [3].

Given the new techniques of sustainable design, Yu et al. [4] provided a comprehensive review of recent research on the use of immersive virtual environments (ImVE) in architectural design collaboration. The study identified, screened, and reviewed 29 journal articles published since 2010, focusing on three aspects: ImVE in the architecture, engineering, and construction (AEC) industry, ImVE for supporting virtual collaboration, and applications of ImVE in design collaboration. The review highlighted the need for future research and technological development in areas such as ImVE support for design collaboration at the early design stage, cognitive research on design collaboration in ImVE, and enhancements to ImVE technologies to incorporate advanced design features.

In this regard, through using artificial intelligence techniques, the study conducted by Jayabalan et al. [5] discussed the use of steel plates in various engineering fields and the importance of considering buckling as a failure mode. It focused on rectangular steel plates with centrally placed circular openings and different support conditions. The study utilized artificial intelligence techniques, including Gene Expression Programming (GEP), Artificial Neural Network (ANN), and Evolutionary Polynomial Regression (EPR), to predict the critical buckling loads of these plates. Datasets from the literature were compiled and used to develop the models [5].

Another study [6] assessed the use of in-filled tubes, specifically steel shell tubes filled with concrete, as a successful configuration for axially loaded members like columns and struts. This configuration offers advantages such as eliminating the need for shuttering, reinforcement bars, and ties, while increasing flexural and axial capacities and enhancing ductility. However, a main disadvantage is the potential for local buckling and decomposition. Previous studies have explored solutions using intermediate stiffeners or shear connectors. This research proposed a different approach using double cold-formed sigma sections as steel shell tubes. Sixteen specimens with varying lengths, cross-section dimensions, and shell thicknesses were tested under concentric and eccentric compression loads. The results recorded ultimate capacities, lateral deformations, and normal strains. Theoretical capacities were calculated using different standards and software, with deviations from experimental results ranging from 13% to 24% [6].

According to the necessity for the development of appropriate standard projects for providing highways with roadside service facilities and increasing efficiency, Samoilov et al. [7] suggested interconnecting space-planning solutions based on a triangular module instead of a square or rectangular one. According to the results of this research, the use of this modular system can help reduce the harmful impact on the environment and effectively utilize renewable energy sources [7].

The research conducted by Mangeli et al. [8] focused on rock-cut architecture, a lesser-known type of vernacular architecture that differs from conventional architectural practices. The study aimed to explore the techniques, designs, and excavation procedures employed in this type of architecture. The research compares and contrasts the techniques, types, and settlement context materials, recognizing three general excavation techniques. The main case study was the Meymand residential complex, the largest rock-cut complex in Iran. The study examined 50 residential units in the oldest part of the village, analyzing techniques and design styles and comparing them [8].

Regarding the importance of public spaces and green areas on the mental and physical health of individuals, with a focus on thermal comfort as a key indicator, Baquero Larrivaand Higueras García [9] examined the outdoor thermal comfort of older adults in Madrid, Spain, and Newcastle upon Tyne, United Kingdom, during the autumn season. The study utilized a mixed methodology involving environmental measurements and surveys conducted on-site. The findings of this research highlighted the need for the design of more comfortable and healthy public spaces that enhance the quality of life for all citizens, aligning with the principles of active aging and healthy cities [9].

In the study conducted by Abouelela [10], the focus was on investigating the relationship between the work environment and job happiness at King Faisal University. The researcher aimed to understand the opinions of faculty and staff members regarding their environmental and functional needs at work, with a specific focus on improving the interior design of workspaces to create happiness in the work environment. Overall, this research emphasized the importance of the work environment in shaping individuals' happiness and quality of life. It provided insights into the factors that contribute to job satisfaction and highlighted the role of organizations in creating a conducive work environment that fosters happiness and encourages innovation and creativity [10]. In a similar study [11], the importance of interior design in academic libraries, specifically in the context of college and university libraries, was discussed. This research indicated the need for academic libraries to evolve from being mere repositories of books to becoming spaces for research and communication [11].

Conflicts of Interest: The author declares no conflict of interest.

References

1. Da Costa, B.B.F.; Silva, C.F.P.; Maciel, A.C.F.; Cusi, H.D.P.; Maquera, G.; Haddad, A.N. Simulation and Analysis of Thermal Insulators Applied to Post-Disaster Temporary Shelters in Tropical Countries. *Designs* **2023**, *7*, 64. [CrossRef]
2. Nezzi, C.; Ruiz-Pastor, L.; Altavilla, S.; Berni, A.; Borgianni, Y. How Sustainability-Related Information Affects the Evaluation of Designs: A Case Study of a Locally Manufactured Mobile Tiny House. *Designs* **2022**, *6*, 57. [CrossRef]
3. Cabeza-Lainez, J.; Almodóvar-Melendo, J.-M.; Revenga-Dominguez, P.; Rodríguez-Cunill, I.; Xu, Y. New Simulation Tool for Architectural Design in the Realm of Solar Radiative Transfer. *Designs* **2022**, *6*, 72. [CrossRef]
4. Yu, R.; Gu, N.; Lee, G.; Khan, A. A Systematic Review of Architectural Design Collaboration in Immersive Virtual Environments. *Designs* **2022**, *6*, 93. [CrossRef]
5. Jayabalan, J.; Dominic, M.; Ebid, A.M.; Soleymani, A.; Onyelowe, K.C.; Jahangir, H. Estimating the Buckling Load of Steel Plates with Center Cut-Outs by ANN, GEP and EPR Techniques. *Designs* **2022**, *6*, 84. [CrossRef]
6. Reda, M.A.; Ebid, A.M.; Ibrahim, S.M.; El-Aghoury, M.A. Strength of Composite Columns Consists of Welded Double CF Sigma-Sections Filled with Concrete—An Experimental Study. *Designs* **2022**, *6*, 82. [CrossRef]
7. Samoilov, K.; Kuspangaliyev, B.; Sadvokasova, G.; Kuanyshbekov, N. Prospects of Triangular Modular Structures for Roadside Service Buildings. *Designs* **2022**, *6*, 90. [CrossRef]
8. Mangeli, M.; Aram, F.; Abouei, R.; Mehdizadeh Saradj, F. A New Look at Excavation Techniques and Design of Rock-Cut Architectures. *Designs* **2022**, *6*, 64. [CrossRef]
9. Baquero Larriva, M.T.; Higueras García, E. Influence of Microclimate on Older Peoples' Outdoor Thermal Comfort and Health during Autumn in Two European Cities. *Designs* **2023**, *7*, 27. [CrossRef]

10. Abouelela, A. The Effectiveness of the Role of Interior Design in Creating Functional and Institutional Happiness for Work Environments: King Faisal University as a Model. *Designs* **2022**, *6*, 45. [CrossRef]
11. Abouelela, A. Towards a Better Interior Design for the Academic Library at College of Education—King Faisal University. *Designs* **2022**, *6*, 47. [CrossRef]

designs

MDPI

Article

The Effectiveness of the Role of Interior Design in Creating Functional and Institutional Happiness for Work Environments: King Faisal University as a Model

Amira Abouelela

Art Education Department, College of Education, King Faisal University, Al-Ahsa 31982, Saudi Arabia; aabouelela@kfu.edu.sa; Tel.: +966-558658227

Abstract: Happiness is a natural human right that all seek to achieve. The quality of people's lives may be directly affected by the quality of their working life, which is affected by the quality of their work environment. This has become the focus of attention of work institutions in society due to its great importance and strong impact on success. The purpose of this study was to investigate the institutional work environment at King Faisal University by surveying faculty and staff members regarding their opinions on meeting their environmental and functional needs at work by improving the interior design of workspaces to create happiness in the work environment. The aim of this study was to reveal the relationship between employees' performance levels and their work environment, in addition to making happiness and quality of life major priorities and creating a stimulating work environment. The researcher used descriptive analysis to analyze the relationship between aspects of work and the levels of job satisfaction and happiness among employees of King Faisal University. The researcher used a five-point Likert scale to measure the responses to the questionnaire items, and reached several conclusions, including that the level of job happiness at King Faisal University is not affected by the variables of gender, social status, or the nature of the job, and that the university provides a work environment that helps achieve job happiness and allows for job innovation and creativity.

Keywords: happiness; job happiness; quality of life; workplace; healthy interior design; office breakout area furniture

Citation: Abouelela, A. The Effectiveness of the Role of Interior Design in Creating Functional and Institutional Happiness for Work Environments: King Faisal University as a Model. *Designs* **2022**, *6*, 45. https://doi.org/10.3390/designs6030045

Academic Editor: Min-Hwi Kim

Received: 7 April 2022
Accepted: 9 May 2022
Published: 16 May 2022

Publisher's Note: MDPI stays neutral with regard to jurisdictional claims in published maps and institutional affiliations.

1. Introduction

Happiness is a word with two connotations. One is experimental and refers to the psychological condition in the present moment and a feeling of positive emotions, such as joy, enthusiasm, love, and hope. In this context, happiness lies in overcoming negative feelings with positive feelings. The second connotation is evaluative; people think about the happiest moments in their lives, as happiness is the aspiration of every human being. According to Mayer and Diner (1995), happiness can be defined as the experience of frequent positive affect, infrequent negative affect, and an overall sense of satisfaction with life as a whole [1]. The British philosopher David Hume said, "People do not pursue professions, acquire antiques, invent inventions, or publish sciences, or contemplate the stars, except to reach happiness ... either by toil and trouble, or money and wealth, or laughter and singing." Happiness is defined by the Oxford English Dictionary (Hawker and Waite, 2007) as "a state of mind or feeling comprising contentment, satisfaction, pleasure or joy." It is further described as the state of a pleasurable content of mind, which results from success or the attainment of what is considered good. Satisfaction is another word that is frequently used interchangeably with happiness.

The Oxford Dictionary defines satisfaction as "the feeling of pleasure that arises when you have the things you want or need or when the things you want to happen" [2]. Happiness is defined as the feeling of permanent and integral satisfaction with life as a whole [3] and is also seen as a positive feeling and a powerful motivator of human behavior [4]. The simple concept is that happiness describes the extent to which a person can measure and judge the quality of his or her life positively [5]. Aristotle defined happiness as life with meaning, purpose, and ends. Several books and studies have been published confirming that happiness helps us be more productive, as it helps us think creatively and work efficiently.

Happiness in the workplace has become very important and is related to the present societal context, crucial to both employees and institutions. People will spend more of their adult life at their job than doing anything else. Work will take up more time than families, friends, and hobbies combined. It would be nicer if that time were spent at a job that actually makes them happy. Happy people and happy institutions can create many good things, such as increased productivity, quality, customer satisfaction, creativity, innovation, adaptation, and flexibility, and decreased loss, work stress, and occupational diseases [6].

Happy employees can quickly solve the problems they face constructively since happiness positively affects human behavior. Happiness is an internal feeling of joy and pleasure that is externally reflected in employees' behavior and mood and appears in their reactions to colleagues. Happy people deal with everyone around them positively and accept what is going on. The efficiency of administrative institutions at universities generally depends on the efficiency of the people within them and their ability to work. The level of happiness of people at work affects their performance and success at the institution to which they belong.

Job happiness is reflected in employees' daily work by a feeling of comfort, calm, and positivity in participating in the work they are assigned to perform. Job happiness also helps employees face and overcome challenges [7]. A work environment is described as happy when employees often experience positive feelings there [8].

Happiness in the workplace is closely related to the development and innovation of today's societies. It is not just a feel-good thing, it is really, according to Plato "for man to be happy" [6]. Workplace happiness refers to "an individual's work and life satisfaction" or "subjective well-being" at the workplace [9,10]. Workplace happiness includes job satisfaction as well as individual measures of happiness involved in work [11]. Happiness at work is the feeling that employees enjoy what they do and are proud of themselves; they enjoy having other people around; thus, they give better performance. It is not an administrative luxury; thus, it is necessary to understand happy practices, which are one of the keys to institutional development and a factor of success for the beneficiaries. In addition, the institution can bring happiness to its employees by focusing on the interior design of the workplace along with the employees and those who deal with them, as all of these elements are linked and affect each other.

Quality of life, or wellbeing, is a concept closely linked to happiness, but it is a self-contained concept. In addition to happiness, it also has an empirical sense and refers to how people evaluate their lives and the source from which they derive their sense of purpose. Quality of life is not only about individuals' feelings but also about their performance on the personal and social levels. Robert Levering, the founder of the Great Place to Work Institute, said that a great workplace is where you feel confident in the organization you work for, proud of what you do, and joy working with the people you work with [12]. The more people are satisfied with their work and life, the more it impacts their happiness at the workplace. Happiness is important and has an impact on the success of any institution, which always strives to maintain happiness within the work environment by knowing and studying the factors that affect employee happiness to enhance it because of its remarkable impact on increasing productivity [13].

Institutional happiness is not a new concept but rather an extension of the concept of internal communication in public relations focused on making employees happy in the work environment; the new concept is happiness in institutional work in which all sectors participate. A creative and innovative environment contributes to employees' happiness, and they are more creative, focused, and able to lead and face challenges. Creative and innovative people are more productive. Research and practical experience have proven that employee happiness should be a priority. Happy employees mean higher productivity, better production quality, better reputation, competitive advantage, and sustainable success, which leads to happy customers. Accordingly, governments, private institutions, and civil society organizations play a role in creating a stimulating, appropriate, and happy environment for employees thus that they can provide services that exceed customers' expectations and requirements.

Entertainment in the work environment motivates employees to come up with great ideas by having fun games to play during their break time. Happiness is a basic requirement that everyone strives to achieve, whether at work or otherwise.

Some surveys have shown that happy employees stay four times longer than dissatisfied employees, and happiness plays a big role in their being 12% more productive than unhappy employees; they enjoy performing their job duties. Employees who are happy at work exhibit higher levels of job-related performance when compared to employees who are unhappy in the same work environment [14].

Interior design related to human health and wellness focuses on all of the design elements used, including materials, lighting, and colors, along with studying air and sound quality, paying attention to green spaces and achieving integration between exterior spaces and the interior environment to achieve the desired result [15]. Interior design greatly affects people's feelings about interior spaces; feeling cramped and trapped will affect the creativity and productivity of the occupants of the space. One of the elements that has a dramatic effect is the height of the ceiling, as high ceilings improve focus and creativity and enhance mood and improve the air quality and space. Functional space, natural light, security, and safety exit [16].

The physical environment has a clear impact on individuals and their health, and both individuals and groups have needs in the work environment, and they must be well understood to provide a safe and comfortable environment. In interior design, some elements can affect mental health, such as color, which can affect people's mood when they are in offices or in workplaces in general [17]. Experienced interior designers understand how design affects mental health and that it is necessary to consider the design criteria of spaces, what requirements employees have for their offices, and whether the current design reduces tension or causes it, and then take into account modifications in the proposed design. Lighting is one of the most important elements of interior design that must be achieved optimally, whether natural or artificial, as lighting affects the feeling of functional comfort and the level of performance of tasks. Appropriate types of lighting must be provided, using materials according to the functional nature of the place [18]. Interior design generally has a role in helping the occupants of a building to feel comfortable and calm through aesthetic and functional aspects, taking into account the physiological and psychological factors of occupants of different buildings with different functions.

2. Literature Reviews

There are many recent literature reviews in the field of academic library space design. In this study, we reviewed several studies that benefited from its most prominent features. It should be pointed out that the studies that were reviewed were published between 2002 and 2018 in several countries, which indicates their temporal and geographic diversity.

In the following, we present a summary of these studies, then we explain what the current study offers by identifying the differences between it and previous studies, and finally note the aspects of previous studies that benefit the current study.

Wasarat and Sharif (2014) noted that happiness in the workplace refers to how satisfied people are with their work and life. Happiness represents the feeling of subjective wellbeing of the individual. Achieving happiness in the workplace is very important for improving productivity. Happy people are productive people, while those who are unhappy may not pay full attention to their tasks. Some scholars also believe that organizations that work to maintain workplace happiness in the long term are able to increase productivity. In order to achieve this, the employees must know the factors that affect happiness in order to effectively promote it in the work environment. The researchers also noted the scarcity of research on employee happiness in the past and presented a conceptual framework for happiness in the workplace [13].

Tosiriwatanapong (2014) positioned wellbeing as one of the success factors in the workplace. Given that people's body, mind, and spirit are linked to the physical environment, a well-designed workplace can increase the wellbeing of employees, which leads to increased productivity of the work organization. The design has a great role in helping employees deal with their feelings within the work environment and can promote positive feelings. The study focused on contemporary health themes, with the objective of finding out the emotional insight of designers regarding the physical environment at the workplace and identifying its characteristics. Working with designers in various specialties, the authors conducted in-depth interviews with seven respondents and carried out an online survey questionnaire with 100 respondents using a convenience sampling method [19].

Akhtar et al. (2014) noted that better interior design was critical to increasing employee productivity in all business areas. They identified five dimensions in interior design: furniture, lighting, noise, temperature, and fixtures. The main objective of the study was to examine the relationship between office interior design, work environment, and employee productivity. They collected data from communications sector employees at the Sahiwal Division in Punjab, and a sample size of 200 was determined to conduct this research. The primary data were collected using a structured questionnaire, and then descriptive statistics analysis, correlation, and regression by SPSS were used to find out the effect of interior design on employee productivity. The results showed that interior design has a positive relationship with employee productivity [20].

In reviewing the literature, we noted that the current study agrees with previous studies on the main topic and its general objective, and it differs in several aspects, as follows.

This research differs from other research in that it focuses on revealing the opinions of King Faisal University employees about the interior design of the work environment with regard to achieving job happiness and their levels of satisfaction, as well as the importance of having places designed for rest during work, either outside the official workplace or not, and it collects and analyzes their proposals and needs that should be taken into account in developing and improving the interior design to achieve career happiness. We designed a questionnaire consisting of a set of questions about job happiness, its importance, the nature of interior design, and design considerations in office rooms, in addition to the availability of places designed for rest in the work environment. Then, opinions based on the questionnaire were collected and analyzed to identify the positive and negative points from the respondents' point of view.

3. Problem Statement

The problem lies in recognizing the importance of job happiness and determining how to create a work environment that brings job happiness to employees by spreading the concept of happiness and satisfaction. This study focused on the current work environment of the administration building at King Faisal University and how the administrators were working to create a happy work environment for employees. The two basic principles are the science of happiness and the science of design, which contribute to the design of the workspace to improve the quality of the internal environment and confirm the role of interior design and its effectiveness in creating a work environment in which the principle of institutional functional happiness is achieved.

4. Study Objectives

The aims of this study are as follows:

- Analyze the opinions of respondents through an electronic questionnaire on the concept of job happiness;

- Measure the effect of interior design on job performance and the work environment and its role in achieving job happiness;
- Discover how to create a happy environment for work through interior design thus that employees can feel happier in work and life;
- Study the relationship between the level of employee performance and the work environment;
- Determine how to make happiness and quality of life a major priority in the work environment at King Faisal University;
- Promote achieving the goal of spreading happiness in the workplace with practical and effective steps.

5. The Importance of This Study

The importance of this study lies in what it presents to higher education institutions about what a happy work environment is and the positive impact it has on university employees, including staff and faculty members. It emphasizes the role of a happy work environment in achieving career happiness and its impact on increasing employee performance and job efficiency, which affects the productivity of the institution. After analyzing the results of the questionnaire, which clarified the current situation at King Faisal University and the hoped-for situation of its employees, some proposals are presented.

6. Study Limitations

Time: The second semester of the 2021–2022 academic year was chosen to implement the study tool. Sample size: the study was limited to a random sample of King Faisal University employees. Location: King Faisal University. Subject: the effectiveness of the role of interior design in creating functional and institutional happiness for the work environment was investigated using King Faisal University as a model.

7. Study Questions

The study sought to answer the following research questions:

- Is the current internal work environment conducive to achieving job happiness?
- What is the effectiveness of the role of interior design in creating functional and institutional happiness for the work environment?
- How can employees be encouraged to reach their highest levels of happiness and quality in the work environment?
- What are the future vision and proposed solutions to achieve functional and institutional happiness in the work environment?

8. Materials and Methods

The researcher used the descriptive analytical method to study and analyze the levels of satisfaction and happiness at an educational institution (King Faisal University) through an electronic questionnaire answered by an exploratory sample of university employees.

9. Study Tool

The appropriate tool was chosen, a questionnaire, to survey employees of King Faisal University on their opinions about job happiness in the institutional work environment, and the required data were collected through their responses to the questionnaire. The questionnaire was divided into five main sections, each consisting of a set of questions. Out of the total sample of university employees (N = 100), 57 responded to the questionnaire.

The five sections of the questionnaire were as follows: (1) administrative spaces (administrative offices for employees); (2) interior space planning for offices; (3) availability of natural and artificial lighting sources; (4) materials and colors; and (5) rest areas within the work environment.

The researcher used the Likert scale, which is widely used in most fields of science to measure tendencies, desires, and attitudes, as well as to measure behaviors and preferences in psychological tests [21]. It is scored from 1 to 5, with 5 indicating "strongly agree" and 1 indicating "strongly disagree," and the scores were used to calculate arithmetic indicators and averages by computer analysis programs to derive results from questionnaires and research.

A 5-point Likert scale was used to measure the responses to the questionnaire items. The scale in this study is shown in Table 1.

Table 1. Five-point Likert scale.

Response	Strongly Disagree	Disagree	Neutral	Agree	Strongly Agree
Degree	1	2	3	4	5

The importance of happiness and quality of life in the work environment can be described as follows:

Delivering Happiness

Delivering Happiness (DH), the leading company in spreading the concept of happiness and satisfaction among employees in the work environment is headquartered in San Francisco, California, USA. It worked with the government of Dubai to develop the Dubai Happiness initiative and provided consultations with many major international companies, such as Facebook, to help create happy work environments for their employees. Mountain View has announced its partnership with DH, which aims to create happy work environments for employees in Egypt and the Middle East. The company aims to use innovative ideas to design workplaces that bring happiness by utilizing modern science and studying the factors that affect the creation of happiness, and applying them to achieve the desired goals, including in the following areas.

Happiness and success

The feeling of happiness is transferable from one person to another. Employees who feel happy in their work environment and while performing their job tasks are great models for peers who do not feel happy or feel less motivated. In addition, happiness in the workplace is directly related to improving employee performance and thus increasing productivity.

Reducing stressin the workplace

Workplace stress is a growing concern as employees face overworking conditions, job insecurity, and low levels of job satisfaction. Workplace stress has been shown to have a detrimental effect on employee health and wellbeing and has a negative impact on workplace productivity [22].

Studies on stress reduction include the following:

- Kagan, Kagan, and Watson conducted a field study with 373 employees of the emergency medical service of a firefighting department that lasted 3 years. They developed a framework to define stress and classify programs to reduce psychological and educational stress. It consisted of seven psychological educational programs, each of which had a relative impact in the short and long term. Improvements were found before and after follow-up. The results of the study supported the value of psychological training programs for preventive mental health in the workplace [23];
- Smith and Sainfort aimed to propose a new way of conceptualizing job design and job stress based on the balance among job elements. They integrated social psychological theories of job design with job stress concepts to develop a model of job balance that addresses how the organization and job design can influence worker health. The model defines how job design can improve "loading" factors on the worker by "balancing" aspects of the job that can produce stress. The implications of this model for enhancing worker health by controlling workplace stress are discussed. The model provides a holistic approach for designing workplaces that balance aspects of production and stress [24].

Enjoying happiness in the work environment helps to reduce the feeling of stress, and on the contrary, unhappiness affects positive feelings and leads to an inability to think creatively and logically.

Happiness and positivity at work

Happy people are more creative and inspiring and have a positive effect on those around them; they can create an enjoyable work environment that helps others integrate and work in it and increase the number of members of the work team [25].

Design standards for administrative buildings

Modules in administrative offices: when designing administrative buildings, it is necessary to rely on modules in the floor plan, in the facades and sections, whether the building is brick, iron, aluminum, glass, or reinforced concrete. The designer chooses the module that gives the best solution for the building for which the interior design is intended, and this module depends on the area of the space, which can be determined by the number of people who work in it and the type of work they do. It is possible to use natural lighting in an office to a large extent, and accordingly, it is possible to determine the appropriate areas for each purpose for which it is used. Administrative buildings must meet some conditions in the design process, such as:

- The type of floor plan (open, closed) as Figure 1 according to the nature of the building;
- The number of employees in the different departments in the building;
- The number of public users of the building;
- Various services in the building (toilets, offices);
- The safety of the building indicated by the presence of escape ladders to be used in emergencies.

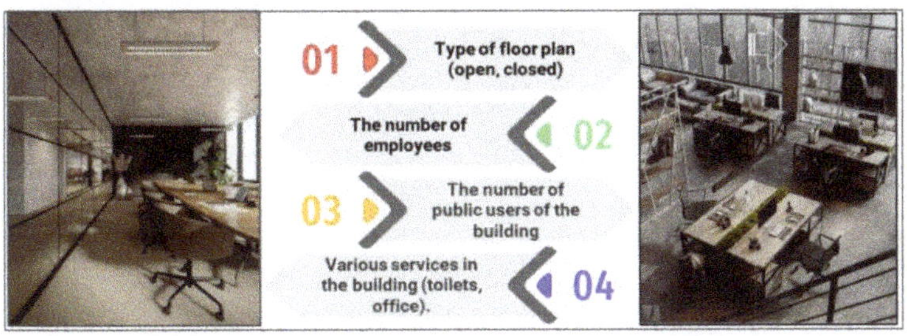

Figure 1. Interior design of open-plan and closed-plan offices. (1) https://www.behance.net/gallery/68740375/Loft-style-furniture-part-5?trackingid=T32PLY3L&mv=email/ (accessed on 15 March 2022); (2) https://www.flickr.com/photos/lemeridienhotels/8427593833/in/photostream (accessed on 10 March 2022).

Foundations and standards for office interior design

Administrative offices are designed based on several criteria:

- The interior design should be appropriate to the nature of the work;
- They should provide a suitable work environment for employees and clients;
- The results of studies of the office space and windows should be used to determine the lighting and the nature of the work;
- There should be an appropriate distribution of the internal space of the administrative office to comply with the requirements of the nature of work and maintain movement corridors;
- The appropriate colors should be chosen for the nature of the work;
- Appropriate materials should be used for ceilings, walls, and floors.

The main elements of office design

- Flooring: it is preferable to use wooden floors, with carpets or rugs in managers' offices;

- Ceilings: it is important to choose materials intended to silence external sounds and to use roof materials that are resistant to moisture and heat;
- Walls: the colors used on the walls must be light because dark colors have an effect on employees' creativity in their work;
- Accessories: office accessories such as lampshades, vases, office sets, portfolios, art-boards, and curtains should be provided.

The design of a good work environment depends on matching the work and the environment with the needs and requirements of the employees shown in Table 2, and these requirements include a set of variables that affect employee performance [12]. Employees generally want to contribute to the best of their ability because they believe in the importance of their work and their duty to accomplish it. However, employees are humans with universal human needs in the workplace; they need to have their contributions valued and their ideas respected. When these human needs are met in the workplace, people do their best, share, and contribute to delivering the massive returns that the most successful companies achieve [26].

Table 2. Needs and requirements of employees in work environment [26].

No	Physiological Needs	Psychological Needs
1.	Thermal comfort	Privacy
2.	Indoor air quality	Personal space
3.	Optical comfort	Physical space
4.	Acoustic comfort	Friendships
5.	Ergonomics	Telecommunications

Creating a productive and comfortable interior atmosphere

The concept of interior design for workspaces is based on the design of collaborative and impactful office spaces. Interior designers can understand the importance of the space by meeting the people who will be using it and learning from their experience of how the design can support them purposefully. With rapid changes, designers need to allow movement or rearrangement to create many types of spaces or ways in which people can work. Movement from one place to another may become increasingly important.

The interior design of the workplace significantly affects human behavior. Employees spend many hours in their offices every week. This can negatively affect their motivation and productivity if they do not feel comfortable and happy in their work environment. Paying attention to the work environment is one of the most important factors in achieving goals. Therefore, the standards of interior design for administrative spaces must be applied by providing good design, appropriate lighting and colors, and office furniture commensurate with the activities of the space, considering the impact, which is reflected in the thoughts and feelings of employees, whether they are happy to belong to the organization. The work environment is related to the creativity of its interior design, which is important for improving its level. The more the work environment is improved, the more the institution progresses [27].

There are four factors involved in designing co-working spaces:

1. Warm up the lighting with openings for natural light;
2. Create an audio spectrum by providing multiple options in terms of power and noise level that can vary according to activity and mood;
3. Create a work environment that feels like home with warm, attractive, and comfortable spaces with a connection to nature and natural materials;
4. Encourage spontaneous activities by designing comfort hubs that allow people to move in and out of the space and encourage them to stop and chat.

Advantages of glass partition walls

Glass partitions between rooms and corridors create more visual distraction than was there before as Figure 2; some see them as not providing the privacy needed in the work environment, and they may be inconvenient and distracting for employees. There are many sophisticated and modern designs and shapes of glass walls used in open spaces, helping to create a feeling of spaciousness in the work environment. Glass walls are characterized by strength and durability, which means the possibility of shattering them is very low, and the interior design of administrative offices may require dividing the spaces using glass panels, which increase transparency between sections, and some can also help bring in daylight and enhance communication and cooperation between employees as shown in Table 3 [28].

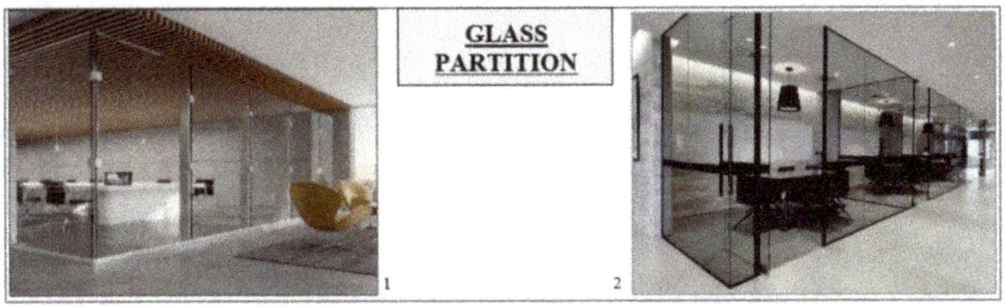

Figure 2. Use of vertical glass partitions to divide office spaces. (1) https://archello.com/product/klarity-freestanding-glass-partition (accessed on 12 February 2022); (2) https://www.indiamart.com/bombayglass-delhi/glass-partition.html (accessed on 12 February 2022).

Table 3. Advantages of glass partition walls.

	Advantages	Illustration
1.	**Expansion of space and a feeling of lightness and spaciousness**	
2.	**100% safe and laminated** Many people believe that glass is fragile and not as strong as traditional walls, but it is safe and laminated, as well as environmentally friendly	
3.	**Possibility of sound isolation and optical transparency** As much sound insulation as traditional walls	
4.	**Space zoning at lowest cost** Glass walls are less expensive than traditional construction, take days to install, and have a more attractive look	
5.	**Increased natural light** Glass walls help to increase the amount of natural light and freshen the air.	

Work Break Environments

Office break furniture is an area separate from the official workplace, provided as a resting place for employees where they can spend time taking breaks during the official workday in a way that does not conflict with their job duties that is away from the offices and the official work atmosphere [29].

Modern work systems require employees to spend much more time in offices in front of computer screens. Health and safety laws require employees to take frequent breaks from screens during work hours. A break area is any open space for employees or visitors separate from the usual work area. It can be a place for employees to relax and have lunch or even hold casual meetings with other employees or clients. Different types of seats can be provided if the appropriate space is available, such as multi-purpose chairs or chairs around a central table; thus that there is a place for meetings or lunches and seating units to make employees feel comfortable and relaxed when taking a break [30]. Relaxation spaces are the most flexible spaces in the office. A rest area for employees should be in a strategic location where it is easy to move from one place to another. These places are intended for employees to feel comfortable and calm during their daily work. One of the rooms in the administrative building could be furnished with comfortable seating units designed in a way that helps people to rest, relax, renew, and restore their functional activity, along with motivational books with short positive phrases to meet people's various needs. Such spaces are suitable for grabbing a bite to eat or drinking a cup of coffee, meeting new customers, or greeting visitors. Giving employees casual areas to collaborate and brainstorm is crucial to their productivity and wellbeing. Break spaces are integral to any energized, creative workplace. A strong Wi-Fi connection is essential for any sub-area, but the technologies that employees need and use to get the job done must be considered. The basic principle in any mixed workspace with collaborative areas is that people should be able to carry on with their work. Technology must be seamless to achieve true collaboration in the workplace.

Design concept for breakout spaces

It is necessary to know the main requirements for the space and focus on the design. Research conducted by the Workplace Intelligence Unit indicates that having many side spaces does not represent an efficient use of space because the design does not match the primary function. Rest areas are often playful spaces that allow offices to show the lighter side of their work-life and bring out the fun side of the organization. When planning the interior design of the work environment using healthy materials, sound, smell, texture, and colors, these spaces can be indoor spaces as the aim is to reduce anxiety and enhance productivity as shown in Table 4 [31].

A happy work environment helps to generate positive feelings and a sense of peace, psychological comfort, and activity. In addition, it encourages participating with others, developing cooperative work, and working without paying attention to the time. When everyone spends time working without putting pressure on each other, it is a model of a happy environment that plans effective work and makes employees feel more productive.

Table 4. Design concept for breakout spaces.

No	Description	Figure
1.	**Quiet work space** Among high-performing employees, 58% say they need more quiet workplaces. A modern office environment should have quiet spaces. Flexible workspaces today need areas of calm that enable employees to be creative and focused on coming up with great ideas [32].	
2.	**Use of technology and charging devices** Technology is important in solving problems and managing files and reports, and modern workplaces rely entirely on technology to achieve work efficiency and reduce time and money problems, and accomplish required tasks [33].	
3.	**Relaxation, meditation** Meditation is a simple and quick method that helps reduce the stress of daily work while bringing a sense of inner peace [34].	

Table 4. *Cont.*

No	Description	Figure
4.	**Informal meetings** Informal meetings can be held to address issues, whether monthly, weekly, daily basis. Large chalkboards can be hung on a wall to present problems and ideas are presented, and share information [35].	
5.	**Brainstorming sessions** Brainstorming combines calm and informal thinking to solve problems, providing an opportunity and encouraging everyone to think innovatively. Interior design for a brainstorming space uses methods that focus on creative design and activities, which helps to evoke ideas by providing the best conditions [36].	

10. Discussion

This section includes the following: a presentation and discussion of the results related to the questionnaire filled out by employees of King Faisal University; data analysis related to the responses about the dimensions of job happiness and the role of interior design in achieving job happiness; and the process of identifying the consistency and compatibility among the respondents' opinions about the questions and clarifying them.

The researcher carried out several procedures, as follows:

In the first step, an electronic questionnaire was designed and prepared to gather opinions on the effectiveness of interior design in creating functional and institutional happiness for work environments, using King Faisal University as a model. The questionnaire was made up of five sections, with each section containing a set of questions or phrases. The Deanship of Scientific Research at King Faisal University was consulted to approve the ethics of the scientific research and agreed to send the questionnaire via e-mail to the university's employees.

In the second step, the researcher carried out the analysis by first determining the goal, which was to measure the extent to which job happiness was achieved in the institutional work environment at King Faisal University by analyzing the responses to the questionnaire by university employees. Next was determining the unit of measurement or counting to enumerate the analysis categories, which was the frequency of responses to each question in each section. The total sample size was N = 100, and 57 responded to the questionnaire.

The following are the results related to the first question: Is the current internal work environment conducive to achieving job happiness?

The statistical analysis of the 57 responses to question 2, shown in the first section of Table 5, indicates that 8% strongly agreed, 23% agreed, 13% were neutral, 8% disagreed, and 5% strongly disagreed as shown in Figure 3. Through these responses, it can be seen that they were aware of job happiness and their need for it in the work environment.

From the statistical analysis of the 57 responses to question 5 in the first section, the questionnaire (Table 5) indicates that 8.8% strongly agreed, 33.3% agreed, 19.3% were neutral, 22.8% disagreed, and 15.8% strongly disagreed as shown in Figure 4. From this analysis, it was found that the number who agreed and disagreed was almost equal. That is, the current internal work environment needs to be developed and improved to achieve job happiness.

The results related to the second question: What is the effectiveness of interior design in creating functional and institutional happiness for the work environment?

The statistical analysis of the responses to question 6 in the first section of the questionnaire (Table 5) indicates that 53.1% strongly agreed, 40.4% agreed, 19.3% were neutral, 1.8% disagreed, and 3.5% strongly disagreed. From this analysis, we can see the importance of interior design in the workplace. The analysis of responses to question 7 indicates that 59.6% strongly agreed, 22.8% agreed, 12.3% were neutral, 3.5% disagreed, and 1.8% strongly disagreed as shown in Figure 5. From this analysis, we found that interior design affects job happiness in the workplace.

Table 5. Frequency of responses in the first section on administrative spaces (administrative offices for employees).

No	First Section Administrative Spaces (Administrative Offices for Employees)	SD	D	N	A	SA
1.	Do you know what functional happiness is?	4	3	12	13	25
2.	Do you feel functional happiness in your work environment?	5	8	13	23	8
3.	Does feeling happy in the work environment reduce the level of anxiety and help enhance productivity?	3	1	1	12	40
4.	Is job happiness a basic requirement for employees in the work environment?	1	3	2	9	42
5.	Is the current internal work environment suitable for achieving job happiness?	9	13	11	19	5
6.	Are you interested in the interior design of the workplace at your institution?	2	1	11	23	20
7.	Does the interior design of the workplace affect the achievement of job happiness?	1	2	7	13	34

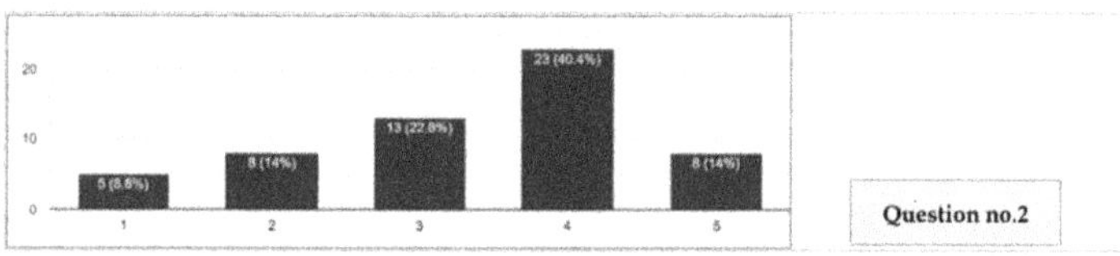

Figure 3. Analysis of responses to question 2 in first section.

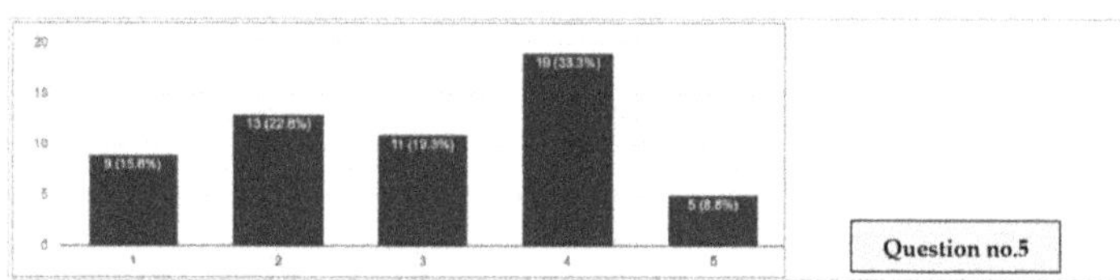

Figure 4. Analysis of responses to first question.

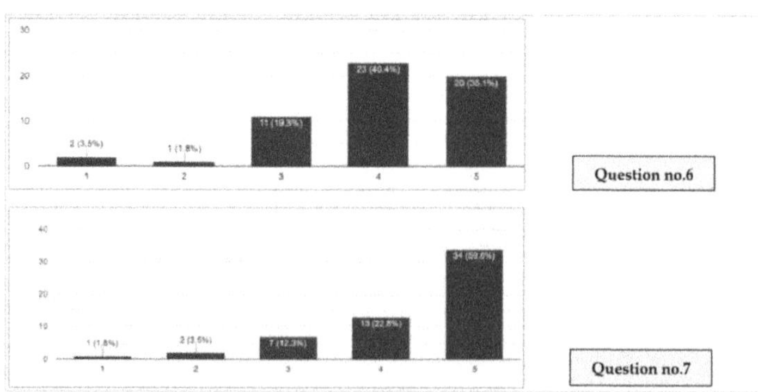

Figure 5. Analysis of responses to questions 6 and 7 in first section.

Table 6 shows the frequency of responses in the second section, office room interior (space planning), which consists of five questions. The analysis of responses to question 1 (Is the interior space of the office commensurate with the nature of the work?) indicates that 40.4% strongly agreed, 31.6% agreed, 19.3% were neutral, 3.5% disagreed, and 5.3% strongly disagreed. The analysis of responses to question 2 (Are the office furniture items available in the room appropriate to their size?) indicates that 29.8% strongly agreed, 31.6% agreed, 24.6% were neutral, 10.5% disagreed, and 3.5% strongly disagreed.

Table 6. Frequency of responses in the second section (Office room interior space planning).

No	Second Section: Office Room Interior Space Planning	SD	D	N	A	SA
1.	Is the interior space of the office commensurate with the nature of the work?	3	2	11	18	23
2.	Are the office furniture items available in the room appropriate to their size?	2	6	14	18	17
3.	Does the office furniture achieve aesthetic and functional aspects?	5	12	14	13	13
4.	Do the dimensions of the available office furniture items make you feel comfortable?	7	6	15	13	16
5.	Do the internal movement paths fit into the space of the office?	5	10	15	14	13

The analysis of responses to question 4 (Do the dimensions of the available office furniture items make you feel comfortable?) indicates that 28.1% strongly agreed, 22.8% agreed, 26.3% were neutral, 10.5% disagreed, and 12.3% strongly disagreed. The analysis of responses to question 5 (Do the internal movement paths fit into the space of the office?) indicates that 22.8% strongly agreed, 24.6% agreed, 26.3% were neutral, 17.5% disagreed, and 8.8% strongly disagreed. The analysis of responses to question 3 (Does the office furniture achieve aesthetic and functional aspects?) indicates that 22.8% strongly agreed, 22.8% agreed, 24.6% were neutral, 21.1% disagreed, and 8.8% strongly disagreed.

Statistical graphs of the data in Table 6 showing the analysis of responses to the second section are shown in Figure 6.

Table 7 shows the responses to the availability of natural and artificial lighting sources, which consists of six questions. The analysis of responses to question 5 (Does your office interior design make you feel private?) indicates that 49.1% strongly agreed, 22.8% agreed, 10.5% were neutral, 7% disagreed, and 10.5% strongly disagreed. The analysis of responses to question 4 (Do you prefer vertical partitions (walls) for the office room to be solid?) indicates that 45.6% strongly agreed, 26.3% agreed, 7% were neutral, 10.5% disagreed, and 10.5% strongly disagreed. The analysis of responses to question 3 (Is the current lighting for the office commensurate with the interior space?) indicates that 36.8% strongly agreed, 31.6% agreed, 14% were neutral, 7% disagreed, and 10.5% strongly disagreed. The analysis of responses to question 1 (Is there a source of natural lighting in your office?) indicates that 33.3% strongly agreed, 21.1% agreed, 8.8% were neutral, 1.8% disagreed, and 35.1% strongly disagreed. The analysis of responses to question 2 (Do you suffer from a lack of natural light sources?) indicates that 31.6% strongly agreed, 1.8% agreed, 10.5% were neutral, 17.5% disagreed, and 38.6% strongly disagreed. The analysis of responses to question no 6 (Do you prefer to work in an open-plan office with eye contact with colleagues?) indicates that 12.3% strongly agreed, 15.8% agreed, 12.3% were neutral, 15.8% disagreed, and 43.9% strongly disagreed.

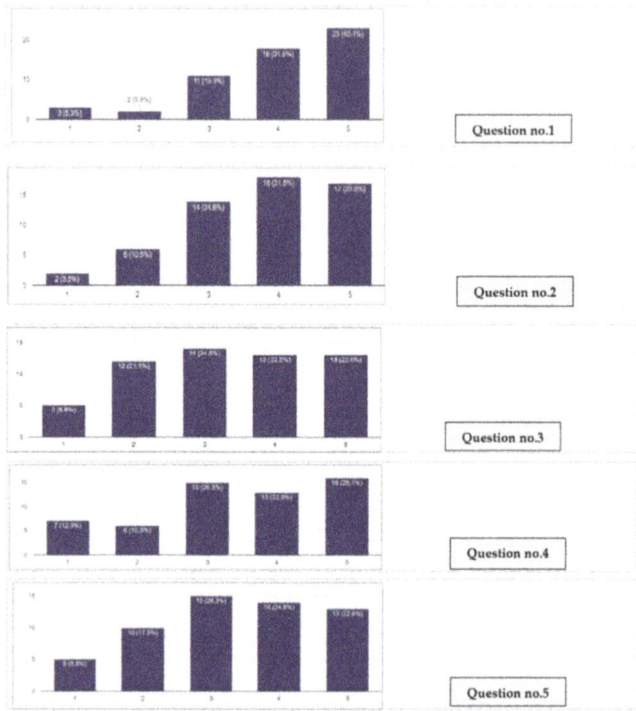

Figure 6. Analysis of responses to the second section.

Table 7. Frequency of responses to the third section on availability of natural and artificial lighting.

No	Third Section: Availability of Natural and Artificial Lighting Sources	SD	D	N	A	SA
1.	Is there a source of natural lighting in your office?	20	1	5	12	19
2.	Do you suffer from a lack of natural light sources?	22	10	6	1	18
3.	Is the current lighting for the office commensurate with the interior space?	6	4	8	18	21
4.	Do you prefer vertical partitions (walls) for the office to be solid?	6	6	5	14	26
5.	Does your office interior design make you feel private?	6	4	6	13	28
6.	Do you prefer to work in an open-plan office with eye contact with colleagues?	25	9	7	9	7

Statistical graphs of the data in Table 7 showing the analysis of responses to the third section are shown in Figure 7.

Table 8 shows the responses on materials and colors, which consists of five questions. The analysis of responses to question 2 (Do the materials used in the interior design elements (floors, walls, ceiling) for the office make you feel comfortable and functional?) indicates that 17.5% strongly agreed, 36.8% agreed, 21.1% were neutral, 14% disagreed, and 10.5% strongly disagreed. The analysis of responses to question 5 (Are the materials used in the office furniture items appropriate from your point of view?) indicates that 17.5% strongly agreed, 36.8% agreed, 24.6% were neutral, 8.8% disagreed, and 12.3% strongly disagreed. The analysis of responses to question 3 (Are the colors used in a way that helps you feel comfortable and positive?) indicates that 14% strongly agreed, 38.6% agreed, 15.8% were neutral, 15.8% disagreed, and 15.8% strongly disagreed. The analysis of responses to question 4 (Does the distribution of colors in the office make you feel lethargic and lazy?) indicates that 14% strongly agreed, 7% agreed, 21.1% were neutral, 19.3% disagreed, and 38.6% strongly disagreed. The analysis of responses to question 1 (Do the colors used in the furniture items make you feel stressed?) indicates that 10.5% strongly agreed, 8.8% agreed, 14% were neutral, 24.6% disagreed, and 42.1% strongly disagreed.

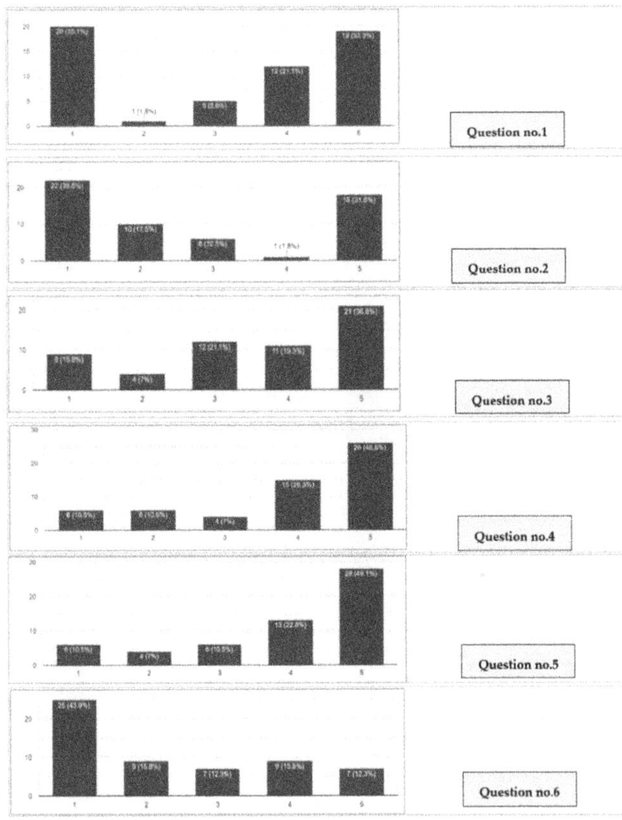

Figure 7. Analysis of responses to the third section.

Table 8. Frequency of responses to fourth section (materials, colors).

No	Fourth Section: Materials, Colors	SD	D	N	A	SA
1.	Do the colors used in the furniture items make you feel stressed?	24	14	8	5	6
2.	Do the materials used in the interior design elements (floors, walls, ceiling) for the office make you feel comfortable and functional?	6	8	12	21	10
3.	Are the colors used in a way that helps you feel comfortable and positive?	9	9	9	22	8
4.	Does the distribution of colors in the office make you feel lethargic and lazy?	22	11	21	4	8
5.	Are the materials used in the office furniture items appropriate from your point of view?	7	5	14	21	10

Statistical graphs of the data in Table 8 showing the analysis of responses to the fourth section are shown in Figure 8.

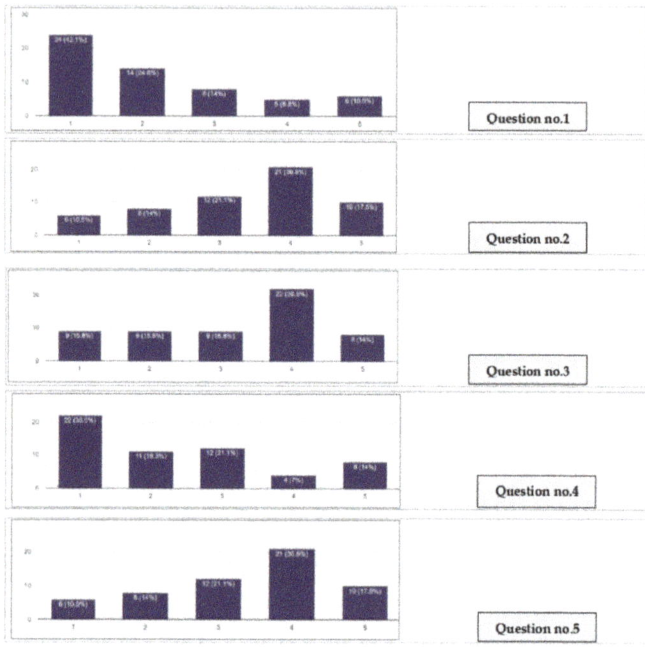

Figure 8. Analysis of responses to the fourth section.

The following are the results related to the third question: how can we help employees reach their highest levels of happiness and quality in the work environment?

The statistical analysis of the 57 responses to the questions in the fifth section of the questionnaire (Table 9), which consists of seven questions about resting places within the work environment, indicates that for question 1, 59.6% strongly agreed, 28.1% agreed, 7% were neutral, 1.8% disagreed, and 3.5% strongly disagreed. From this analysis, we can see the importance of providing time to take breaks from the daily workload within the institution.

Table 9. Frequency of responses to the fifth section (Rest areas within the work environment).

No	Fifth Section: Rest Areas within the Work Environment	SD	D	N	A	SA
1.	Do you feel that you need to take breaks during your daily work?	2	1	4	16	34
2.	Do you prefer to take time off alone?	8	7	11	13	18
3.	Would you rather take a break in your office (formal work area)?	12	13	9	7	16
4.	Would you rather take a break outside your office (formal work area)?	10	7	10	14	16
5.	Are there places to rest from work at your institution?	34	5	11	6	1
6.	Do you feel that having resting places at your institution is important?	1	3	5	6	42
7.	Does the presence of resting places in your institution help achieve job happiness and strengthen relationships between you and co-workers?	6	2	6	10	33

Figure 9 shows the analysis of responses to question 2, about spending time alone, indicates that 31.6% strongly agreed, 22.8% agreed, 19.3% were neutral, 12.3% disagreed, and 14% strongly disagreed. This shows that it was necessary to provide suitable places for rest individually or collectively according to the desires of the employees.

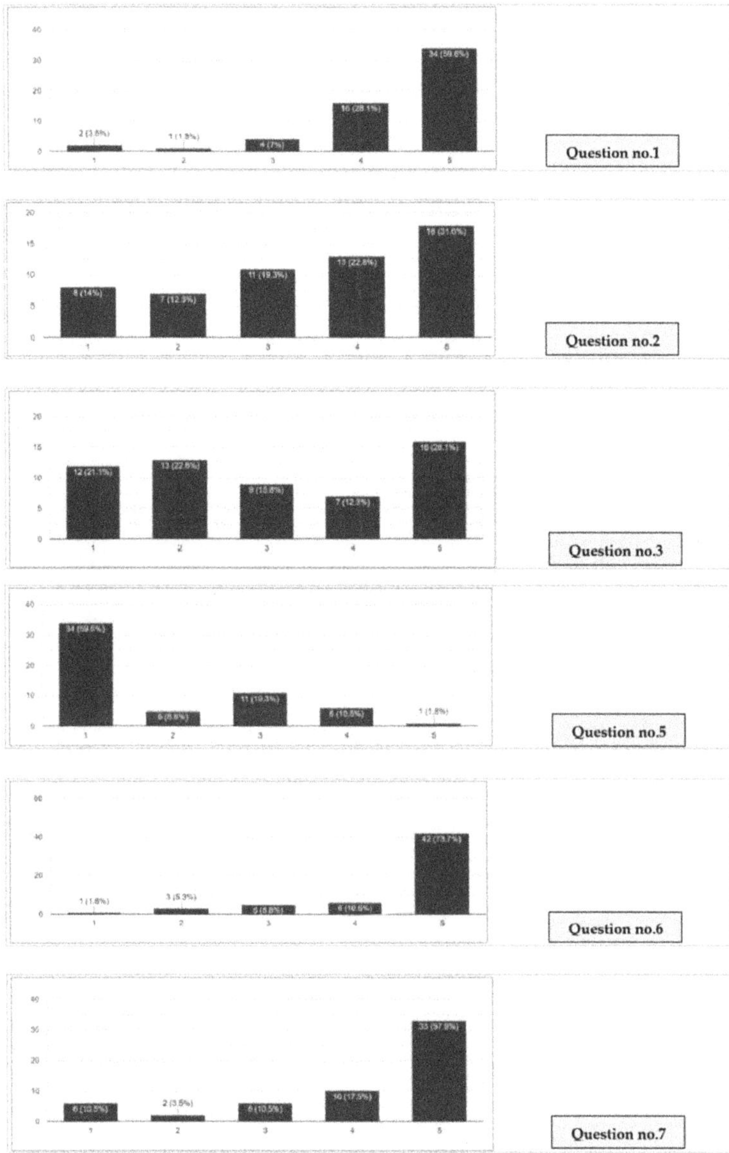

Figure 9. Analysis of responses to the fifth section.

The analysis of responses to question 3, about employees' desire to spend rest time in their offices (official workplaces), shows similar values: 28.1% strongly agreed, 12.3% agreed, 15.8% were neutral, 22.8% disagreed, and 21.1% strongly disagreed.

The analysis of responses to question 5, about the availability of resting places within the institution, shows that 1.8% strongly agreed, 10.5% agreed, 19.3% were neutral, 8.8%

disagreed, and 59.6% strongly disagreed. These percentages indicate that there were no resting places within the institution; thus, there needs to be development and modification in order to provide places to rest during work.

The analysis of responses to question 6 concerning the importance of having resting places at the institution indicate that 73.7% strongly agreed, 10.5% agreed, 8.8% were neutral, 5.3% disagreed, and 1.8% strongly disagreed. These percentages indicate the need to provide resting places and that they have an impact on employee comfort and happiness.

The analysis of responses to question 7, about the presence of resting places and their role in achieving job happiness, indicates that 57.9% strongly agreed, 17.5% agreed, 10.5% were neutral, 3.5% disagreed, and 10.5% strongly disagreed. The researcher believes that providing such spaces would help achieve happiness and quality in the institutional work environment.

The following are the results related to the fourth question: What are the future vision and proposed solutions to achieve functional and institutional happiness of the work environment?

The researcher presented suggestions about achieving job happiness in the institutional work environment to the questionnaire respondents, and their suggestions included the following:

- Provide an attractive and safe work environment in which the interior design elements are available while providing well-equipped rest areas with the appropriate furniture that facilitates creativity and innovation;
- Provide a comfortable work environment for King Faisal University employees that suits them and their preferences, as some prefer to be in a closed office where they can focus on work and avoid noise, while others prefer to work in an open-plan office that contributes to cooperation and visual and intellectual communication among the staff;
- Provide additional workspaces outside the offices that enable collaborative or individual work and generation of ideas while strengthening social bonds and exchanging opinions;
- Activate happiness days at the university and its faculties;
- Achieve job happiness through interior design treatments that are commensurate with the nature of the job and the space to improve motivation;
- Provide natural lighting sources within the institution for psychological comfort during work;
- Pay careful attention to the selection of colors that can help to spread positive and stimulating energy in the workplace;
- Provide the appropriate thermal climate in terms of heat and cold, which has an impact on the performance of workers and is reflected on their efficiency at work.

11. Recommendations

- Provide an internal work environment with an interior design that helps employees feel functional happiness;
- Improve employee productivity through air quality, lighting, and temperature, as these are among the factors that positively affect health and enhance performance in a comfortable work environment;
- Understand that innovation in interior design has a positive impact on the level of the work environment in general and motivates employees to work in particular;
- It is necessary to allow employees to get away from their computers; thus, they can have job comfort;
- Encourage researchers and interior architecture designers to put forward creative proposals to redesign the workplace to raise the level of happiness;
- Increase the interest of university-level institutions in the role of interior design in the work environment and to work on their wellbeing and make it a happy environment that stimulates activity, generates constructive energy, and stimulates positivity among employees;
- Provide safety and prevention standards in the work environment;

– Provide a positive and happy work environment that encourages positive communication among workers.

Funding: This research was funded by the Deanship of Scientific Research at King Faisal University, grant number NA00092.

Institutional Review Board Statement: Not applicable.

Informed Consent Statement: Not applicable.

Data Availability Statement: Not applicable.

Acknowledgments: The author acknowledges the Deanship of Scientific Research at King Faisal University for financial support under the Nasher Track (grant no. NA00092).

Conflicts of Interest: The author declare no conflict of interest.

References

1. Myers, D.G.; Diener, E. Who is happy? *Psychol. Sci.* **1995**, *6*, 10–19. [CrossRef]
2. Mcgonagle, C. Happiness in the Workplace: An Appreciative Inquiry. Ph.D. Thesis, Dublin City University, Dublin, Ireland, 2015, *unpublished*.
3. Bekhet, A.K.; Zauszniewski, J.A.; Nakhla, W.E. Happiness: Theoretical and empirical considerations. In *Nursing Forum*; Blackwell Publishing Inc.: Malden, MA, USA, 2008; pp. 12–23.
4. Argyle, M. *The Psychology of Happiness*; Routledge: Abingdon, UK, 2013.
5. Veenhoven, R. *Data-Book of Happiness: A Complementary Reference Work to 'Conditions of Happiness' by the Same Author*; Springer: Berlin/Heidelberg, Germany, 2013.
6. Duari, P.; Sia, S.K. Importance of happiness at workplace. *Indian J. Posit. Psychol.* **2013**, *4*, 453.
7. Abu Life, S. The Role of Organizational Balance and Job Happiness in Enhancing Job Engagement: A Field Study. *Alex. Univ. Adm. Sci.* **2021**, *58*, 133–184.
8. Sahin, F.; Yenel, K.; Kiliç, S. Investigation of Teachers' Views on a Happy Work Environment by PERMA Model = Mutlu Bir Is Ortamina Iliskin Ögretmen Görüslerinin PERMA Modeline Göre Incelenmesi. *Educ. Adm. Theory Pract.* **2019**, *25*, 773–804.
9. Bhattacharjee, D.; Bhattcharjee, M. Measuring happiness at workplace. *ASBM J. Manag.* **2010**, *3*, 112.
10. Carleton, E.L. Happiness at Work: Using Positive Psychology Interventions to Increase Worker Well-Being. Master's Thesis, Saint Mary's University, Halifax, NS, Canada, 2009.
11. Fisher, C.D. Happiness at work. *Int. J. Manag. Rev.* **2010**, *12*, 384–412. [CrossRef]
12. Butler, T.D.; Armstrong, C.; Ellinger, A.; Franke, G. Employer trustworthiness, worker pride, and camaraderie as a source of competitive advantage: Evidence from great places to work. *J. Strategy Manag.* **2016**, *9*, 322–343. [CrossRef]
13. Wesarat, P.; Sharif, M.Y.; Abdul, M.A.H. A conceptual framework of happiness at the workplace. *Asian Soc. Sci.* **2014**, *11*, 78–88. [CrossRef]
14. Cropanzano, R.; Wright, T.A. When a "happy" worker is really a "productive" worker: A review and further refinement of the happy-productive worker thesis. *Consult. Psychol. J. Pract. Res.* **2001**, *53*, 182. [CrossRef]
15. Smith, M.; Wallace, M. An analysis of key issues in spa management: Viewpoints from international industry professionals. *Int. J. Spa Wellness* **2019**, *2*, 119–134. [CrossRef]
16. The University of Minnesota. Ceiling Height Can Affect How a Person Thinks, Feels and Acts. Science Daily. 2007. Available online: www.sciencedaily.com/releases/2007/04/070424155539.htm (accessed on 24 January 2022).
17. Payne, D. Node. An Interior Design Approach to Improving Mental Health. Master's Thesis, University of Manitoba, Winnipeg, MB, Canada, 2020.
18. Montgomery, K.F. Understanding the Relationship between the Design of the Workplace Environment and Wellness. Ph.D. Thesis, Texas Tech University, Lubbock, TX, USA, 2004.
19. Tosiriwatanapong, S. How Workplace Design Affects Emotional Wellbeing of People Who Work in the Design Firm. Ph.D. Thesis, Thammasat University, Bangkok, Thailand, 2014.
20. Akhtar, N.; Ali, S.; Salman, M.; Ijaz, A. Interior Design and its Impact on of Employees' Productivity in Telecom Sector, Pakistan. *J. Asian Bus. Strategy* **2014**, *4*, 74–82.
21. National Thesaurus for Author Names Identifier, NTA. Available online: https://ntaugcnet.co.in/tag/likert-scale-in-research-methodology/ (accessed on 13 December 2016).
22. Bickford, M. *Stress in the Workplace: A General Overview of the Causes, the Effects, and the Solutions*; Canadian Mental Health Association Newfoundland and Labrador Division: St. John's, NL, Canada, 2005.
23. Kagan, N.I.; Watson, M.G. Stress reduction in the workplace: The effectiveness of psychoeducational programs. *J. Couns. Psychol.* **1995**, *42*, 71. [CrossRef]
24. Smith, M.J.; Sainfort, P.C. A balance theory of job design for stress reduction. *Int. J. Ind. Ergon.* **1989**, *4*, 67–79. [CrossRef]
25. Ching, F.D.K.; Binggeli, C. *Interior Design Illustrated*; John Wiley & Sons: Hoboken, NJ, USA, 2018.

26. The Ergonomics Unit. *Office Wise—A Guide to Health & Safety in The Office*; Work Safe Victoria: Geelong, VIC, Australia, 2006.
27. El-Hawary, F.; Fahmy, S.; Ghoneim, O. Standards of interior design and furniture design for workgroup spaces for an environment stimulating creativity and innovation. *J. Archit. Arts Humanit.* **2022**, *7*, 473–492. [CrossRef]
28. El-Zeiny, R.M.A. The interior design of workplace and its impact on employees' performance: A case study of the private sector corporations in Egypt. *Procedia-Soc. Behav. Sci.* **2012**, *35*, 746–756. [CrossRef]
29. Cobaleda-Cordero, A.; Babapour, M.; Karlsson, M. Flexible office, flexible working? A post-relocation study on how and why university employees use a combi-office for their activities at hand. *Int. J. Hum. Factors Ergon.* **2020**, *7*, 26–54. [CrossRef]
30. Jancey, J.M.; McGann, S.; Creagh, R.; Blackford, K.D.; Howat, P.; Tye, M. Workplace building design and office-based workers' activity: A study of a natural experiment. *Aust. N. Z. J. Public Health* **2016**, *40*, 78–82. [CrossRef]
31. Salama, A.M.; Courtney, L. The impact of the spatial qualities of the workplace on architects' job satisfaction. *ArchNet-IJAR Int. J. Archit. Res.* **2013**, *7*, 52–64.
32. Issa. The impact of "Design for Happiness" on the quality of life in the workplace. *J. Archit. Arts Humanit.* **2021**, *6*, 1775–1793.
33. Ben-Rahmoun, S. The internal work environment and its impact on the job performance of administrators: A field study in the faculties of the University of Batna. *J. Hum. Soc. Sci.* **2013**, *8*, 203–221.
34. Haapakangas, A.; Hongisto, V.; Varjo, J.; Lahtinen, M. Benefits of quiet workspaces in open-plan offices–Evidence from two office relocations. *J. Environ. Psychol.* **2018**, *56*, 63–75. [CrossRef]
35. Knoll, Inc. *Creating Collaborative Spaces That Work: A Performance-Based Approach to Successful Planning*; Knoll Workplace Research: New York, NY, USA, 2013.
36. Bonnardel, N.; Didier, J. Brainstorming variants to favor creative design. *Appl. Ergon.* **2020**, *83*, 102987. [CrossRef] [PubMed]

designs

MDPI

Article

Towards a Better Interior Design for the Academic Library at College of Education—King Faisal University

Amira Abouelela

Art Education Department, College of Education, King Faisal University, Al-Ahsa 31982, Saudi Arabia; aabouelela@kfu.edu.sa; Tel.: +966-558658227

Abstract: Academic libraries include college and university libraries. Interior design is an important way to bring about the required change in transforming academic libraries from a repository of books into places for research and communication. While interior design plays a major role in providing an innovative functional design in academic libraries within universities, the problem of the study lies in the importance of academic libraries and their main role in university education institutions in encouraging and supporting scientific research among students and faculty members. According to the role of interior design in designing academic libraries, the researcher finds that there are problems with interior design in the academic library at the College of Education. It needs to be re-designed to improve functionality and aesthetics and enhance the internal space for users to achieve effective use of the internal library space. The study aims to create a better design of interior environments in academic libraries in order to feature creativity and innovation. The research method is descriptive–analytical to describe and analyze the current interior design of the academic library at the College of Education at King Faisal University and collect real information about the library and the problems of interior design. Then, it was presented through the opinion poll tool to the beneficiaries to know their opinions about the current design of the library and to benefit from them in the design proposal of the academic library according to the correct design considerations. Then, the opinion poll tool was displayed again according to the design proposal to know the views of the beneficiaries again to contribute to providing an appropriate research environment for students and faculty members.

Keywords: interior design for libraries; library buildings; academic library; university libraries

Citation: Abouelela, A. Towards a Better Interior Design for the Academic Library at College of Education—King Faisal University. *Designs* **2022**, *6*, 47. https://doi.org/10.3390/designs6030047

Academic Editor: Álvaro Ramírez-Gómez

Received: 21 March 2022
Accepted: 16 May 2022
Published: 18 May 2022

1. Introduction

The academic library is an integral part of the formal education system which provides time-bound education from primary school level to university level. An academic library works as a base for teaching, learning, research, etc.

In addition, it is an essential part of higher education institutions (universities, colleges, and institutes) and is one of the cultural institutions having an important scientific role in supporting curricula according to different disciplines and supporting research by undergraduates, graduate students, and faculty members; this role is no less important and necessary than any other role that can be played by any other scientific institution.

Academic libraries are an important part of their academic institution, and the vision, mission, and goals of the academic library are linked depending on their academic institution's mission, vision, and goals. The academic library is a basic supporter of the academic and research needs of faculty members and students [1].

Academic libraries are considered sources of knowledge and research that create opportunities for learning and research for students. It is necessary to know the needs of the users of academic libraries so that the academic library can provide these services efficiently to its users and increase users' satisfaction through quality management.

In addition, the design standards for academic libraries in higher education institutions are directed towards advancing their role in providing suitable places for learning and research and achieving the university institution's missions. In addition, they are directed towards the continuous improvement in the internal space of libraries in design and functionality in proportion to their area to be suitable for students' study and learning as a common work and learning space [2]. The role of interior design in academic libraries is an important role by establishing proper planning for the internal departments of the library according to their functions through a functional and aesthetic design that fulfills the requirements of the internal space of the library and is commensurate with the requirements of the users, students, and faculty members. The interior design is based on the foundations of standards for designing academic libraries in universities; planning, designing, and renovating academic libraries involve many design criteria for the needs of users in the library. The development of the traditional design of the academic library starts with reshaping gathering, studying, and reading spaces into more collaborative and tech-enriched spaces by re-planning the reading and study seating units into a design conducive to collaborative learning. The design is the most important part of the academic library as it can affect the users; it must be attractive aesthetically, suitable, and functional to attract the patrons to come to the library. In addition to the flexibility, the library's interior design must be more suitable to be applied to the academic library nowadays due to the new technological discoveries.

When planning the academic library space, it should have a variety of spaces to suit all the diverse needs of students, as the needs of the individual student differ from the needs of groups according to some of the activities in which the students participate. Four different types of space must be planned within the academic library: spaces of collaboration, sanctuary, interaction, and community. These four spaces provide students with diverse environments that suit their needs during their different times.

2. Literature Review

There are many recent literature reviews in the field of academic library space design. This study will review several studies that have benefited from its most prominent features. The researcher would like to point out that the studies that will be reviewed came in the period between 2002 and 2018 and included several countries, which indicates their temporal and geographical diversity.

In the following, we present a presentation of these studies; then, we explain what the current study offers by identifying the difference between the current study from the previous studies; and finally, we explore the aspects of benefiting from previous studies in the current study.

Rizzo (2002) provides some useful tips for four types of spaces: highly active and engaging shared spaces; interactive, collaborative spaces for teamwork or individual work; quieter spaces such as reading and study rooms; and places for contemplation out of the way for quiet and deep thought. According to the researcher, a good and successful academic library design would be by achieving a design balance between these four types of internal spaces for the library and to be characterized by design flexibility according to the needs of users, whether at exam times or group collaboration spaces that suit a larger number of students while working on a group project [3].

Beard and Dale (2010) provide five categories of different user spaces based on researchers' observations of UK higher education institutional libraries. They are individual information gathering, group collaborative work in open space, silent individual study, small-group collaborative work, and teaching and structured learning. Each of the five categories is designed to support an educational focus. The researchers suggested that applying learning points from others innovatively and flexibly, keeping in mind its own needs, can be a good library design [4].

Cunningham (2012) provides a scheme useful in clarifying the needs of users to be taken into account in the design of the library space that architects can benefit from and use. This schema consists of four levels: the lowest level of the pyramid refers to access and connections (including location, zones, collection, information, and network), the next level is used and activities (including reading, writing, collaborating, furniture, and tools), and the next level is sociability (which includes communal, social, quiet, noisy, independent, and group). The higher level of the pyramid refers to comfort and image (which includes the ambiance and sense of scholarship) indicates the highest-level attribute of comfort and feel for an ideal learning space [5].

Jochumsen et al. (2012) propose a four-space model that has been used in public libraries in Nordic countries. It provides the desired outcome of library spaces at a conceptual level. The four are inspiration space; learning space; meeting space, and performative space. "The four spaces are not to be seen as concrete 'rooms' in a physical sense, but rather as possibilities that can be fulfilled both in the physical library and in cyberspace. In an ideal library, these four spaces will support each other, and thereby support the library's objectives." According to the model, the library's overall objective is to support four goals: experience; involvement; empowerment, and innovation. These could also be overlapping functions that interact in the library space physically and virtually [6].

Narum (2013) presents a proposal consisting of four questions to be considered when designing future learning spaces. The first question is, what do we want learners to become? The second question is, what are the experiences for this to happen? The third question is, what are the spaces that enable these experiences? The fourth question, how do we know? The researcher suggested focusing on the idea of transformation so that it would be easier to see how investments in physical spaces made a difference in how students experience learning. These questions are not directly related to libraries but are closely related to the mission of academic libraries in supporting the design of learning and research spaces for learners [7].

Clugston (2013) provides the interior designer's perspective on the principles of new library design. The library's interior design should include flexible and multifunctional spaces that can be permanently rearranged to suit the different functions within the library ensuring efficient use of the interior space. The design of the spaces must also be commensurate with different learning styles and needs in addition to providing sufficient spaces that ensure comfort for learners and are commensurate with their numbers, providing them with cooperative and social spaces that suit the needs of learners, and providing information in clear and direct communication [8].

Ling, Fan, and Boya (2018) provide the importance of the role of physical academic library facilities and their internal impact on library space and readership attraction. Since academic libraries are part of higher education activities, researchers have empirically analyzed the importance of various library design features that help enhance student learning satisfaction. They found that the lighting environment, the acoustic environment, as well as the location of the library building were the main determinants that influenced the use of the university library by students in a major university in China, and they made their suggestions for modifying the design of the university library: a library to suit the educational needs of students, rebuilding the standing and function of university libraries in the age of digital information and transforming them from a provider of reference material into actual spaces for learning on campus [9].

McCabe, JP (2003) stresses the importance of studying the psychological effects of color. Additionally, what is caused by dark colors and bright colors, and emphasizes the necessity of adopting a contemporary approach in choosing the colors and materials used in the design of libraries to create an attractive, friendly, and stimulating image for the beneficiaries. Colors should be combined selectively and based on a neutral background, color may appear on elements such as selected walls and partitions, and the choice of fabric for furniture, colors, and finishes should be appropriately chosen to enhance the attractiveness of the spaces for the specific age group of the user while harmonizing seamlessly with the overall concept of the space as an attractive and stimulating place to visit [10].

By reviewing literature reviews, we note that the current study agrees with previous studies on the main topic and its general objective is that it differs from it in several aspects that this study seeks to address, namely:

- The research problem included the actual reality of the interior design of the academic library at the College of Education at King Faisal University from the point of view of the beneficiaries of the faculty members and students.
- This study used the descriptive-analytical method in collecting real information about the library and the problems of the interior design of its interior space.
- An opinion poll tool was used by one sample of faculty and students to collect accurate information and opinions about the current interior design of the library from the point of view of the beneficiaries to know their opinions and needs to take them into account and to benefit from them in the development of and improvement in the academic library.
- The researcher presented a proposal for the modern interior design of the academic library that is designed by the needs of the beneficiaries and fits with the requirements of the times while solving the design problems existing in the current design and benefiting from the views of the beneficiaries to meet their needs and aspirations.
- The researcher used the second opinion poll tool for the same sample of faculty members to view the interior design proposed by the researcher for the academic library to reveal their opinions and responses.
- The researcher benefited from previous studies on the points related to the design standards of libraries and the instructions for dividing the internal spaces in the library to achieve the effective use of the internal space and design flexibility according to the levels of the beneficiaries' needs and their internal activities that must be taken into account in the design of the library space.

3. Problem Statement

The problem of the study lies in the importance of academic libraries and their main role in university education institutions in encouraging and supporting scientific research among students and faculty members. According to the role of interior design in designing academic libraries, the researcher finds that there are problems with interior design in the academic library at the College of Education at King Faisal University. It needs to be re-designed to improve functionality and aesthetics and enhance the internal space for users to achieve effective use of the internal library space.

Design problems of the current interior design of the academic library:

The researcher will display a set of current pictures of the academic library that illustrate the design problems in it shown in Table 1.

Table 1. The problems of interior design in the academic library.

No.	Problems	Figures
1.	-The need for more items of furniture and interior equipment, such as reception desk, shelves, and computers, according to the number of users and internal departments. -The internal lighting is not commensurate with the library space, which affects the efficiency of the internal use of the library by users.	

Table 1. *Cont.*

No.	Problems	Figures
2.	-Failure to exploit the internal space of the library with the optimal distribution of used furniture units and their diversity in proportion to the internal activities.	
3.	-The current interior design of the academic library is not an attractive learning environment for students. -The dimensions of the used book storage units are not suitable as they are huge and have great depth and were not used in an optimal functional way.	
4.	-The internal space of the academic library, in its current form, does not fit design and functionality with the needs of the beneficiaries. -The furniture used is heavy and not flexible or easy to move.	
5.	-The use of heavyweight wooden partitions in the internal division affects the narrowness of the internal space and the movement paths according to the services provided within the library.	
6.	-The lack of proportional wooden vertical dividers in dividing the interior spaces of the library functionally and aesthetically.	
7.	-There is no visual contact between users or a sense of familiarity due to the solid wood vertical partitions.	

4. Study Objectives

The current study aims to develop the interior design of the academic library at the College of Education at King Faisal University to support its main objectives, which are:

- The research aims to provide the needs of users within the academic library as places for collections and places for reading, discussions, learning, and research.
- Providing a stimulating interior design environment for graduate students, faculty members, or undergraduates for scientific research.
- Achieving the design standards of the academic library functionally and aesthetically to achieve the functional requirements.
- Providing research educational resources for graduate students, faculty members, or undergraduates.

5. The Importance of Study

- The importance of the study is due to the importance of the topic it raises about academic libraries and their role in encouraging and supporting scientific research among students and faculty members.
- Academic libraries are the center of academic activities, providing many services to their students and faculty members to facilitate the teaching, learning, and research process.
- Providing a modern and comprehensive set of information sources that are closely related to the academic curricula, academic programs, and scientific research conducted at the university.
- Efficient interior design plays a major role in the design of the library's interior space and has a significant role in the efficient use of the library by users.

6. Study Limits

- Time Limits: The second semester of the academic year 2021–2022.
- Human limits: Faculty members, undergraduate, and graduate students of King Faisal University.
- Spatial Limits: Academic Library College of Education at King Faisal University.
- Subject Limits: Developing the interior design of the academic library at the College of Education in student departments at King Faisal University to support its main objectives of providing an appropriate and effective research environment for users.

7. Study Questions

The research seeks to answer the following research questions:

- What are the necessary design standards for designing academic libraries in universities?
- What is the reality of the academic library at the College of Education at King Faisal University in the shade of the design standards that need to be developed?
- What is the proposed vision for designing a better interior environment in the academic library at the College of Education at King Faisal University in light of the necessary design standards?

8. Materials and Methods

The research process was based on the descriptive-analytical approach, and the relevant literature was presented, and the description of the interior design of the academic library at the College of Education, King Faisal University, in which the study was conducted, was presented, and real information was collected about the library and its interior design problems, and used in the application of the correct design, followed by the proposal for the design by the researcher for the academic library, which he hopes will contribute to providing a suitable research environment for students and faculty members by the standards of the interior design of the academic library to achieve efficient internal use and create a friendly place for library users.

8.1. Study Tool

- The researcher used an opinion poll tool consisting of 22 questions directed to the beneficiaries of the faculty members and students about the current interior design of the academic library to reveal the opinions and needs of the users. The response of the beneficiaries on the card is (yes or no) to a sample (consisting of 67 users of faculty members and students).
- Then, this tool (an opinion poll) was distributed again to the beneficiaries to know their opinions about the design proposal for the academic library submitted by the researcher so that the beneficiaries' response to the card would be (yes or no) and the sample consisted of 67 users of faculty members and students.

8.2. Types of Libraries

There are four major categories of libraries as Figure 1:

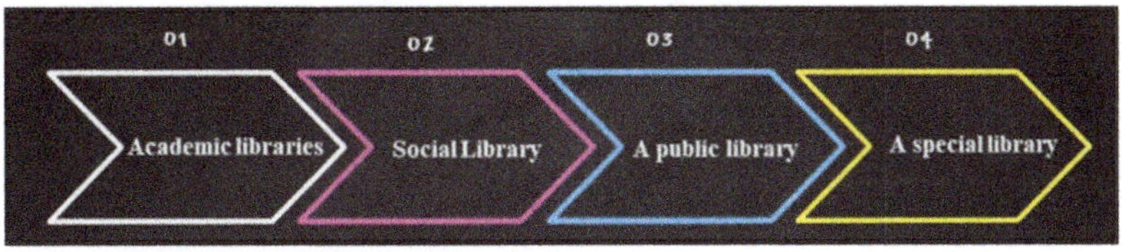

Figure 1. The diagram shows the types of libraries.

- Academic libraries:

The academic library is part of the politics, part of the culture, and part of the response of its parent institution; whatever affects higher education, affects academic libraries [6]. Academic libraries are the libraries that are located in universities, colleges, and institutes and are characterized by their important role in supporting their affiliated institution to meet educational needs and support studies at the undergraduate and graduate levels by supporting curricula, scientific research, and research by faculty members and students at the university [11].

Components of the academic library space:

- Lecture halls.
- Collaborative learning and meetings.
- User seating units.
- Workstation spaces.
- Viewing and listening rooms [12].

Modern Academic Libraries:

In the twenty-first century, academic libraries are in a technological transformation from what they were before, where information, scientific resources, and books are accessed digitally via the internet, containing books, references, e-books, seating units for collaborative or individual learning, and meetings [13].

- Social Library:

This type of library has an essential role in society, as libraries are a source of knowledge and culture in addition to the services provided by the library, as it opens the way for learning and helps in eradicating illiteracy and education in addition to helping to form ideas through reading and perusal, which enhances the progress of society culturally, scientifically, and creatively.

- Public Library:

This is a library that is available for all people and is considered a local gateway to knowledge that contributes to the cultural development of the person and social groups [14].

- Special Library:

A special library is a library that specializes in providing information sources on a specific topic, meaning that it serves a targeted group of users and helps specialists in providing the services they need. Examples of private libraries (are corporate libraries, government libraries, medical libraries and others) [15].

Types of Academic Libraries as Figure 2 [16].

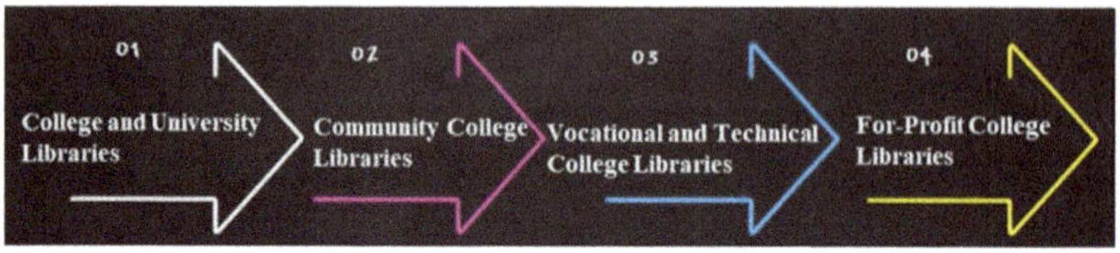

Figure 2. The diagram shows the types of academic libraries.

Main places of the Academic Library:

- The main reading room:

This is a vital and very important area in terms of movement and activity, and its area is determined by the number of visitors, and the following conditions are required: that it be in the heart of the library, that it be close to the area of the book galleries and have a direct relationship with it, provided with good lighting suitable for reading.

- The circulation section:

The circulation section is a section attached to the main reading section that contains books, references, letters, and other educational materials organized in a way that allows and helps to control the process of borrowing and retrieval in an orderly and accurate manner.

- Department of Scientific Journals:

This section is considered one of the important and main sections in the libraries, on which the library depends mainly to respond in meet the continuous needs in providing information and modern research resources in the field of specialization for researchers and scholars.

- Retreats:

These are separate rooms equipped with cupboards and shelves for books and references. They also contain cupboards belonging to the researchers to keep their papers. The spacers between these rooms should be made of sound-insulated glass (transparent dividers) to provide natural light with an artificial light source as well as furniture items suitable for each room. These rooms are supported by auxiliary rooms such as photocopying, printing, and computers.

- Library Management:

The library management is considered the main body responsible for the library by performing the various functions according to the internal departments, and its interior design must be commensurate with the work volume and the number of employees [17].

9. Discussion

- The results related to the first question:
- What are the necessary design standards for designing academic libraries in universities?

To answer this question, the researcher presented the design and planning considerations for academic libraries that should be taken into consideration when designing, planning, and developing academic libraries.

Planning considerations of the academic library:

There are scientific terms related to the description, planning, and design of the academic library, such as Figure 3.

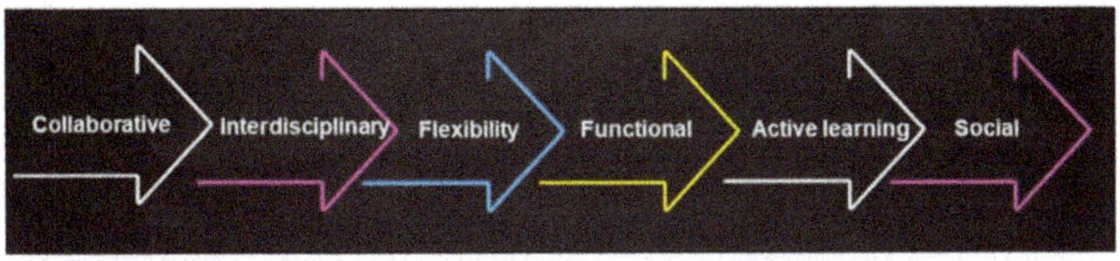

Figure 3. The diagram shows scientific terms for planning and design of the academic library.

- The layout and interior design of the library should be commensurate with its nature, functional sections, the number of visitors, and their needs.
- The academic library should be designed on a functional basis.
- The need to focus on the effective interior design of the spaces designated for students and faculty users to support learning and provide for their needs.
- The main study and reading areas should be close to the bookshelves.
- The interior design of the academic library must be able to expand and develop in the future [18].
- The physical space plays an important role in helping the academic library meet the needs of users. The movement paths that link the departments and some of them must be taken into account to help in accommodating a large number of students and faculty members through the library space daily.
- The designer should facilitate entry and exit and eliminate congestion and conflict in the movement paths between visitors and employees by allocating a special entrance to the administration.
- All the major factors that affect the use of academic library space must be taken into account.
- Providing technological services that the academic library should provide to its students and users, ideal places to provide computers and other information technology.
- Achieving calm in the library through the use of tightly closed windows, which provide sufficient calm.

The attributes of the Modern Academic Library:

- Full of natural light.
- Shelves are easy to reach.
- Technology is accessible to patrons and visible to staff.
- Possibility to sit for cooperation and independent work.
- Seating for collaboration and independent work.
- Distinct zones for different uses [19].

Academic Library Lighting Standards:

Natural lighting:

This is the lighting that comes from the sun, and its amount in the interior space depends on the number and area of wall openings such as windows and skylights. It is physiologically most appropriate for humans, but it changes and varies with time and season. The natural lighting provides visual and psychological comfort for humans and helps with activity and vitality. Natural lighting is the best in academic libraries, as it is an essential component of architecture and light is one of the most important elements of interior design. Relying on modern technologies, engineers have tried to control the harmful part of the sun's rays and use its heat. Natural daylight helps reduce energy consumption and can be a source of glare, so adjustable window coverings should be provided to allow direct sunlight into the reading areas.

Artificial Lighting:

This is the illumination received from electric lamps, and it is based on three main ways to illuminate the space:

- General lighting: This is the lighting homogeneously used throughout the library.

- Thematic lighting: This is lighting certain areas of the internal space of the library to help illuminate certain sections and functions.
- Focused lighting: This is a type of objective lighting that focuses on specific functions in the library [20].

The internal environment of libraries contains various elements, the most important of which is lighting, because most of the activities that take place in libraries are directly related to sight. The quality of the internal environment of the library depends on the quality of lighting and space adequacy in addition to its layout and the quality and diversity of furniture.

In addition, the process of providing appropriate lighting for reading and work halls depends on the proper use of natural and artificial lighting, simultaneous use, or each one alone, according to what the need requires. Reading and searching are very important in libraries. It requires the provision of good lighting which is appropriate aesthetically and functionally for the success of the library's interior design in general. The tasks within the library depend mainly on high-quality lighting, as technology has introduced new and advanced types of lighting requirements, as it affects the interior design of the library significantly and can also improve its efficiency.

- Bookcases should be provided with adequate lighting so that users can access references and books simply and without visual disturbance.
- The general lighting design of the library should be suitable for many activities.
- Artificial lighting is used in case of insufficient natural lighting using fluorescent lamps, and the lighting from electric lamps must be indirect so as not to cause luster.
- Plastic paper curtains were used for the windows to block direct sunlight.

Challenges with Academic Libraries:

- Noise:

Noise in libraries, in general, is a constant concern for library users, students, faculty, and administrators. It is one of the common problems, and by using some solutions such as installing sound insulation techniques, noise can be controlled and modified. [21]

- Power and capacity outlets:

This is possessing sufficient "strength". Having sufficient power is required for beneficiaries to charge their devices. Spaces without sufficient power for the users result in them being underused by the students.

- Sharing space with learning partners:

Allocate space for learning partners on campus. Some of these partners required more space than initially anticipated.

- Technology and its impact:

Patrons come to the library to use a variety of modern technologies. Technology provides a tool for academic library services. It also works on shaping and defining how services are provided to beneficiaries.

- The results related to the second question:
- What is the reality of the academic library at the College of Education at King Faisal University in the shade of the design standards that need to be developed?

To answer this question, the researcher used an opinion poll tool by designing a card consisting of 22 questions directed to the beneficiaries of the faculty and students about the current interior design of the academic library to reveal their opinions and needs through their responses to an opinion poll, and 67 responded shown in Figure 4, and as shown in Table A1; two were excluded because they were not valid.

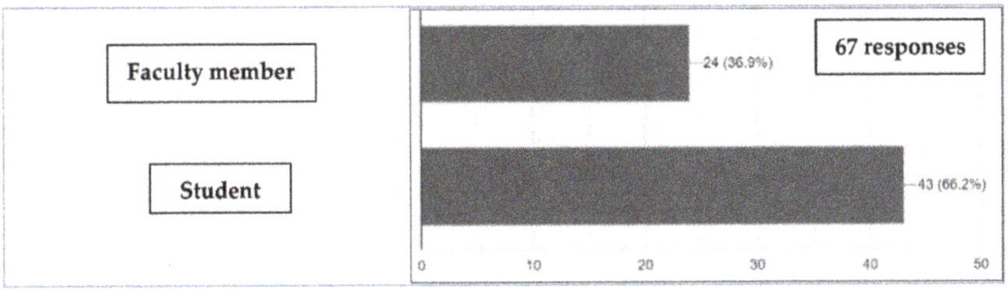

Figure 4. The distribution of the sample between the responses of faculty members and students.

Opinion poll result:

From the previous figure, we find that the sample consists of 67 beneficiaries, of whom 24 are faculty members (36.9%) and 43 are students (66.2%). Two were excluded because they were not valid.

From Table A1, it is clear that 87.7% of the beneficiaries' responses emphasized the role of the interior design of academic libraries in providing an attractive and effective learning environment for the beneficiaries, and 83.1% emphasized the importance of developing the current interior design of the academic library by developing furniture items for the different needs of cooperative and individual education, and reaching a design atmosphere that achieves visual comfort, as the current interior space of the academic library does not fit the design and functionality with the needs of the beneficiaries. A total of 80% of responses confirmed that the interior design of the library does not fit the requirements of the times and 73.8% responded that the floor plan of the library was not properly planned. In addition, 76.9% confirm that the current interior design does not achieve visual coherence and a sense of familiarity.

In addition, the distribution of book and reference storage units is not proportional to the library space, as it is concentrated in some aspects, causing narrow movement paths in some parts and widening in other parts. A total of 86.2% of beneficiaries expressed their opinion in changing the colors used. They also confirmed the lack of equipment for electronic search sources by 83.1%, and 81.5% would like to remove the vertical wooden dividers used to divide the internal spaces of the library, which affects the narrowness of space and the feeling of lack of visual communication between the beneficiaries. Some of the views of the beneficiaries were that the current internal environment of the academic library provides for the different needs of cooperative or individual education by 40%.

The proposed design of the Academic Library of the College of Education—King Faisal University.

Academic library shape

When determining the basic shape of the academic library at the College of Education, we find that its shape is square, whereas library engineers and consultants agree that the most functional form of the library is the square or a series of squares. Architects sometimes see the square shape as too simple and uninteresting: however, long, narrow spaces and round rooms do not allow for effective planning of shelving ranges or good visual control by staff. The square shape is also easy to divide and rearrange regardless of whether the library is large or small. Successful academic library design relies on large, flexible spaces that can be easily rearranged as the needs of the library change from time to time; shelving and book storage must be planned to accommodate them [22].

Library Zoning:

The primary goal of effective library design and space planning is that the facility must respond to the needs of students and faculty members. The floor plan of the library shows its interior design through the distribution of the different functional sections and their relationship to each other with a study of the internal movement paths between the functional sections as Figure 5.

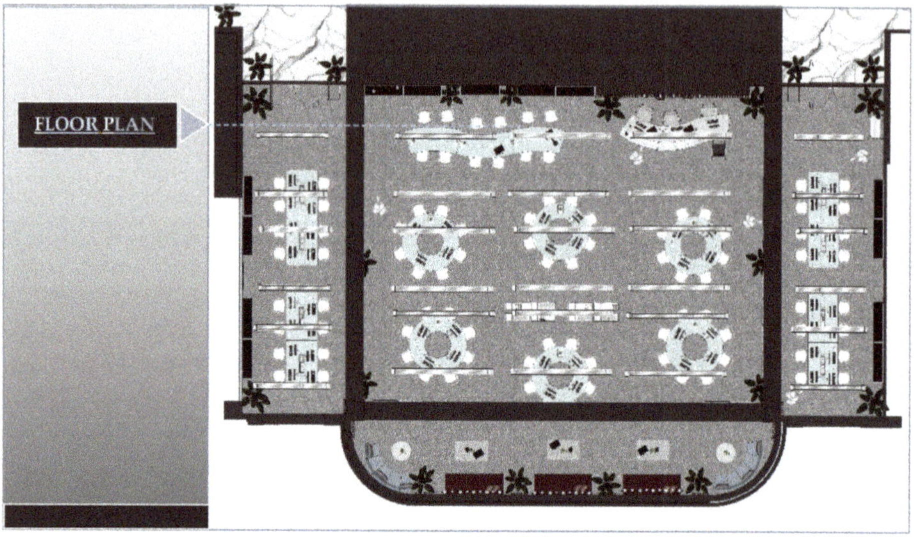

Figure 5. Floor plan of the academic library.

The proposed floor plan of the academic library:

The layout of the floor plan of the library space comes in all shapes and sizes, and the primary goal of effective library interior design and space planning is that an academic library should provide design and functional requirements of users and respond to the needs of its students and faculty while defining the interior departments, their spaces, and their needs [23].

Library equipment

This term includes roughly the items and vocabulary of library furniture from seating units, tables, storage units, and shelves for books and references, supplies, and tools with which the library operates as Figure 6 [24].

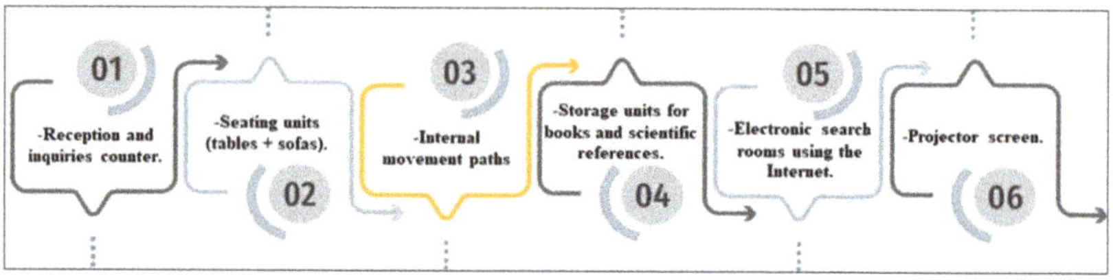

Figure 6. The diagram shows the components of the academic library.

It consists of:

1. Reception and inquiries counter.
2. Seating units (tables + sofas).
3. Internal movement paths.
4. Storage units for books and scientific references.
5. Electronic search rooms using the internet.
6. Projector screen.

Internal Perspective of the Library as Figure 7.

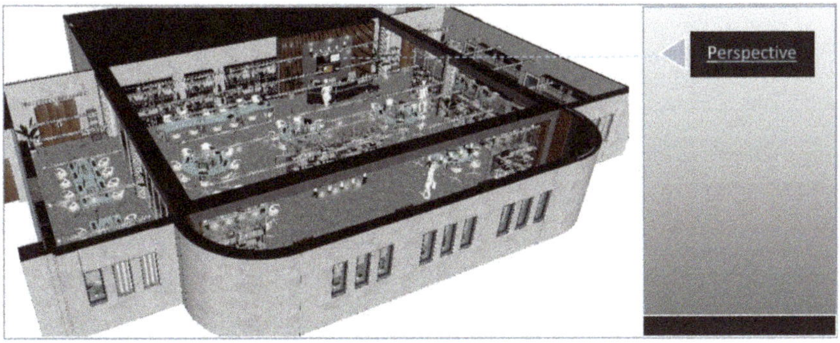

Figure 7. The perspective of the academic library.

Design of Reception Desk:

The proposed interior design of the academic library of the College of Education at King Faisal University as Figure 8. provides quiet and conducive areas for study as well as common areas for university study and collaboration and provides reading areas, discussion areas, learning, and research areas, and facilities for accessing electronic resources. The proposed finishes were carefully shaped, and there was a conscious decision to continue coordinating the interior design elements, the various furniture elements, the materials used, and the colors suggested in the design. Bookcases are designed as a different combination of shelves that allows users to store and display books in a smooth and easily accessible way as well as computer desks, tables, and chairs to get the right balance in distributing different furniture items according to functional requirements and according to capacity for different activities. It also designed integrated spaces for student discussions and presentations, providing spaces designed for small group meetings as Figure 9.

Figure 8. The proposed space for the design of the reception space in the library.

Figure 9. The perspective of reception and inquiries counter.

The reception must be directly connected to the entrance with its connection to the library's internal monitoring network as Figure 10; it is wide enough and the largest number of students passing at any time is counted. The reception counter as Figure 11. is designed to inquire about data, books, or references needed by researchers, as well as for external or internal borrowing. It is designed to suit everyone in addition to people with special needs. The back of the information counter is designed with a cladding of wooden slats, 3.5 m high and 6 m wide to cover the back of the counter.

Figure 10. Internal perspective shows reception and seating units.

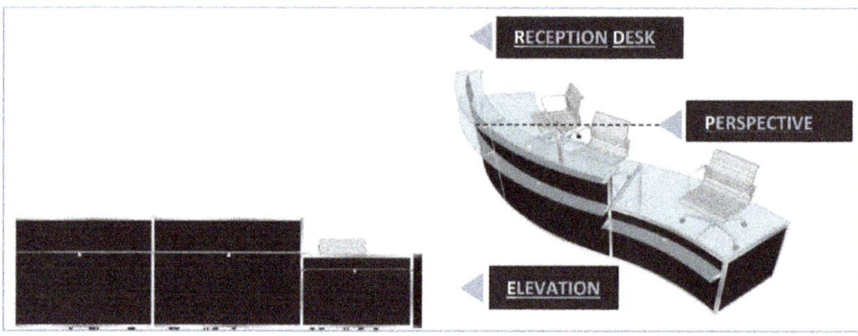

Figure 11. Reception desk design.

The proposed new colors in the design.

The colors of the academic library trended to neutral tones to create a mobile library to view, research, and learn, in addition to the possibility of clarity of architectural details.

The appropriate light gray color for the walls was chosen, with the use of neutral colors in the interior space, and white color in the chairs, and burgundy color in the seating units (sofas); it is also preferable to use warm colors in the reading space. The columns are clad with straight brown wooden ribs with a dark gray color for the carpet of the library flooring instead of the dark blue color. Transparent glass partitions were used to divide the interior spaces according to the design instead of opaque wooden partitions, which caused the narrowness of the interior space and the inefficiency of lighting, the failure to achieve visual communication and a feeling of spaciousness, and the expansion of the library space.

Digital Resource Department:

Digital Resource Department: This section is mainly responsible for offering computer-based library services. It has 32 computers for the library students and faculty members. The use of computers throughout the academic library needs to be planned in the library's structure and systems simultaneously with the interior design to ensure continued energy savings at computer locations as Figures 12 and 13.

Figure 12. Internal perspective of an online search room before modification.

Figure 13. Internal perspective of an online search room suitable for 16 researchers.

The space was re-designed as Figure 14. by changing the furniture to modern furniture (glass tables equipped with computers and internet networks) with the replacement of the wooden partition with a glass partition that helps privacy in addition to achieving the internal visual connection to the library while allowing interior lighting and linking the functional spaces to each other and adding some storage units for books or tools for users. A glass partition that was installed in place of the wooden partition, with dimensions of 5.65 cm in width and 2.65 cm in height, was placed to separate the space for searching online and borrowing books and reading them. It achieves transparency and privacy for researchers. The glass partitions are considered a demountable wall so the spaces can be reconfigured over time. The concept of transparency is an approach to showcasing learning activities taking place in the academic library through open concepts, technology, and furnishings, and limiting physical barriers that might otherwise obstruct a user's open view.

Figure 14. The internal perspective of an online search room consists of two glass tables.

Standards for selecting Academic Library Furniture

Academic libraries have very particular requirements that all their furnishings must meet. The selection and arrangement of furniture within the library is an important issue that needs to consider the human factors that influence how people interact within the environment. It is also relevant in the interior design of the library which demonstrates the concept of acceptable personal space to feel comfortable, through how close people are to sitting or standing next to each other. There are a few important factors to consider whenever you are designing furniture items for an academic library. Reading desks and tables are the important components of library furniture. Library furniture consists of many types of racks to display all kinds of books that students could pick and read themselves. It is necessary to know the appropriate standards for choosing furniture in the academic library. The furniture must be stable and durable, characterized by efficiency, comfort, and resistance, and made of good materials; furniture and design play a huge role in determining how students interact with a library through the following:

1. The design concept is based on balancing the needs of the users, students, and faculty members, and the ability to make the most of the designed space.
2. The interior design of academic libraries should provide an appropriate study for all internal functional departments in an innovative conception that helps in the effective use of the library.
3. Appropriate furniture in the library can make a huge difference to the attractiveness and functionality of library spaces.
4. The durability and reusability of furniture items.
5. The furniture item contributes to creating a balanced environment.
6. The need for natural light and artificial light to ensure that the area is properly lit, thereby avoiding eye strain [25].

Flooring:

High-efficiency materials have been used for permanent use, and floors can be covered with insulating materials such as carpets and rugs to reduce noise by absorbing sound in reading spaces.

Search and Reading Hall:

Reading hall standards:

- It should be in the heart of the library as Figure 15.
- It should be in the quietest area of the library.
- It is preferable to take into account the entry of natural light.
- Preferably near the entrance.
- The paths of movement must not conflict.

Figure 15. The interior design of the reading space before modification.

The reading and borrowing space have been modified by removing the wooden divider and adding storage units for books with a modern design that matches the interior design and used furniture pieces as Figure 16. In addition to providing the reading part with rectangular glass tables, the distribution of internal lighting was taken into account, which achieves comfort for users as Figures 17 and 18.

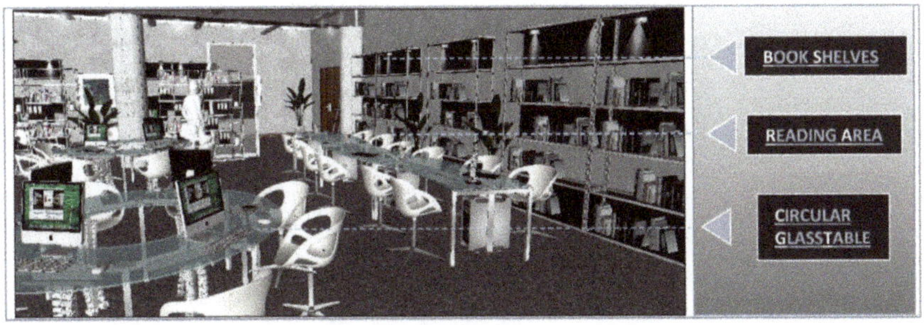

Figure 16. The interior design of the reading space contains modern glass tables and storage units for books and internal borrowing.

Figure 17. The interior design of the reading hall space contains tables and computers for online search and storage units for books and internal borrowing.

Figure 18. Shows the space for reading and research.

Library Bookcases:

Library bookcases are a great addition to any working environment, providing storage and display space while helping to keep your workspaces clean, tidy, and organized.

There are some considerations in distributing them within the library:

- The dimensions of the aisles between the shelves should be considered, which should reach 85 cm in libraries that many users come to. The shelves were distributed in the center of the hall to avoid the sun's rays and surrounded the reading areas where the shelves are divided according to the topics.
- It should be taken into account that the maximum height of the bookshelves is 175 cm.
- The bookcase units in this form were distributed in the parts of the library, where they were placed in the middle of the library hall according to the design standards of libraries. They were also attached to the glass partitions and shelves were placed in

front and behind the glass partition to become double shelves and were distributed numerically in proportion to the internal dimensions of the library as Figures 19 and 20. providing book storage and scientific resources according to the different disciplines and internal departments of the library.

Figure 19. The interior design shows the glass partition, book storage units, and interior paths.

Figure 20. Storage units for books.

Providing appropriate furniture in the interior space of an academic library makes a difference in the attractiveness and function of academic library spaces. This in turn increases the attraction for students to use the spaces more fully; a variety of seating units that help support teamwork is therefore essential in designing an effective collaborative seating space. Various seating and work surfaces must be adapted to meet the needs of different group activities and be appropriate to the size of the group with technical support as Figures 21 and 22 [26].

Figure 21. The current picture shows the space for reading and borrowing.

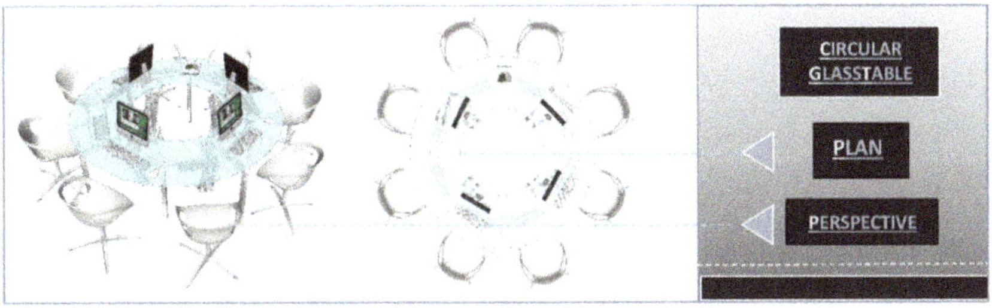

Figure 22. The library contains 6 modern glass circular tables, each table accommodates 8 people, and the total number of round tables accommodates 48 people.

Library Break Areas:

Seating space is an essential element in the design of the academic library. The library should have a variety of seating units such as benches, reading benches, and reading tables to choose from around the approach to learning. The reading area in the design model proposed for the academic library is also carefully designed so that the visitor can choose an individual reading area or a common area [27].

Figure 23 shows the proposed design of break areas, where the seating units were designed as contemporary furniture to allow users to sit and read or study in a quiet atmosphere that enjoys privacy, in addition to providing a shared social space, and allowing students to spend time away from desks. The book storage units were re-placed in a better design. The walls were also designed with some wooden cladding. In addition to providing natural light through windows, effective use of daylight can reduce energy consumption as Figure 24. The library is not only the nerve center of academic learning but also a good place to spend one's leisure time. Plastic paper blinds were used for the openings to block direct sunlight.

Figure 23. Shows the internal resting area of the library.

Figure 24. The picture shows the proposed design of break areas.

Movement paths within the library:

Internal movement paths in academic library design relate to three paths of movement

1—Reader movement:

Which is considered one of the most important movements within the library and must be connected to all services without any complexity.

2—Staff movement:

Since the administration is one of the most important spaces within the library, and also because the employee provides the services required for readers.

3—The movement of books, references, and various services of the library:

It must be a hidden movement, that is, none of the visitors feel it, and it usually has a back entrance [28].

- The results related to the third question:
- What is the proposed vision for designing a better interior environment in the academic library at the College of Education at King Faisal University in light of the necessary design standards?

To answer this question, the researcher distributed this tool (opinion poll) consisting of 22 questions again to the beneficiaries to reveal their opinions about the design proposal of the academic library of the researcher, and 65 responded, as shown in Table A2.

Opinion poll result:

From the previous figure, we find that the sample consists of 67 beneficiaries, of whom 24 are faculty members (35.8%) and 43 are students (64.2%) as Figure 25.

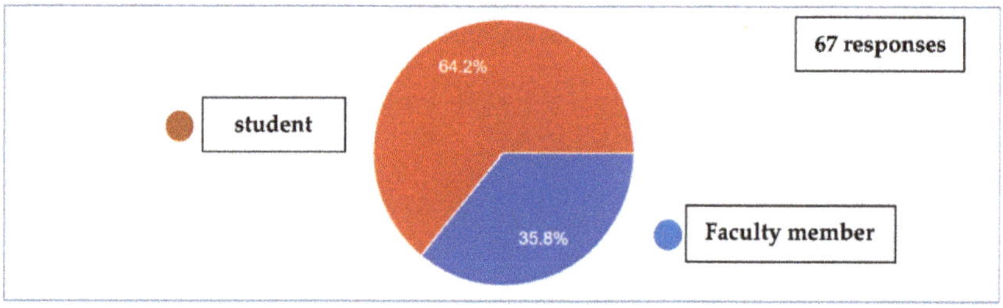

Figure 25. The distribution of the sample between the responses of faculty members and students.

The proposed design of the academic library was discussed by presenting it to the beneficiaries of the faculty and students, which was well received and satisfied through their opinions on the opinion poll. This is evident in Table A2. By analyzing the opinion poll of the beneficiaries' responses about the proposed design of the academic library from the researcher, where 67 beneficiaries responded, we find that there is general satisfaction with the new proposed design of the library, and many questions were approved by 100%, which was in response to the views and aspirations of the beneficiaries from faculty members and students on the development of the interior design of the academic library in line with the requirements of the times and their needs, and the modification of the floor plan design and functionality to achieve the efficiency of internal use, which has a positive impact on the beneficiaries.

In addition to providing the library with various and appropriate items of furnishing and interior fixtures, the efficient use of natural and artificial lighting be considered was taken into account. Some opinions, which constitute a small percentage of 9%, were about the inappropriateness of the color scheme in helping to feel energetic and the desire to learn.

Project Description:

The developmental design proposal of the academic library of the College of Education in the student departments at King Faisal University is an important objective that aims to develop the interior design of the library through a modern and innovative design vision commensurate with the nature of the interior space and its requirements, in addition to the internal space of the library and how to design it with the most appropriate design solutions, including the realization of services and departments. It should be available in the library to achieve the aesthetic and functional aspect. This is reflected in the actual performance of students and faculty members in general and all library visitors from faculty to university members as Figure 26.

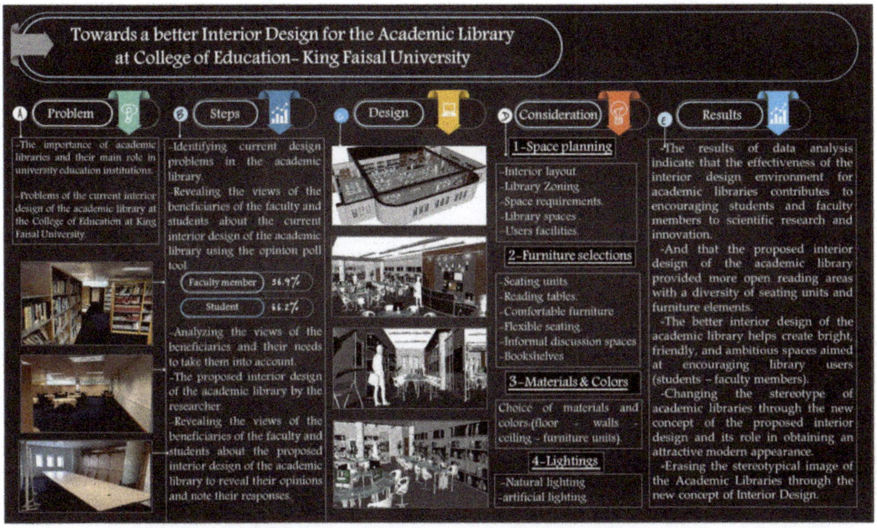

Figure 26. The framework explains the steps of the research.

10. Conclusions

- The results of data analysis in Table A2 indicate that the effectiveness of the interior design environment for academic libraries contributes to encouraging students and faculty members towards scientific research and innovation.
- Additionally, they indicate that the proposed interior design of the academic library provided more open reading areas with a diversity of seating units and furniture elements.
- The better interior design of the academic library helps create bright, friendly, and ambitious spaces aimed at encouraging library users (students and faculty members).
- The design helps in changing the stereotype of academic libraries through the new concept of the proposed interior design and its role in obtaining an attractive modern appearance.
- The design helps to show the important role of interior designers to create attractive and appropriate academic libraries.
- We erased the stereotypical image of academic libraries through the new concept of interior design.

11. Recommendations

- Libraries are good homes for introducing and providing new and experimental technology to students by designing the interior functional spaces that require providing the technology that students need.
- The interior design must be designed to accommodate the academic needs of users.
- The need to pay attention to the design of internal interaction spaces where library users interact with the resources and services provided.
- The necessity of applying modern technology in the interior design of academic libraries to reach more effective and efficient use spaces.

- The need to continuously poll stakeholders about the problems and challenges facing them in the ability to continuously improve, by interviewing the users and conducting questionnaires.

Funding: This research was funded by Deanship of Scientific Research at King Faisal University, grant number NA00082.

Institutional Review Board Statement: Not applicable.

Informed Consent Statement: Not applicable.

Data Availability Statement: Not applicable.

Acknowledgments: The author acknowledges the Deanship of Scientific Research at King Faisal University for the financial support under Nasher Track (Grant No. NA00082).

Conflicts of Interest: The authors declare no conflict of interest.

Appendix A

Objective: Information gathering about the current interior design of the Academic Library.

Description: The results of the responses of the sample of the faculty members and students in an opinion poll tool consisting of 22 questions about the current interior design of the academic library, College of Education, King Faisal University to reveal their opinions and needs.

Table A1. Opinion poll on the current interior design of the Academic Library, College of Education, King Faisal University from the point of view of the beneficiaries.

No.	Questions	Yes	No
1.	-Does the interior design of academic libraries have a major role in providing an attractive and effective learning environment for the beneficiaries?	57	8
2.	-Is the interior design of the library commensurate with the requirements of the times?	13	52
3.	-Was the layout of the academic library properly planned?	17	48
4.	-Does the interior design of the academic library help you to encourage you to spend better and more time in it?	22	43
5.	-Is the interior space of the academic library in its current form compatible in design and functionality with the needs of the beneficiaries?	15	50
6.	-Have the internal movement paths been taken into account in the floor plan of the academic library?	22	44
7.	-Does the current internal working environment of the academic library provide for the different needs of cooperative or individual education?	26	39
8.	-Are the appropriate interior furnishing items available in the academic library?	14	52
9.	-Does the current interior design achieve visual communication between the beneficiaries and a sense of familiarity?	15	50
10.	-Are the items of office furniture currently available in the academic library commensurate with its area?	8	57
11.	-Are the office furniture elements characterized by the aesthetic and functional aspects?	8	57
12.	-Is the furniture used flexible and easy to move and rearrange?	15	50
13.	-Were a variety of seating units used to suit the functional needs of the beneficiaries?	10	55
14.	-Did the lighting sources vary between functional and aesthetic lighting?	7	58

Table A1. *Cont.*

No.	Questions	Yes	No
15	-Were the natural light sources in the internal library space used efficiently?	9	56
16	-Are the wooden vertical dividers appropriate for dividing the internal spaces of the library?	12	52
17.	-Is the distribution of book and reference storage units proportional to the library space?	13	52
18	-Are the units for storing and keeping books and references appropriate from the design aspect and dimensions with the rest of the design elements?	13	52
19.	-Is there a reception desk that provides services and inquiries for students and faculty members?	26	41
20.	-Do you prefer developing the current interior design of the academic library?	54	11
21.	-Do the colors used help to feel active and desire to learn?	10	55
22.	-Does the current interior design have electronic research sources?	11	54

Appendix B

Objective: Information gathering about the proposed interior design for the academic library.

Description: The results of the responses of the sample of the faculty members and students in an opinion poll tool consisting of 22 questions about the proposed interior design for the academic library, College of Education, King Faisal University to reveal their opinions and needs.

Table A2. Opinion poll on the proposed interior design for the academic library, College of Education, King Faisal University from the point of view of the beneficiaries.

No.	Questions	Yes	No
1.	-Does the interior design of academic libraries have a major role in providing an attractive and effective learning environment for the beneficiaries?	67	0
2.	-Is the proposed interior design of the academic library commensurate with the requirements of the times?	67	0
3.	-Was the layout of the academic library properly planned?	66	1
4.	-Does the proposed interior design for the academic library help you encourage you to spend better and more time there?	66	1
5.	-Does the interior space of the academic library fit in design and functionality with the needs of the beneficiaries?	66	1
6.	-Have the internal movement paths been taken into account in the floor plan of the academic library?	66	1
7.	-Does the internal working environment of the academic library design proposal provide the different needs of cooperative or individual education?	66	1
8.	-Do the proposed design have the appropriate interior furnishing elements in the academic library?	63	4
9.	-Does the proposed interior design achieve visual communication between the beneficiaries and a sense of familiarity?	67	0
10.	-Are the office furniture elements in the design proposal for the academic library commensurate with its area?	64	3
11.	-Are the office furniture elements characterized by the aesthetic and functional aspects?	67	0
12.	-Is the furniture used flexible and easy to move and rearrange?	64	3
13.	-Were a variety of seating units used to suit the functional needs of the beneficiaries?	61	6
14.	-Did the lighting sources vary between functional and aesthetic lighting?	66	1

Table A2. *Cont.*

No.	Questions	Yes	No
15	-Were the library's internal natural lighting sources used efficiently?	65	2
16	-Are the vertical glass partitions suitable for dividing the internal spaces of the library?	67	0
17.	-Is the distribution of book and reference storage units proportional to the library space?	66	1
18	-Are the units for storing and keeping books and references appropriate from the design aspect and dimensions with the rest of the design elements?	65	2
19.	-Is the reception desk appropriate by design to provide services and inquiries to students and faculty members?	67	0
20.	-Is the proposed interior design for the academic library commensurate with your future aspirations?	65	2
21.	-Do the colors used in the proposed design help to feel active and desire to learn?	61	6
22.	-Does the proposed interior design of the academic library have electronic research sources?	67	0

References

1. Obille, K.L.B. An evaluation of standards for academic libraries in the Philippines. *J. Philipp. Librariansh.* **2007**, *27*, 109–150.
2. Sufar, S.; Talib, A.; Hambali, H. Towards a better design: Physical interior environments of public libraries in peninsular Malaysia. *Procedia-Soc. Behav. Sci.* **2012**, *42*, 131–143. [CrossRef]
3. Rizzo, J.C. Finding your place in the information age library. *New Libr. World* **2002**, *103*, 457–466. [CrossRef]
4. Beard, J.; Dale, P. Library design, learning spaces, and academic literacy. *New Libr. World* **2010**, *111*, 480–492. [CrossRef]
5. Cunningham, H.V.; Tabur, S. Learning Space Attributes: Reflections on Academic Library Design and Its Use. *J. Learn. Spaces* **2012**, *1*, n2.
6. Jochumsen, H.; Rasmussen, C.H.; Skot-Hansen, D. The four spaces—A new model for the public library. *New Libr. World* **2012**, *113*, 586–597. [CrossRef]
7. Narum, J.L. A Guide: Planning for Assessing 21st Century Spaces for 21st Century Learners, Learning Spaces Collaboratory. *Learn. Spaces Collab.* **2013**, 1–20. Available online: https://scholar.harvard.edu/files/anu/files/lscguide_pennlibraries.pdf?m=1526310820 (accessed on 20 March 2022).
8. Clugston, V. The interior designer's view. In *Better Library and Learning Space: Projects, Trends and Ideas*; Facet Publishing: London, UK, 2013; pp. 250–257.
9. Li, L.H.; Wu, F.; Su, B. Impacts of library space on learning satisfaction—An empirical study of university library design in Guangzhou, China. *J. Acad. Librariansh.* **2018**, *44*, 724–737. [CrossRef]
10. Mccabe, G.B. *Planning the Modern Public Library Building*; Libraries Unlimited: Westport, CT, USA, 2003.
11. Williams, C.; Budd, J.M. The Academic Library: Its Context, Its Purpose, and Its Operation. Englewood, Colo.: Libraries Unlimited, 1998. 372p. $58 cloth (ISBN 156308614X); $35 paper (ISBN 1563084570). LC 97-35962. *Coll. Res. Libr.* **1998**, *59*, 483–484. [CrossRef]
12. Zaugg, H.; Graham, C.R.; Lim, C.P.; Wang, T. Current and Future Directions of Blended Learning and Teaching in Asia. In *Blended Learning for Inclusive and Quality Higher Education in Asia*; Springer: Singapore, 2021; pp. 301–327.
13. Staines, G. *Universal Design: A Practical Guide to Creating and Re-Creating Interiors of Academic Libraries for Teaching, Learning, and Research*; Elsevier: Amsterdam, The Netherlands, 2012.
14. Pace, A. 21st century library systems. *J. Libr. Adm.* **2009**, *49*, 641–650. [CrossRef]
15. Gueye, B. World Library and Information Congress: 69th IFLA General Conference and Council. 1996. Available online: http://webdoc.sub.gwdg.de/ebook/aw/2003/ifla/vortraege/iv/ifla69/prog03.htm (accessed on 20 March 2022).
16. Gupta, J. Types of Libraries. Available online: https://www.kuk.ac.in/userfiles/file/distance_education/Year-2011-2012/B_Lib%20lecture%201.pdf (accessed on 25 February 2022).
17. Lance, K.C. Community college libraries and librarians and student success: A North Carolina study. *Community Jr. Coll. Libr.* **2016**, *22*, 103–124.
18. Satija, M.P.; Singh, P.; Chander, H. The Roadmap for Indian Academic Libraries in the Twentyfirst Century. *Indian J.* **2017**, *7*, 1–12.
19. Calvert, P. The Next Library Leadership: Attributes of Academic and Public Library Directors. *Electron. Libr.* **2004**, *22*, 191–192. [CrossRef]
20. Price, C.; Pierce, V. The Modern Academic Library: Space to Learn. *South Carol. Libr.* **2018**, *3*, 3.
21. Karaseva, L.V. Traditions, and innovations in lighting of libraries. In *IOP Conference Series: Materials Science and Engineering*; IOP Publishing: Bristol, UK, 2019; p. 033015.

22. Lange, J.; Miller-Nesbitt, A.; Severson, S. Reducing noise in the academic library: The effectiveness of installing noise meters. *Libr. Hi Tech* **2016**, *34*, 45–63. [CrossRef]
23. Brown, C.R. *Interior Design for Libraries: Drawing on Function & Appeal*; American Library Association: Chicago, IL, USA, 2002.
24. Choy, F.C.; Goh, S.N. A framework for planning academic library spaces. *Libr. Manag.* **2016**, *37*, 13–28. [CrossRef]
25. Uddin, M.J.; Hasan, M.N. Use of information technology in library service: A study on some selected libraries in Rajshahi District of Bangladesh. In *International Conference on Asian Digital Libraries*; Springer: Berlin/Heidelberg, Germany, 2011; pp. 385–389.
26. Halim, S.; Tulistyantoro, L.; Wulandari, D. Library Interior Design for Digital Native Generation. *KnE Life Sci.* **2020**, *5*, 85–93.
27. Goswami, S. Furniture for Modern Libraries: A Study. *Int. J. Inf. Dissem. Technol.* **2021**, *11*, 177–180. [CrossRef]
28. Sato, T.; Kishimoto, T.; Yamada, T. Relationship between visitors' movement path, staying activity, and spatial structure in the library as a "third place": Focusing on Yamanashi prefectural library. In Proceedings of the 11th International Space Syntax Symposium, SSS 2017, Lisbon, Portugal, 3–7 July 2017; pp. 26.1–26.16.

designs

MDPI

Article

How Sustainability-Related Information Affects the Evaluation of Designs: A Case Study of a Locally Manufactured Mobile Tiny House

Chiara Nezzi, Laura Ruiz-Pastor, Stefania Altavilla, Aurora Berni and Yuri Borgianni *

Faculty of Science and Technology, Free University of Bozen-Bolzano, 39100 Bolzano, Italy; chiara.nezzi@unibz.it (C.N.); lauramaria.ruizpastor@unibz.it (L.R.-P.); stefania.altavilla@unibz.it (S.A.); aurora.berni@natec.unibz.it (A.B.)

* Correspondence: yuri.borgianni@unibz.it

Abstract: Sustainability-related information affects people's choices and evaluation. The literature has made significant efforts to understand the best ways of delivering this kind of information to shape consumer behavior. However, while most studies have focused on packaged products and direct information provided through eco-labels, preferences could be formed differently in other design domains. The paper investigates the effect of the perceived amount of indirect information on the evaluation of an architectural artefact. A sample of 172 participants visited a locally produced mobile tiny house, made with a considerable amount of sustainable materials. The same participants answered a questionnaire about their perceived knowledge, quality, appropriateness and sustainability of the tiny house. The general level of knowledge of the tiny house was used as a proxy of the amount of indirect information received. Although the knowledge of the tiny house was generally low, ratings regarding the other dimensions were overall extremely positive. In particular, no evident relation was found between knowledge of the tiny house and sustainability, while the latter is significantly linked to quality aspects. These outcomes deviate from the evidence from other studies; this might be due to indirect vs. direct information and the peculiarity of the study carried out in the field of buildings. The gathered demographic and background data of the participants make it possible to highlight the role played by gender and age in affecting the evaluations, but the absence of a significant impact of experience in the field, education and origin. The results are compared with findings related to the evaluation of sustainable products and green buildings in particular.

Keywords: sustainability; indirect information; awareness; buildings; eco-design; consumer behavior; background

Citation: Nezzi, C.; Ruiz-Pastor, L.; Altavilla, S.; Berni, A.; Borgianni, Y. How Sustainability-Related Information Affects the Evaluation of Designs: A Case Study of a Locally Manufactured Mobile Tiny House. *Designs* **2022**, *6*, 57. https://doi.org/10.3390/designs6030057

Academic Editor: Farshid Aram

Received: 28 April 2022
Accepted: 14 June 2022
Published: 16 June 2022

1. Introduction

The success of sustainable products or solutions in the market is strongly affected by consumers' behavioural attitudes and individual environmental sensitivity [1]. Design for sustainability supports the development of sustainability-related product features, so to favour these pro-environmental preferences and attitudes [2]. Therefore, a strong contribution is expected by designers and product developers, who can follow eco-design directions to guide people towards more sustainable choices. Otherwise said, the design process should be steered to meet the goal of ecologising product ideas without falling into mere greenwashing [3].

As a result, the foci of design for sustainability include the development of sustainable features at the product level that can arouse awareness at an individual or collective level. However, tailored design methods to affect those behaviours are not mature yet, as witnessed by the fact that the systematic integration of people's behaviour into design processes for sustainable products has only recently been outlined [4].

Among the first attempts to affect consumer behaviour, priming [5,6] is an experimental design technique adaptable to a sustainability-related context. The use of visual primers could indeed facilitate positive outcomes from the consumers' decision-making processes regarding the evaluation or choice of sustainable products over standard alternatives. MacDonald and She [7] proposed a model to enhance a proper consumer-driven approach and provide designers with sustainability-oriented recommendations to exploit the full potential of the use of sustainable information. A similar approach was adopted in [8], where Shades of Green were proposed to help designers, researchers and industries in structuring their sustainability communication.

It should also be highlighted that social awareness in terms of sustainability could be enhanced not only by priming and steering people's preferences, but also by spreading knowledge about sustainability. According to [9], a proper diffusion of knowledge about sustainability issues could indeed affect consumers' perception of their individual impact on the environment, possibly raising their willingness to make more informed and ecological choices. As a consequence, considering sustainability information in design processes is critical to the success of sustainable products on the market [10]. The impact of elicited sustainability information on consumers' thoughts and choices has been widely studied by many design scholars. Shared conclusions can be summarised as follows.

First, clear sustainable information can positively affect people's sustainable behaviours, e.g., buying a sustainable product [11,12]. Consumers tend to evaluate sustainable products more favourably if a sustainable solution or advertisement is clearly presented [13]. According to [14], consumers are more prone to contribute actively to sustainable development with such clear and informed sustainability knowledge. These conclusions are consistent with [15], which stated that transparent sustainable product features touch the social and moral responsibility of consumers, hence involving them in the greening of consumption.

Second, making sustainability-related product features apparent results in a more eco-conscious product evaluation [16]. The intentional omission of key sustainability information could result in an inaccurate communication of the environmental benefits of a product, leading to its failure on the market. Declared or implicit omission should only be adopted for customary product choices, which sustainable products do not usually range within, e.g., [17].

Third, the consideration of sustainability-related performance in the early stages of the design process commonly leads to a raised social and moral awareness about sustainability issues [18].

However, it is worth stressing that the provision of sustainability-related information is controlled in most studies, especially in the design field. This situation is not common in real-world scenarios, especially in some domains such as the building industry (see Section 2 for more details). Designers might face situations in which the evaluation of their developed products is affected by uncontrolled circumstances and exchanges of information, e.g., [19], including sustainability-related performance. The paper swivels on this core aspect and originally investigates the impact of the uncontrolled provision of knowledge on people's evaluation, along with a number of factors that are more traditionally considered and manipulated. The experiment in the present study regards a prototype of a green building as a paradigm of a newly developed product benefitting from people's evaluation. In the experiment, information about the prototype's features was provided to participants in an uneven, uncontrolled and indirect way.

The specific objectives and research questions are presented in Section 2, following a literature review. The subsequent sections are as follows. After an overview of the methodology (Section 3), the results are shown in Section 4. While Section 5 includes discussions, the conclusions are drawn in Section 6.

2. Literature Review

2.1. Forms of Sustainable Information and Other Factors Affecting Preference and Choice of Products

As reported in the literature, sustainability information can be presented and communicated mainly in direct and indirect forms.

As regards direct forms, the sustainable features of a product can be made visible through visual cues, the use of which has been deeply investigated in the literature regarding their actual effectiveness and impact on consumers' perception, e.g., [20]. A typical example of direct forms is represented by eco-labels, which represent an intuitive and easy-to-construct way of sharing knowledge visually. They constitute a functioning logic to indicate the enhanced environmental performance of sustainable products over their standardised versions. As a result, their contribution to sustainable development has been extensively addressed by researchers [21]. Studies on the impact of sustainability labelling on purchase intentions and the quality perception of products can be found in [22,23]. D'Souza and colleagues [24] developed a framework of cognitive perspectives of sustainable products to possibly increase the impact and effectiveness of unclear or difficult to understand eco-labels. Indeed, sustainability information should always be coupled with direct and empathic effects on consumers' attitudes and choices. In this regard, an eye-tracking experiment was conducted in [25] to detect the amount of visual attention given by participants to different eco-labels. They showed that labels generated a perception of compliance with sustainable aspects in most participants. Therefore, it is imperative for eco-labels to connect with consumers' individual environmental perception. An effective improvement of their use could focus on enhancing their visibility and their diffusion in consumers' habitual purchasing habits [20].

Indirect or diffused forms of sharing knowledge could be adopted too—placing eco-labels on packaging cannot apply to all forms and categories of designs. In-depth studies about this typology of sustainability-related information sharing are fewer. An example is presented in [3], where participants were provided with a set of information about different typologies of the same product, which did not directly refer to their environmental impact. Therefore, the sustainability-related characteristics should have been inferred here. Another example is [26], which deals with the communication of sustainable information related to a university campus by managing sustainability communication based on a combination of marketing strategies and surveys' submissions. Similarly, Saber and Weber [27] investigated the sustainability communication in supermarkets and retailers.

Beyond the presentation of information through design, it has already been highlighted in Section 1 that collective and individual factors play a role too.

As for collective and social factors, research has paid great attention to cultural aspects; cases in point are [28,29]. A more complex network of contextual and social factors affecting design evaluation is illustrated in [30].

In relation to individual aspects, a large amount of literature has focused on the background and demographic data of the evaluators, e.g., [31–33]. Liu and colleagues [34] show the psychological factors affecting the choice of green buildings. Great emphasis is given here to environmental attitude and general knowledge. In relation to personal knowledge and understanding, the effects on product evaluation are also studied in terms of familiarity [35], expectations [36,37], and, more broadly, values [38]. Here, it is to be highlighted that the concept of knowledge differs from mere information acquisition, e.g., [39]. In most of the studies illustrated above with regard to forms of sustainability-related communication, it is hard to infer if the provided information has been properly processed by individuals.

2.2. Perception and Evaluation in the Field of Constructions and the Built Environment

The previous subsection has shown that forms of information delivery might differ and that the effect of its elaboration is hard to study. Most research is conducted in simulated laboratory settings and the validity of results might be challenged in real-world scenarios. It can be also hypothesised that different design domains have typical ways of exposing potential consumers to products and offering information. In the present paper, particular attention is paid to the building industry. This domain lends itself to the development of designs where direct (sustainability-related) information is hardly included in the representation of the products, which is seldom considered according to the previous subsection.

Table 1 summarises how previous literature has dealt with the evaluation of buildings and the built environment (markedly urban spaces) where sustainable aspects were involved. The columns of Table 1 report the following aspects, respectively.

- The considered source.
- How participants have interacted with the designs so as to compare the case studies with real-world situations and to infer how they could shape their evaluation.
- The methods used to extract perception and evaluation data.
- The size of the sample and characteristics of participants.
- Additional critical information about the way the studies have been conducted and other relevant data, markedly about knowledge, collective and individual aspects.

Table 1. Summary of extant literature contributions dealing with the evaluation of sustainable designs within buildings and the built environment.

Source	Representation and Interaction Mode	Investigation Method	Participants' Sample	Research Protocol and Relevant Data
[40]	Real user experience collected in traditional and modern buildings in Cameroon	Questionnaire	1750 questionnaires were answered depending on the residents' geographical area	- the participants were aware of the characteristics of the residential area - data about socio-demographic questions and residential experience was collected
[41]	Rendering and 3D models of a future neighborhood redevelopment project	Questionnaire	269 respondents selected in a specific neighborhood	- the participants were not aware of the characteristics of the future buildings - data about socio-demographic questions and residential experience was collected - preferences evaluation was based on an 11-point rating scale
[42]	Rendering and picture of a post-industrial landscape renovation	Questionnaire	450 residents randomly selected	- the participants were aware of the characteristics of the residential area - data about socio-demographic questions and residential experience was collected
[34]	Experiences with green buildings	Online survey	342 residents (valid answers)	- the participants reported an experience a posteriori; no evaluation of a new design - data about background and some demographic factors were collected
[43]	Real-world experiments at the Department of Civil Engineering of the University of Aveiro, Portugal.	Questionnaire	150 random users among students, researchers, professors, and administrative staff	- direct questionnaires were administered with enquiries about water consumption behaviour and preferences - the participants were not aware of the characteristics of sustainable water consumption behaviours - no socio-demographic or background questions were asked

Table 1. *Cont.*

Source	Representation and Interaction Mode	Investigation Method	Participants' Sample	Research Protocol and Relevant Data
[44]	User experience of life in a temporary house	Questionnaire and interviews	32 families interviewed and 181 questionnaires collected	- the participants were aware of the characteristics of the buildings; - no socio-demographic or background analysis was performed - interview and questionnaire made to understand the importance of the space and the sustainability of the used material
[45]	Rendering and pictures of 32 scenario of built environment design	Questionnaire	752 respondents divided into three different groups of target population that included building users/building owners, road users	- the participants were not aware of the characteristics of the buildings - socio-demographic and background information was collected
[46]	Two real buildings for demonstration scopes	Interviews	61 participants with different degrees of experience	- the participants were aware of the characteristics of the buildings - 32 in-depth, semi-structured interviews with building professionals were conducted - 29 shorter interviews with building users were conducted
[47]	Pictures of six selected urban streets in the city of Seoul	Interviews	Six experts in public space, transportation, and behaviour	- the participants were not aware of the characteristics of the urban streets - no socio-demographic or background analysis was performed - sustainability, amenity, placeness and accessibility of the urban streets based on open questions were evaluated
[48]	Five scenarios (electricity production; vegetable; green roof implementation and rainwater harvesting) of future development of a residential area in Barcelona	Questionnaire	60 respondents selected among residents, experts and public institutions	- the participants were not aware of the characteristics of the residential area - no socio-demographic or background questions were asked - respondents used a 5-point ranking of the scenario considering sustainability, environmental, economic and social indicators
[49]	Rendering and pictures of 160 wood constructions	Questionnaire	159 respondents selected among wood construction users	- the participants were aware of the characteristics of the building - background and demographic questions were asked - open questions were added - sustainability and economic criteria were evaluated

2.3. Literature Gap and Objectives

The collected literature shows that:

- Most of the designs were evaluated by experts or people with a significant awareness of the contextual factors related to buildings and urban spaces. The evaluation of ordinary people is rarely dealt with, while their views are critical when it comes to preferences, choices and purchases.
- In a few cases, real-world situations were studied; among them, very peculiar aspects were focused upon, see [43]. Conversely, the experience found with designs, especially if they present innovative features, is much more reliable if actual artefacts are involved [50]. In the case of green buildings or sustainable-oriented architectural interventions, it is assumed that new characteristics are included to fulfil the requirements of increased sustainability.
- The knowledge of the presented designs is dissimilar across participants of different studies but poorly considered as a factor affecting evaluation and acceptance in those contributions with participants having different degrees of awareness.
- A large number of studies consider demographic and background data, which have proven critical to evaluations.

In this context, an appropriate study for the evaluation of a green building should include:

- The use of a real building or realistic prototype even though this may be inconvenient because of the size of artefacts in the construction industry.
- The involvement of ordinary people.
- The consideration of people's knowledge and its effect on evaluations and perception, where information is provided in an indirect way, which is more realistic in a real-case scenario.
- The consideration of people's background and demographic data, along with their effect on evaluations and perception.

The present study is organised consistently with the above requirements, where the consideration of bullets 1 to 3 represents an original aspect, since they are not found contextually in the extant literature.

The objectives of the paper are to study the following:

1. Ordinary people's overall perception and evaluation of a real green building;
2. To what extent the (perceived) knowledge of the properties of a sustainable building affects evaluation and perceived sustainability, which contribute to product desirability and choice as shown above;
3. The effect of background and demographic data on perception and evaluation;
4. The interplay between multiple evaluation criteria to get more insight into the evaluation phenomenon.

3. Methodology and Context of the Study

3.1. Research Design

The study was conducted within the project "Tiny FOP MOB—A Real World Laboratory made of wood and hemp travelling through the Vinschgau Valley". The acronym FOP MOB (in German, FOrschungs- und Praxis-MOBil) recalls the main objective of this research project, namely, to design and develop a mobile Real-world Laboratory (RwL). The concept of the RwL has gained traction in the last decade and is breaking into the design domain too [51]. The building prototype (hereinafter Tiny FOP MOB) is therefore a mobile tiny house that was designed and constructed to represent a sustainable example in the building sector because of the materials used and their local origin (see below).

The project scopes align with the pursuance of the research objectives and the prototype has been used as a case study to address the objectives. The research procedure is shown in Figure 1, where the major steps are indicated along with the paper's sections where these steps are described. In particular, an experiment involving volunteer participants was designed and conducted, made more apparent in the following subsections.

Participants were asked to visit the prototype and provide feedback by means of a questionnaire. Experiments were initially approved by the statistical office of the project leader (EURAC Research) and the ethical commission of the Free University of Bozen-Bolzano, Bolzano, Italy. Both these institutions are located in South Tyrol, a bilingual (German and Italian) region in Italy. Questionnaires' outcomes were subsequently collected to conduct a statistical analysis, intended to correlate variables of interest related to the evaluation of perception of the Tiny FOP MOB and relevant according to research objectives. A correlation analysis was carried out to capture the links across a large number of variables; regressions were not used, as the goal was not to predict evaluations based on potential causal factors. Correlation functions (Spearman, Somer's D) were chosen based on the involved variables—details are to be found in Section 4. As aforementioned, results were critically discussed (Section 5) with a focus on the effects of perceived knowledge, background and demographical data on the evaluation of the prototype, consistent with the research objectives.

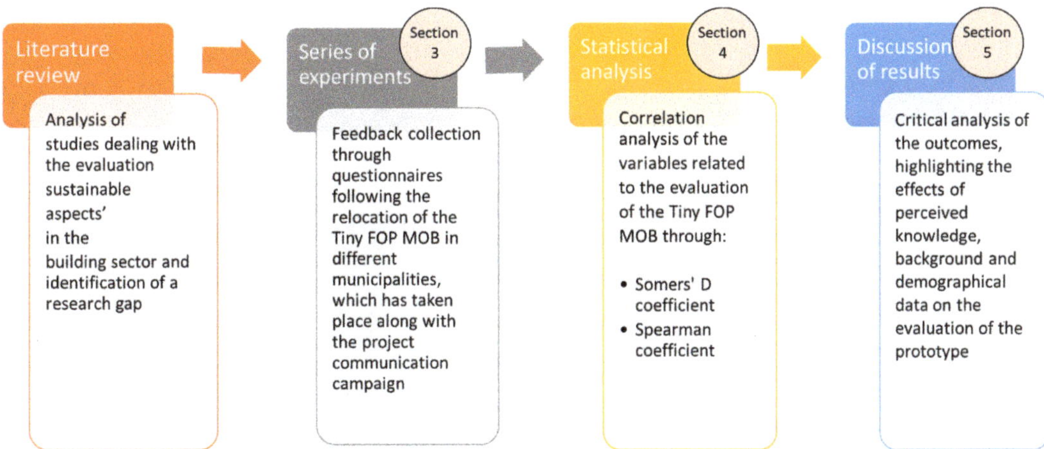

Figure 1. Research procedure schematised into steps and the following sections mostly devoted to illustrating such steps.

3.2. *Product and Characteristics Thereof*

The Tiny FOP MOB was designed and manufactured by two local companies of South Tyrol, Italy, more specifically of the Vinschgau Valley. The prototype was produced with natural materials, e.g., a mixture of hemp, lime and water for the building blocks, wood for external coatings and furniture elements. Sustainable peculiarities of the Tiny FOP MOB include the materials utilised, which are almost zero-kilometre, but also the entire process of development and construction. The environmental impact of the building blocks is CO_2 negative based on preliminary estimates. As a result, the Tiny FOP MOB was supposed to stimulate people to think about sustainability in the building sector and provided a concrete example thereby. Its use as a RwL further stressed its sustainable orientation from a social viewpoint.

The Tiny FOP MOB is mounted on a trailer, and, thanks to this feature, it could be transported in different towns of the Vinschgau Valley (see the next subsection). The prototype is illustrated in Figure 2 (outside) and Figure 3 (inside).

Figure 2. The Tiny FOP MOB prototype (exterior view).

Figure 3. The Tiny FOP MOB prototype (interior view).

3.3. *Relevant Information about the Project*

The mobility of the Tiny FOP MOB mirrors its scope to work as an RwL and matches its ability to spread knowledge about sustainability and related research outside major towns in South Tyrol, where research facilities are located. Consequently, a shared travel calendar was established in order to stop the Tiny FOP MOB in five pilot locations in the Vinschgau Valley, following consultations with local authorities. The prototype was moved to the localities of Schlanders, Mals, Latsch, Prad am Stilfserjoch and Graun im Vinschgau in the period of July–November, 2021. Additional tests were performed in spring 2022 in Schlanders where the Tiny FOP MOB is currently parked at the time of writing (June 2022). The prototype route throughout the Vinschgau Valley, as well as key information about the project's objectives, were described in newspaper articles by the local press, on the project website and in informative materials distributed in the above towns. Materials were made available in the surroundings of the prototype (in the form of informative posters) in line with the instructions reported in the next subsection. A number of initiatives about sustainability with specific reference to the building sector were organised too. It follows that participants' awareness of the project, the design requirements of the Tiny FOP MOB and, mainly, the delivery of information about sustainability-related aspects, could not be controlled and was uneven across participants. These circumstances justify the research objective (see Section 2) of understanding the impact of sustainability information shared at a social level, in indirect forms, and to different extents, across participants. In other words, the sustainability information addressed in this work can be indicated as diffuse and indirect, in contrast with the direct form provided, for instance, by eco-labels.

3.4. Participants and Relevant Aspects of the Experimental Procedure

The sample of convenience for this study was formed by 172 participants who took part in the experiment. Participants were directly recruited on the location sites of the prototype. All adults represented potential participants irrespective of their background, education, gender, reason for being in the Vinschgau Valley, e.g., residence, tourism, work. Participation was voluntary; all passers-by in the five localities in Vinschgau Valley were invited to visit the Tiny FOP MOB and answer the questionnaire (see below). Participants just had to confirm being at least 18 years old due to ethical and legal issues. After recruitment, participants were informed of the following guidelines and rules of conduct.

- Specific information about product peculiarities would have been given only after the visit unless this was explicitly requested to the experimenters before or during the visit; as well, paper-based or online informative material about the project and the Tiny FOP MOB was given to participants based on their requests.
- A scheduled timetable was not planned by providing participants with the chance to observe the prototype as long as they wanted and needed. If the Tiny FOP MOB was free at the time of recruitment, the visit could take place immediately.
- A limitation on the number of simultaneous visitors was imposed, together with the rule of wearing a mask inside the Tiny FOP MOB, due to the COVID-19 pandemic situation at the time of the experiments.

Background, personal and demographic data was also requested, which included:

- Gender;
- Age range (options: 18–30, 31–40, 41–50, 51–60, 61–70, 71+);
- Origin, markedly if the participant lived in a South Tyrolean municipality;
- Education (options: primary school, secondary school, high school, second-level vocational school, University degree, Ph.D.);
- Job.

Participants were left free to skip some answers if they believed the provided data could violate their privacy. Apart from missing information, the final sample was constituted by:

- 63 men and 79 women;
- 39 people aged 18–30; 24 people aged 31–40; 22 people aged 41–50; 38 people aged 51–60; 9 people aged 61–70; 8 people aged 71 or older;
- 84 South Tyroleans and 59 people whose residency was outside South Tyrol;
- 2 people with primary school; 10 people with secondary school; 44 people with high school; 17 people with vocational school; 54 people with an University degree; 15 people with a Ph.D. degree;
- 14 people working as architects, engineers, urban planners, entrepreneurs or managers in the building or wood industry, who could be considered experts in the field; 129 non-experts.

3.5. Questionnaires and Extracted Variables

As evident from Table 1, questionnaires have turned out to be the most used method to acquire information about people's evaluations, especially in the building industry. As aforementioned, questionnaires were also used in the present research.

The questionnaires were developed and used during the whole duration of the experiments in German and in Italian. Participants could choose to fill in questionnaires either paper-based or online. The main purpose of the questionnaire was to investigate comfort, perception of the Tiny FOP MOB and participants' opinions concerning global challenges beyond personal and background data. The total time needed to complete the questionnaire was approximately 15 min; no data potentially leading to the identification of respondents was collected.

With regard to the addressed objective of the paper, this subsection, along with the paper as a whole, presents the questionnaire's part (translated in English) concerning the perception of the quality and sustainability of the Tiny FOP MOB. The Institution the authors belong to was in charge of the design of this part of the questionnaire and analysed the corresponding results. This part of the questionnaire included 13 different statements (see Table 1) to be evaluated by means of a Likert scale ranging from 1 ("Totally disagree") to 5 ("Totally agree"). The statements in Table 2 are reported alongside the name of the corresponding variables that are manipulated in the next sections.

Table 2. Questionnaire's list of statements and variable names related to knowledge, perception, quality and sustainability.

Statement	Variable	Source or Precedent for Using the Variable
1. I knew the Tiny FOP MOB and the associated project before the visit.	Knowledge	[10]
2. The Tiny FOP MOB is a good quality product.	Quality	[10]
3. The Tiny FOP MOB represents a product to be preferred over other types and competing products.	Preference	[10]
4. The Tiny FOP MOB has many advantages over other types and competing products.	Advantages	[10]
5. The Tiny FOP MOB has no disadvantages compared to other types or competing products.	Lack of disadvantages	[10]
6. The Tiny FOP MOB is a creative and original product.	Creativity	[10]
7. The Tiny FOP MOB could be a branded product of South Tyrol.	Brand	[52]
8. I would willingly stay in the Tiny FOP MOB for a shorter or longer period of time.	Staying	[53]
9. The Tiny FOP MOB is a suitable building module to live in permanently.	Living	[53]
10. The Tiny FOP MOB is a suitable building module for organising small conferences/seminars.	Seminars	[53]
11. The Tiny FOP MOB is a building module that is suitable as a workplace.	Workplace	[53]
12. The Tiny FOP MOB is a suitable building module to spend the holidays in.	Holidays	[53]
13. The Tiny FOP MOB is a sustainable product.	Sustainability	[54,55]

The first statement directly enquires about the different levels of knowledge about the product and the project itself. The variable "Knowledge" has provided the authors with an indication of whether participants were aware of the project's characteristics and scopes, but without specific details about how they acquired this knowledge due to the project's limitations. This kind of information is strictly related to the need to pursue the second objective mentioned in Section 2.3.

The questions concerning quality (2 to 6) were taken from an already available list of statements [10], which survey the quality perception in the domain of sustainable products. These questions address a number of aspects aimed at forming choice and preference of sustainable products over alternatives. These questions target primarily the first objective mentioned in Section 2.3.

The questionnaire is integrated with bespoke statements related to the project and the investigated product, i.e.,

- Brand identity (7), as a measure of the product's capability to fit and represent the South Tyrolean territory, similar to [52];

- Suitability for different purposes (8 to 12) in order to assess the perceived capability of prototype to serve some of the scopes it could potentially fulfil, as in [53];
- Perception of sustainability (question 13), which is closely targeted in the present work because of its dealing with sustainable design. Precedents of investigating the perception of designs' sustainability can be found in [54,55].

The variables 7 to 13 can complement the quality perception in the pursuance of the paper's objectives, markedly the fourth one, reported in Section 2.3.

4. Results

4.1. Objective 1: Overall Perception and Evaluation of a Green Building

Beyond the participants' perception of knowledge, the questionnaire was intended to reveal a number of aspects related to the evaluation of the Tiny FOP MOB. Consistently with the objectives, ordinary people were involved in the evaluation, as the general description of the sample and the fact they were passers-by should suggest.

Table 3 includes the number and the frequency (in percentage terms) of answers for each variable. The median is also reported. Some participants skipped some questions; therefore, the totals of the rows slightly differ. A statistical analysis was performed with the software SPSS, PASW Statistics version 26 (IBM Corporation) by using data of each participant—the same software application was used for all the statistical analyses. Regarding the distribution of the results, the median and mode of the variable [Knowledge] is 1, which means that most of the participants seemingly had little to no information about the project before and during their participation in the experiment. As for the rest of the variables, the values are predominantly mainly between 3 and 5. While all these variables denote an overall positive perception of the Tiny FOP MOB, the high values attributed to [Sustainability] and [Creativity] are particularly noteworthy.

Table 3. Number of answers for each variable of the perception questionnaire reported along with percentages and medians.

Variable	Totally Disagree (1)	%	Somehow Disagree (2)	%	Indifferent (3)	%	Fairly Agree (4)	%	Totally Agree (5)	%	Median
1. Knowledge	91	55.5	12	7.3	30	18.3	13	7.9	18	11	1
2. Quality	1	0.6	2	1.9	6	3.7	76	46.9	77	46.1	4
3. Preference	2	1.2	5	4.3	45	28	68	42.2	41	24.6	4
4. Advantages	1	0.6	1	0.6	50	31.3	64	40	44	27.5	4
5. Lack of disadvantages	5	3.2	11	7.1	63	40.4	60	38.5	17	10.9	3
6. Creativity	2	1.2	5	3	8	4.8	55	33.1	96	57.8	5
7. Brand	3	1.8	9	5.5	27	16.6	61	37.4	63	38.7	4
8. Staying	3	1.9	10	6.3	31	19.4	64	40	52	32.5	4
9. Living	5	3	14	8.4	38	22.9	53	31.9	56	33.7	4
10. Seminars	4	2.4	9	5.5	17	10.4	70	42.7	64	39	4
11. Workplace	2	1.2	5	3	15	9	66	39.8	78	47	4
12. Holidays	1	0.6	5	3.1	19	11.7	49	30.1	89	54.6	5
13. Sustainability	2	1.2	0	0	3	1.9	36	22.2	121	74.7	5

4.2. Objective 2: Effect of the Perceived Knowledge on the Evaluation of the Prototype

Based on the paper's objectives, particular focus has been given to the relationship between the previous knowledge about the Tiny FOP MOB project [Knowledge] and the perception of the tiny house after the visit. This objective has been addressed by calculating the Spearman's correlation between [Knowledge] and the other variables. Here, this association function has been chosen because of its suitability for ordered variables.

Given the focus of the research, specific attention was paid to the effect of [Knowledge] on [Sustainability]. However, it emerged that the correlation between the variables [Knowledge] and [Sustainability] was 0.068. Hence, this correlation was almost absent according to Landis and Koch's [56] interpretation and non-significant ($p = 0.391$).

The correlations between [Knowledge] and the other variables were also weak and non-significant, as for example [Knowledge] and [Preference] (0.129).

4.3. Objective 3: Effect of Background and Demographic Data on the Evaluation of the Prototype

The individual and demographic data described in Section 3.4 were used as background variables. They were correlated with the perception variables described in Section 3.5 in order to study how the background of the participants affected the results.

The statistical functions used to analyze the correlations between these groups of variables were Somers' D coefficient and Spearman correlation, due to the nature of the variables. Somers' D was used to compare ordinal and nominal variables (gender, origin and job), while Spearman correlation has been used for pairs of ordered variables in compliance with [57]. The significant correlations found between the background and the evaluation variables can be seen in Table 4. The third column indicates the category or the property associated to the background variable for which larger values of the corresponding evaluation variables were obtained.

Table 4. Number of answers for each variable of the perception questionnaire reported along with percentages and medians.

Background Variable	Evaluation Variable	Direction (Increasing)	Strength of Correlation	p-Value
Gender	3 Preference	Woman	0.228	0.011
	6 Creativity		0.175	0.035
	7 Brand		0.237	0.008
	8 Staying		0.239	0.009
	9 Living		0.244	0.007
	10 Seminars		0.280	0.001
	11 Workplace		0.221	0.011
	12 Holidays		0.263	0.002
	13 Sustainability		0.147	0.045
Age	5 Lack of disadvantages	Younger	0.308	0.0003
	7 Brand		0.248	0.003
	12 Holidays		0.269	0.001
	13 Sustainability		0.237	0.005

The gender of participants displayed the largest number of significant relationships with the variables. In all the nine cases for which significant correlations were found, women tended to assign higher values to evaluation variables. On the other hand, origin, job and education were not significantly associated with any of the evaluation variables, and, as such, are not present in Table 4. The age of the participants was significant for four of the variables studied: lack of disadvantages, brand, holidays and sustainability. In all these cases, younger people tended to provide higher evaluations to the Tiny FOP MOB.

4.4. Objective 4: Interplay among Evaluation Criteria

In order to compare the evaluation variables, a Spearman's Correlation analysis was performed, as they were all order variables. The results of the analysis are presented in Table 5, which includes the magnitude and the significance of the correlation. In this respect, the option was adopted of the SPSS software to flag the correlations at the p-value thresholds 0.01 (**) or 0.05 (*), as a common rule of thumb.

Table 5. Correlations among evaluation variables.

	2 Quality	**3 Preference**	**4 Advantages**	**5 Lack of Disadvantages**	**6 Creativity**	**7 Brand**	**8 Staying**	**9 Living**	**10 Seminars**	**11 Workplace**	**12 Holidays**	**13 Sustainability**
2 Quality	-											
3 Preference	0.516 **	-										
4 Advantages	0.422 **	0.717 **	-									
5 Lack of disadvantages	0.282 **	0.478 **	0.418 **	-								
6 Creativity	0.310 **	0.355 **	0.351 **	0.357 **	-							
7 Brand	0.175 *	0.383 **	0.341 **	0.242 **	0.406 **	-						
8 Staying	0.343	0.287	0.227	0.290	0.344	0.136	-					
9 Living	0.233 **	0.367 **	0.221 **	0.319 **	0.295 **	0.381 **	0.497 **	-				
10 Seminars	0.229 **	0.254 **	0.152	0.154	0.351 **	0.240 **	0.370 **	0.444 **	-			
11 Workplace	0.183 *	0.252 **	0.110	0.187 *	0.217 **	0.265 **	0.366 **	0.504 **	0.493 **	-		
12 Holidays	0.129	0.240 **	0.149	0.178 *	0.209 **	0.394 **	0.416 **	0.613 **	0.334 **	0.350 **	-	
13 Sustainability	0.340 **	0.414 **	0.316 **	0.236 **	0.475 **	0.267 **	0.297 **	0.305 **	0.194 *	0.304 **	0.243 **	-

No correlations across all the variables presented negative values showing the absence of inverse relations.

Through Figures 4–6, it was possible to gain insight into some of the strong correlations emerging in the analysis. These figures show, by means of the size of bubbles, the number of co-occurrences of values attributed to the considered pairs of variables. In particular, the highest correlation could be found between [Advantages] and [Preference] (0.717) (Figure 4), which led us to consider these quality dimensions possibly redundant, although they are, in principle, distinct.

The correlations of [Sustainability] with other variables were significant in many cases. The measures of association between [Sustainability] and the other evaluation variables varied between 0.194 ([Seminars]) and 0.475 ([Creativity]). The perception of sustainability was significantly correlated with all the other variables, especially with [Preference] (Spearman correlation 0.414) and [Creativity] (Spearman correlation 0.475) (Figures 5 and 6). Regarding the association between the variables concerning quality and appropriateness, they mainly exhibited slight to moderate correlation.

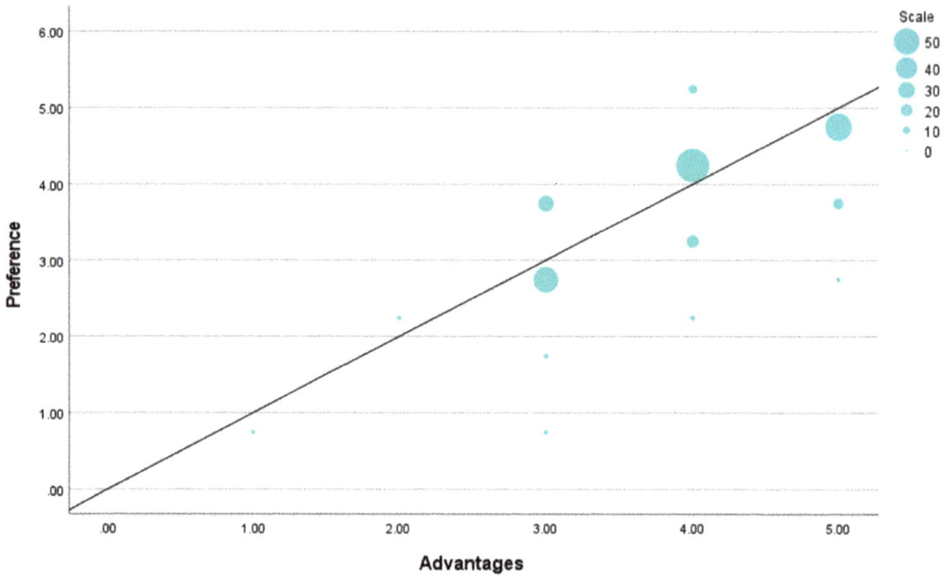

Figure 4. Relationship between Preference and Advantages.

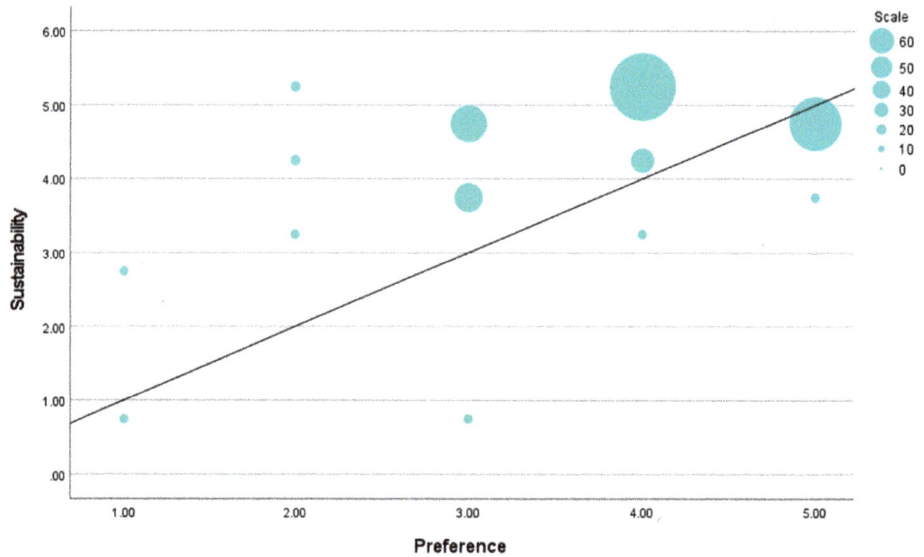

Figure 5. Relationship between Sustainability and Preference.

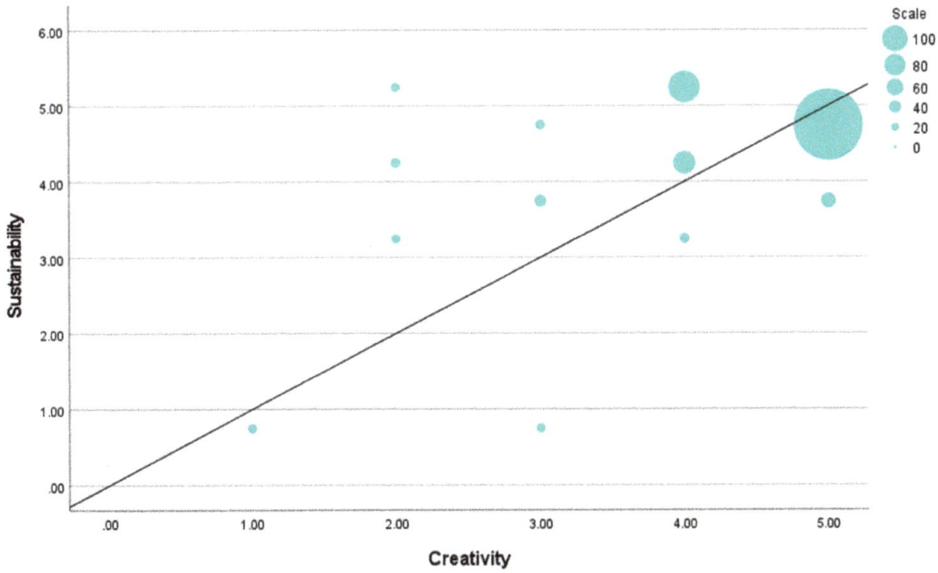

Figure 6. Relationship between Sustainability and Creativity.

5. Discussions

The present section is devoted to critically discussing the outcomes of the study and suggesting their interpretation and implications. The original aspects of the study and the requirements for a thorough investigation of a green building, recalled in Section 2.3, prevent the comparison with very similar studies. Therefore, comparisons of the findings are largely affected by a large number of contextual factors and differences in the settings of the studies.

5.1. Objective 1: Overall Perception and Evaluation of a Green Building

As regards the first stated objective, the aim was to study ordinary people's overall perception of a physical building implementing sustainable characteristics. The results showed that the evaluation of the Tiny FOP MOB prototype was very positive across a considerable number of evaluation criteria ranging from identification of advantages to appropriateness, creativity and sustainability.

These results support the positive attitude towards green buildings across several factors that were found in [34], where participants were surveyed after living and having experienced green buildings. However, post-occupancy evaluations of green buildings do not always show high satisfaction levels, e.g., in [58]. Likely, the characteristics of what makes a building greener or more sustainable might play a role in this respect and more research is therefore needed. When switching the attention to sustainable products and other forms of designs, evaluations differ substantially as well, e.g., in [10].

With reference to the attainment of this objective, it must be nevertheless noted that the sample of participants was random, but the volunteer participation could make it poorly representative of "ordinary people". It could be hypothesised here that those willing to participate had an aprioristic positive view of the presented design, which justifies the high evaluations shown in Table 3. This might explain why the knowledge of the product acquired by participants did not affect their perception of sustainability and quality (see Section 5.3).

5.2. Objective 2: Effect of the Perceived Knowledge on the Evaluation of the Prototype

In relation to this focal research objective, the results conflict with past research findings (see the abundant literature presented in Sections 1 and 2.1) where the role of knowledge and awareness of the sustainable properties of products has largely affected the perception of sustainability and preferences. The possible causes of misalignment with previous studies are discussed below along with pointing out some peculiarities and limitations of the study.

- The aforementioned peculiar contextual factors, such as unevenness of information given to participants, knowledge possibly coming from different sources and in different modalities, etc., are candidates to explain the divergence of the presented results from previous work.
- While all the analysed answers are inherently subjective, this might particularly apply to knowledge, as each participant could have evaluated differently the amount of information processed and the lack of necessary knowledge to assess the prototype in a fully aware manner. In other words, the impossibility of verifying the metrics used by participants to provide [Knowledge] values represents an important limitation of the paper, besides being a difference with respect to most previous literature.
- An additional hypothesis is that a tiny house, clearly built with natural materials, might be considered per se a sustainable product; hence, details about the project, such as the planned use of the Tiny FOP MOB for an RwL or the origin of materials, could have poorly oriented evaluations. In other terms and with a closer look at the design research, the product considered could lend itself to effective indirect communication of sustainable aspects.

5.3. Objective 3: Effect of Background and Demographic Data on the Evaluation of the Prototype

With regard to the role played by factors concerning the participants in the experiment, few significant correlations were found. The effect of demographic factors emerged overall as greater than the impact of background factors, such as experience in the field and education.

On the one hand, the fact that gender and age affected the evaluation of a green building is in line with [59], where experience was targeted. Additional similarities with [59] include the fact that correlations with occupation were not found and that more positive evaluations were typically given by women and younger people. Nevertheless, while the factors significantly affected by age and gender differed, the results of the present study showed that some impacted factors were shared here, notably including the perception of sustainability.

On the other hand, a major dissimilarity of the present study with respect to past literature and a counterintuitive aspect concerns the undetected effect of background variables. Different results can be found markedly in [34,58].

As an additional factor considered despite the absence of significant correlations, people's origin cannot be considered as a factor underlying strong cultural differences (all the participants were European with Italians, Germans and Austrians constituting nearly the whole sample). Therefore, it would be inappropriate to compare the achieved results with studies treating the effects of cultural aspects.

5.4. Objective 4: Interplay among Evaluation Criteria

The variety of questions and the considerable number of participants allow further reflections beyond the relation between knowledge and perceived sustainability. As [Sustainability] has proved to be largely correlated with quality and appropriateness variables, one might conclude that sustainability is increasingly considered as a necessary design requirement, at least for a subset of people. A consequence might be that the frequent design need to find a trade-off between quality and sustainability is nowadays alleviated. The data gathered in this work conversely suggest that sustainability is a prerequisite for quality and positive experiences. The strong relation between [Creativity], a plainly important dimension in the design field, and [Sustainability] has already been underlined. This link is confirmed by abundant literature stating that people more prone to novelty are also expected to welcome changes driven by sustainability requirements [60].

As far as quality variables are concerned, the degrees of correlation have resulted much larger than in other studies where different evaluation criteria were compared against each other, for instance, in [10]. More research is needed in this area in order to standardize evaluation procedures for sustainable products and use a comprehensive and non-redundant number of questions for characterizing the participants' experience.

6. Conclusions

The paper has investigated the relationship between the perceived knowledge and sustainability of a sustainable product along with other evaluation criteria. From a methodological point of view, the objective was pursued through a questionnaire, which included the quality evaluation of a tiny house prototype, followed by correlation analysis. As aforementioned, sustainable information was provided in an indirect and diffuse form, differently from most cases presented in the literature, which deal with a direct form of sustainability-related information, e.g., eco-labels (see Section 2).

In relation to the four objectives declared in the present study, the most significant findings are listed below:

1. The tiny house received consistently positive evaluations concerning its perceived quality, creativity, appropriateness and sustainability. The majority of evaluators were randomly selected volunteers with a limited number of experts in the field. As such, the involved sample could be considered as well representative of a group of ordinary people despite the participants' likely intrinsic interest towards the product and their probable sustainable attitude.
2. Prior knowledge about the tiny house and the project within which it was designed, developed and built played no evident role in the evaluations.
3. People's background did not affect evaluations significantly either. In contrast, some evaluation variables were affected by gender and age, where women and younger people overall rated the tiny house better in terms of sustainability and other factors.
4. The chosen evaluation criteria were shown to be significantly correlated with a remarkable association between perceived sustainability vs. preference and creativity.

Future work includes the attempt to overcome the limitations mentioned in Section 5, as well as to launch new experiments to understand the role of contextual factors and methodological choices. In particular, an experiment in a laboratory environment is ongoing where the visit of the physical prototype is substituted by a virtual tour and the provision of information is controlled. Possible differences in the results will be used to assess the extent to which user experience changes in a virtual environment (which was assumed in the present paper) and the effect of changing the way information is supplied to participants.

Author Contributions: Conceptualization, C.N., S.A. and Y.B.; methodology, S.A. and Y.B.; validation, A.B. and L.R.-P.; formal analysis, C.N. and A.B.; investigation, C.N., A.B., L.R.-P. and Y.B.; data curation, S.A. and L.R.-P.; writing—original draft preparation, C.N., A.B. and L.R.-P.; writing—review and editing, S.A. and Y.B.; supervision, Y.B.; project administration, Y.B.; funding acquisition, Y.B. All authors have read and agreed to the published version of the manuscript.

Funding: The work is fully supported by the project "Tiny FOP MOB—A Real World Laboratory made of wood and hemp travelling through the Vinschgau Valley", funded by the European Regional Development Fund (ERDF) Investment for Growth and Jobs Programme 2014–2020, 5th call in the Axis 1 "Research and Innovation" of the Autonomous Province of Bolzano-South Tyrol, grant FESR1161.

Data Availability Statement: Not applicable.

Acknowledgments: The authors would like to thank the project partners, and markedly the colleagues from the Center of Advanced Studies at Eurac Research, for their contribution to the present research and the collection of data.

Conflicts of Interest: The authors declare no conflict of interest.

References

1. She, J.; MacDonald, E.F. Exploring the Effects of a Product's Sustainability Triggers on Pro-Environmental Decision-Making. *J. Mech. Des.* **2017**, *140*, 011102. [CrossRef]
2. MacDonald, E.; Whitefoot, K.; Allison, J.T.; Papalambros, P.Y.; Gonzalez, R. An Investigation of Sustainability, Preference, and Profitability in Design Optimization. In Proceedings of the ASME 2010 International Design Engineering Technical Conferences and Computers and Information in Engineering Conference, Montreal, QC, Canada, 15–18 August 2010; pp. 715–728. [CrossRef]
3. Goucher-Lambert, K.; Cagan, J. The Impact of Sustainability on Consumer Preference Judgments of Product Attributes. *J. Mech. Des.* **2015**, *137*, 081401. [CrossRef]
4. Balikci, A.; Borgianni, Y.; Maccioni, L.; Nezzi, C. A Framework of Unsustainable Behaviors to Support Product Eco-Design. *Sustainability* **2021**, *13*, 11394. [CrossRef]
5. Mandel, N.; Johnson, E. When Web Pages Influence Choice: Effects of Visual Primes on Experts and Novices. *J. Consum. Res.* **2002**, *29*, 235–245. [CrossRef]
6. She, J.; MacDonald, E. Priming Designers to Communicate Sustainability. *J. Mech. Des.* **2013**, *136*, 011001. [CrossRef]
7. MacDonald, E.F.; She, J. Seven cognitive concepts for successful eco-design. *J. Clean. Prod.* **2015**, *92*, 23–36. [CrossRef]
8. Turunen, L.L.M.; Halme, M. Communicating actionable sustainability information to consumers: The Shades of Green instrument for fashion. *J. Clean. Prod.* **2021**, *297*, 126605. [CrossRef]
9. Leire, C.; Thidell, A. Product-related environmental information to guide consumer purchases—A review and analysis of research on perceptions, understanding and use among Nordic consumers. *J. Clean. Prod.* **2005**, *13*, 1061–1070. [CrossRef]
10. Maccioni, L.; Borgianni, Y.; Basso, D. Value Perception of Green Products: An Exploratory Study Combining Conscious Answers and Unconscious Behavioral Aspects. *Sustainability* **2019**, *11*, 1226. [CrossRef]
11. Shao, J.; Li, W.; Aneye, C.; Fang, W. Facilitating mechanism of green products purchasing with a premium price—Moderating by sustainability-related information. *Corp. Soc. Responsib. Environ. Manag.* **2021**, *29*, 686–700. [CrossRef]
12. Chiu, M.-C.; Tu, Y.-L.; Kao, M.-C. Applying deep learning image recognition technology to promote environmentally sustainable behavior. *Sustain. Prod. Consum.* **2022**, *31*, 736–749. [CrossRef]
13. Cho, Y.-N. Different Shades of Green Consciousness: The Interplay of Sustainability Labeling and Environmental Impact on Product Evaluations. *J. Bus. Ethics* **2014**, *128*, 73–82. [CrossRef]
14. Stöckigt, G.; Schiebener, J.; Brand, M. Providing sustainability information in shopping situations contributes to sustainable decision making: An empirical study with choice-based conjoint analyses. *J. Retail. Consum. Serv.* **2018**, *43*, 188–199. [CrossRef]
15. Goucher-Lambert, K.; Moss, J.; Cagan, J. Inside the Mind: Using Neuroimaging to Understand Moral Product Preference Judgments Involving Sustainability. *J. Mech. Des.* **2017**, *139*, 041103. [CrossRef]
16. Cho, Y.-N.; Soster, R.L.; Burton, S. Enhancing Environmentally Conscious Consumption through Standardized Sustainability Information. *J. Consum. Aff.* **2017**, *52*, 393–414. [CrossRef]
17. Wang, J.; Shen, M.; Chu, M. Why is green consumption easier said than done? Exploring the green consumption attitude-intention gap in China with behavioral reasoning theory. *Clean. Responsible Consum.* **2021**, *2*, 100015. [CrossRef]
18. Woodhouse, A.; Davis, J.; Pénicaud, C.; Östergren, K. Sustainability checklist in support of the design of food processing. *Sustain. Prod. Consum.* **2018**, *16*, 110–120. [CrossRef]
19. Raghupathi, D.; Yannou, B.; Farel, R.; Poirson, E. Customer sentiment appraisal from user-generated product reviews: A domain independent heuristic algorithm. *Int. J. Interact. Des. Manuf.* **2015**, *9*, 201–211. [CrossRef]
20. Song, L.; Lim, Y.; Chang, P.; Guo, Y.; Zhang, M.; Wang, X.; Yu, X.; Lehto, M.R.; Cai, H. Ecolabel's role in informing sustainable consumption: A naturalistic decision making study using eye tracking glasses. *J. Clean. Prod.* **2019**, *218*, 685–695. [CrossRef]
21. Iraldo, F.; Griesshammer, R.; Kahlenborn, W. The future of ecolabels. *Int. J. Life Cycle Assess.* **2020**, *25*, 833–839. [CrossRef]
22. Chekima, B.; Wafa, S.A.W.S.K.; Igau, O.A.; Chekima, S.; Sondoh, S.L., Jr. Examining green consumerism motivational drivers: Does premium price and demographics matter to green purchasing? *J. Clean. Prod.* **2016**, *112*, 3436–3450. [CrossRef]
23. De Andrade Silva, A.R.; Bioto, A.S.; Efraim, P.; de Castilho Queiroz, G. Impact of sustainability labeling in the perception of sensory quality and purchase intention of chocolate consumers. *J. Clean. Prod.* **2017**, *141*, 11–21. [CrossRef]
24. D'Souza, C.; Taghian, M.; Lamb, P. An empirical study on the influence of environmental labels on consumers. *Corp. Commun. Int. J.* **2006**, *11*, 162–173. [CrossRef]
25. Pérez-Belis, V.; Agost, M.J.; Vergara, M. Consumers' visual attention and emotional perception of sustainable product information: Case study of furniture. In Proceedings of the 7th International Conference on Kansei Engineering and Emotion Research 2018, Kuching, Malaysia, 19–22 March 2018; Springer: Singapore, 2018; pp. 239–248. [CrossRef]
26. Franz-Balsen, A.; Heinrichs, H. Managing sustainability communication on campus: Experiences from Lüneburg. *Int. J. Sustain. High. Educ.* **2007**, *8*, 431–445. [CrossRef]

27. Saber, M.; Weber, A. How do supermarkets and discounters communicate about sustainability? A comparative analysis of sustainability reports and in-store communication. *Int. J. Retail Distrib. Manag.* **2019**, *47*, 1181–1202. [CrossRef]
28. Sauer, J.; Sonderegger, A.; Álvarez, M.A.H. The influence of cultural background of test participants and test facilitators in online product evaluation. *Int. J. Human-Comput. Stud.* **2018**, *111*, 92–100. [CrossRef]
29. Lin, C.-L.; Chen, S.-J.; Hsiao, W.-H.; Lin, R. Cultural ergonomics in interactional and experiential design: Conceptual framework and case study of the Taiwanese twin cup. *Appl. Ergon.* **2016**, *52*, 242–252. [CrossRef]
30. Carbon, C.C. Psychology of design. *Des. Sci.* **2019**, *5*, E26. [CrossRef]
31. Park, S.J.; Choi, S.; Kim, E.J. The relationships between socio-demographic variables and concerns about environmental sustainability. *Corp. Soc. Responsib. Environ. Manag.* **2012**, *19*, 343–354. [CrossRef]
32. Edinger-Schons, L.M.; Sipilä, J.; Sen, S.; Mende, G.; Wieseke, J. Are Two Reasons Better Than One? The Role of Appeal Type in Consumer Responses to Sustainable Products. *J. Consum. Psychol.* **2018**, *28*, 644–664. [CrossRef]
33. Liu, C.-H. Conducting qualitative and quantitative analyses of sustainable behaviour. *J. Retail. Consum. Serv.* **2021**, *60*, 102474. [CrossRef]
34. Liu, Y.; Hong, Z.; Zhu, J.; Yan, J.; Qi, J.; Liu, P. Promoting green residential buildings: Residents' environmental attitude, subjective knowledge, and social trust matter. *Energy Policy* **2018**, *112*, 152–161. [CrossRef]
35. Blanco, C.F.; Sarasa, R.G.; Sanclemente, C.O. Effects of visual and textual information in online product presentations: Looking for the best combination in website design. *Eur. J. Inf. Syst.* **2010**, *19*, 668–686. [CrossRef]
36. Pettersson, I. Travelling from fascination to new meanings: Understanding user expectations through a case study of autonomous cars. *Int. J. Des.* **2017**, *11*, 11.
37. Becattini, N.; Borgianni, Y.; Cascini, G.; Rotini, F. Investigating users' reactions to surprising products. *Des. Stud.* **2020**, *69*, 100946. [CrossRef]
38. Puska, P. Does Organic Food Consumption Signal Prosociality? An Application of Schwartz's Value Theory. *J. Food Prod. Mark.* **2018**, *25*, 207–231. [CrossRef]
39. Zins, C. Conceptual approaches for defining data, information, and knowledge. *J. Am. Soc. Inf. Sci. Technol.* **2007**, *58*, 479–493. [CrossRef]
40. Nematchoua, M.K.; Tchinda, R.; Orosa, J.A. Thermal comfort and energy consumption in modern versus traditional buildings in Cameroon: A questionnaire-based statistical study. *Appl. Energy* **2014**, *114*, 687–699. [CrossRef]
41. Verso, V.R.M.L.; Fregonara, E.; Caffaro, F.; Morisano, C.; Peiretti, G.M. Daylighting as the Driving Force of the Design Process: From the Results of a Survey to the Implementation into an Advanced Daylighting Project. *J. Daylighting* **2014**, *1*, 36–55. [CrossRef]
42. Loures, L.; Panagopoulos, T.; Burley, J.B. Assessing user preferences on post-industrial redevelopment. *Environ. Plan. B Plan. Des.* **2015**, *43*, 871–892. [CrossRef]
43. Meireles, I.; Sousa, V.; Adeyeye, K.; Silva-Afonso, A. User preferences and water use savings owing to washbasin taps retrofit: A case study of the DECivil building of the University of Aveiro. *Environ. Sci. Pollut. Res.* **2017**, *25*, 19217–19227. [CrossRef] [PubMed]
44. Hosseini, S.A.; Yazdani, R.; de la Fuente, A. Multi-objective interior design optimization method based on sustainability concepts for post-disaster temporary housing units. *Build. Environ.* **2020**, *173*, 106742. [CrossRef]
45. Adeel, A.; Notteboom, B.; Yasar, A.; Scheerlinck, K.; Stevens, J. Sustainable Streetscape and Built Environment Designs around BRT Stations: A Stated Choice Experiment Using 3D Visualizations. *Sustainability* **2021**, *13*, 6594. [CrossRef]
46. Blasberg, L.A. Business Models for Smart Sustainability: A Critical Perspective on Smart Homes and Sustainability Transitions. In *Business Models for Sustainability Transitions*; Palgrave Macmillan: Cham, Switzerland, 2021; pp. 273–302. [CrossRef]
47. Seo, M.; Kim, S. A Comparative Evaluation of Utility Value Based on User Preferences for Urban Streets: The Case of Seoul, Korea. *Sustainability* **2021**, *13*, 5073. [CrossRef]
48. Toboso-Chavero, S.; Madrid-López, C.; Durany, X.G.; Villalba, G. Incorporating user preferences in rooftop food-energy-water production through integrated sustainability assessment. *Environ. Res. Commun.* **2021**, *3*, 065001. [CrossRef]
49. Švajlenka, J.; Kozlovská, M. Perception of the Efficiency and Sustainability of Wooden Building. In *Efficient and Sustainable Wood-Based Constructions*; Springer: Berlin/Heidelberg, Germany, 2022; pp. 35–57.
50. Berni, A.; Borgianni, Y. Making Order in User Experience Research to Support Its Application in Design and Beyond. *Appl. Sci.* **2021**, *11*, 6981. [CrossRef]
51. Pregernig, M.; Rhodius, R.; Winkel, G. Design Junctions in Real-World Laboratories: Analyzing Experiences Gained from the Project Knowledge Dialogue Northern Black Forest. *GAIA-Ecol. Perspect. Sci. Soc.* **2018**, *27*, 32–38. [CrossRef]
52. Xiao, N.; Lee, S.H.M. Brand identity fit in co-branding: The moderating role of CB identification and consumer coping. *Eur. J. Mark.* **2014**, *48*, 1239–1254. [CrossRef]
53. Giacalone, D.; Jaeger, S. Better the devil you know? How product familiarity affects usage versatility of foods and beverages. *J. Econ. Psychol.* **2016**, *55*, 120–138. [CrossRef]
54. López-Forniés, I.; Sierra-Pérez, J.; Boschmonart-Rives, J.; Gabarrell, X. Metric for measuring the effectiveness of an eco-ideation process. *J. Clean. Prod.* **2017**, *162*, 865–874. [CrossRef]
55. Kim, Y.; Oh, K. Effects of Perceived Sustainability Level of Sportswear Product on Purchase Intention: Exploring the Roles of Perceived Skepticism and Perceived Brand Reputation. *Sustainability* **2020**, *12*, 8650. [CrossRef]

56. Landis, J.R.; Koch, G.G. An Application of Hierarchical Kappa-type Statistics in the Assessment of Majority Agreement among Multiple Observers. *Biometrics* **1977**, *33*, 363–374. [CrossRef] [PubMed]
57. Khamis, H. Measures of association: How to choose? *J. Diagn. Med. Sonogr.* **2008**, *24*, 155–162. [CrossRef]
58. Deuble, M.P.; de Dear, R.J. Green occupants for green buildings: The missing link? *Build. Environ.* **2012**, *56*, 21–27. [CrossRef]
59. Kim, M.J.; Oh, M.W.; Kim, J.T. A method for evaluating the performance of green buildings with a focus on user experience. *Energy Build.* **2013**, *66*, 203–210. [CrossRef]
60. Earl, P.E. Lifestyle changes and the lifestyle selection process. *J. Bioecon.* **2016**, *19*, 97–114. [CrossRef]

designs

MDPI

Article

A New Look at Excavation Techniques and Design of Rock-Cut Architectures

Mohammad Mangeli [1,*], Farshid Aram [1,*], Reza Abouei [2] and Fatemeh Mehdizadeh Saradj [3]

1 Faculty of Architecture, Urbanism, and Art, Urmia University, Urmia 5756151818, Iran
2 Faculty of Art and Architecture, Yazd University, Yazd 8915818411, Iran; r.abouei@yazd.ac.ir
3 Faculty of Architecture and Urban Studies, Iran University of Science and Technology, Tehran 1311416846, Iran; mehdizadeh@iust.ac.ir
* Correspondence: m.mangeli@urmia.ac.ir (M.M.); f.aram@urmia.ac.ir (F.A.)

Abstract: Rock-cut architecture is an essential yet little-known type of vernacular architecture whose nature is different from what we understand by the term "architecture". This research seeks to answer the question of which technics, designs and digging procedures have been applied in this type of vernacular architecture. Out of the 300 rock-cut buildings and complexes found in Iran, nearly 70 were functionally assessed. Ten complexes were found to serve residential functions in different climates. Technique, type and the settlement context material were also briefly compared and contrasted, and thus, three general excavation techniques were recognized. The largest rock-cut residential complex in Iran, Meymand, was selected as the main case study. Fifty residential units in the oldest part of the village in two regions on both sides of the main valley were studied in terms of technique and design style. They were also compared and contrasted. The quantitative data obtained in this section were compared and contrasted using the descriptive statistical method. Although the rock-cut buildings are dispersed throughout Iran, three main techniques were employed to excavate them. Most of them were excavated using two or three of the said techniques. Application of the fine technique and the size of the particles constituting the settlement context are the two factors determining the quality of excavation. The findings in the area of special climatic design standards and technologies not only give us a better insight into rock-cut architecture but also contribute to setting some standards for design and construction of rock-cut buildings in the present era.

Keywords: rock-cut architecture; Meymand; vernacular architecture; architectural framework; earth-shelter; cave dwelling; underground architecture

Citation: Mangeli, M.; Aram, F.; Abouei, R.; Mehdizadeh Saradj, F. A New Look at Excavation Techniques and Design of Rock-Cut Architectures. *Designs* **2022**, *6*, 64. https://doi.org/10.3390/designs6040064

Academic Editor: Tiago Pinto Ribeiro

Received: 17 May 2022
Accepted: 5 July 2022
Published: 8 July 2022

1. Introduction

Building space through hollowing out solid rock in the absence of common building materials is an extraordinary architecture technique whose study demands special attention [1–3]. Rock-cut architecture, as a type of earth-shelter architecture, enjoys considerable diversity, as particularly the Iranian plateau and generally the world host its manifestations, which have taken different forms and functions due to climate considerations [4–6].

Common building materials and techniques are not used in rock-cut architecture [7]. Instead, space is built through hollowing out a natural solid context either in the form of a flat horizontal land under which the rock-cut buildings are excavated or the steep surface of foothills or gently sloping to upright hills, which provide the best context for rock-cut architecture [8–10]. The context should mainly be softer than granite, so that it could be excavated with simple tools, and harder than soil, so that it would not fall apart [11].

It is a hard task to understand the techniques used in this singular type of underground architecture [12] since there is a lack of historical documentary evidence about the excavation processes of rock-cut buildings. Moreover, it is no longer practiced in Iran [13]. The only way to find some evidence about the excavation technology of rock-cut architecture is to refer to interviews held with informed people and physical analysis of the architecture.

It is also useful to compare and contrast the architectural styles of rock-cut buildings in different regions.

2. Statement of the Problem

One of the most significant challenges in understanding rock-cut architecture is to answer the fundamental question of how the buildings were designed and excavated [14–16] and what process was performed to build them. Due to lack of historical evidence and documents, the answer to this question requires understanding the process in which the background and the excavation technique were selected [17]. In other words, one should think as similarly as possible to the men who excavated the buildings in the past. Therefore, the main question is what techniques could be used to excavate an architectural space with the least tools. The answer could partly remove the ambiguity about the nature of rock-cut architecture [18]. Unfortunately, the Iranian works of rock-cut architecture are in danger of destruction due to different reasons [19], and thus, it is necessary to gain a fundamental insight into the manner in which they were formed [20].

3. Materials and Methods

The four-year research is a case study of rock-cut architecture in the world heritage site of Meymand village. Excavation techniques and design patterns of the buildings were assessed on the basis of background and historical evidence found in the architectural body of the village. Then, the excavation process of the buildings was sketched out on the basis of the settlement context. Two general sketches were drawn that were very similar to each other and were only different in terms of the steepness of the surface of the excavation context. Later, building plans were recognized and classified into two broad types: linear plan and radial plan.

Out of the 300 rock-cut buildings and complexes found in Iran, nearly 70 were functionally assessed. The existence of this architecture throughout Iran shows the widespread use of this local architecture type. Ten complexes were found to serve residential functions in different climates (Figure 1). The general classification of context and excavation of these assemblies were determined based on the geological zoning of Iran [21]. The slope of the construction site of this building was also recorded in Table 1 based on field observations. Then, by selecting 3 buildings from each complex, calculating the amount of excavation, the area of the building and the area of the surfaces, the percentage of each of the three main digging methods in this building was estimated (Table 1).

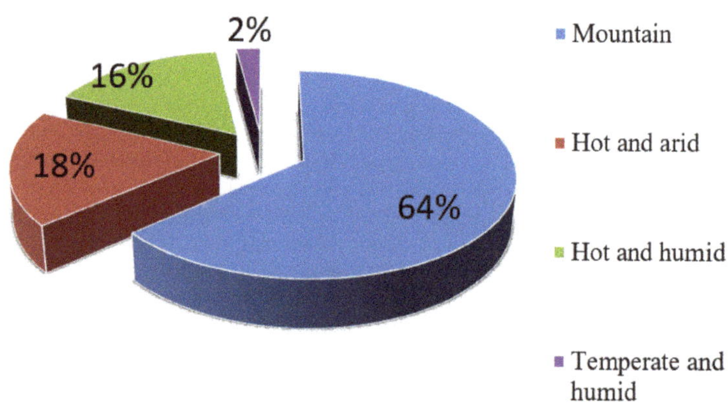

Figure 1. Climate ratio of Iranian rock-cut regions.

Table 1. Technical characteristics of well-known Iranian rock-cut residential complexes.

Name.	Main Excavation Technique	Climate	Context Rock	Context Steepness	Mass Excavation Technique (%)	Steady Excavation Technique (%)	Fine Excavation Technique (%)
Meymand	Rough out	Cold mountain	Pyroclastic	Steep hill and upright wall	8	88	4
Kandovan	Fine	Cold mountain	Pyroclastic	Upright wall	4	44	52
Hilevar	Rough shaping	Cold mountain	Pyroclastic	Steep hill	12	66	22
Savar	Rough out	Cold mountain	Pyroclastic	Steep hill	52	40	8
Arzanfoud	Rough shaping	Cold mountain	Metamorphic	Steep hill	20	75	5
Bafran	Rough shaping	Hot and arid	Sedimentary	Upright wall	30	65	5
Kouhpayeh	Rough shaping	Cold mountain	Sedimentary	Upright wall	30	60	10
Khorrambid	Rough out	Cold mountain	Sedimentary	Steep hill and upright wall	45	40	15
Abazar	Rough out	Cold mountain	Pyroclastic	Upright wall	45	40	15
Viand	Rough shaping	Cold mountain	Pyroclastic	Steep hill and upright wall	30	60	10

Technique, type (Figure 2) and the settlement context material (Figure 3) were also briefly compared and contrasted, and thus, three general excavation techniques were recognized. The largest rock-cut residential complex in Iran, Meymand, was selected as the main case study. A total of 50 residential units in the oldest part of the village in two regions on both sides of the main valley were studied in terms of technique and design style. They were also compared and contrasted. Archaeological evidence shows that the tradition of excavation of this building existed until recent decades [20].

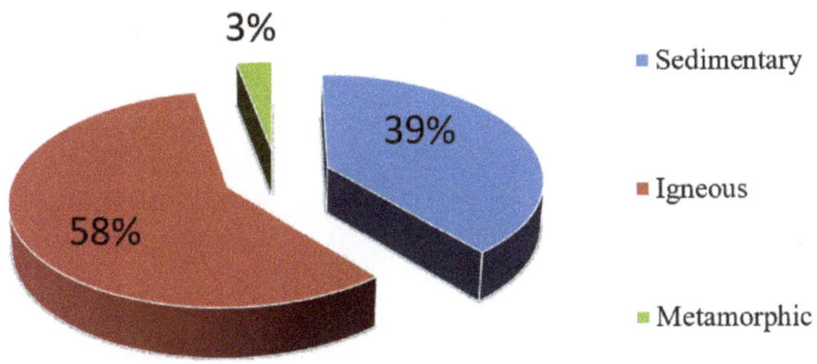

Figure 2. Context-type ratio of Iranian rock-cut buildings.

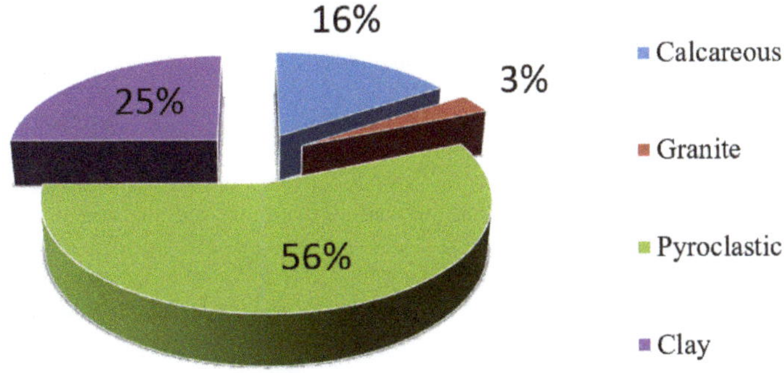

Figure 3. Context-material ratio of Iranian rock-cut buildings.

Animal husbandry and living with livestock have formed the main structure of society and livelihood of the inhabitants in the rock-cut architecture complexes of Iran. Agriculture and horticulture were also seen to a limited extent due to the lack of water around these complexes. The preference of the people of these areas due to climatic issues and access to a suitable context for excavation (mountains and rocky hills) has been the reason for using this type of architecture [19].

The quantitative data obtained in this section were compared and contrasted using the descriptive statistical method. It was found out that the rock-cut architecture of Meymand village follows a certain process and a general technical standard.

Meymand is a historic rock-cut village located in eastern Shahr-e Babak, in the northwest of Kerman province, Iran (Figure 4). Its latitude is 11°13′ N, and its longitude is 55°22′ E. It is 2220 m above the sea. Meymand is located in the southern slope of the Masahim dormant volcano. The village is cut into volcaniclastic rock known as pumice. The five-story village consists of almost 360 houses [22]. Meymand region generally enjoys a mountain climate, but its neighboring plains have a semiarid climate. The history of its formation is unknown; however, Shahr-e Babak was built in the Sassanid era [23]. Archeological excavations at Meymand castle and carbon-14 tests on its remnants date it back to the Parthian era [24]. Earthenware found in Maymand bears significant resemblance to that of the Sassanid era (500–200 A.D.) [25]. Rock carvings around Maymand demonstrate that men lived in Meymand region several millennia ago [26].

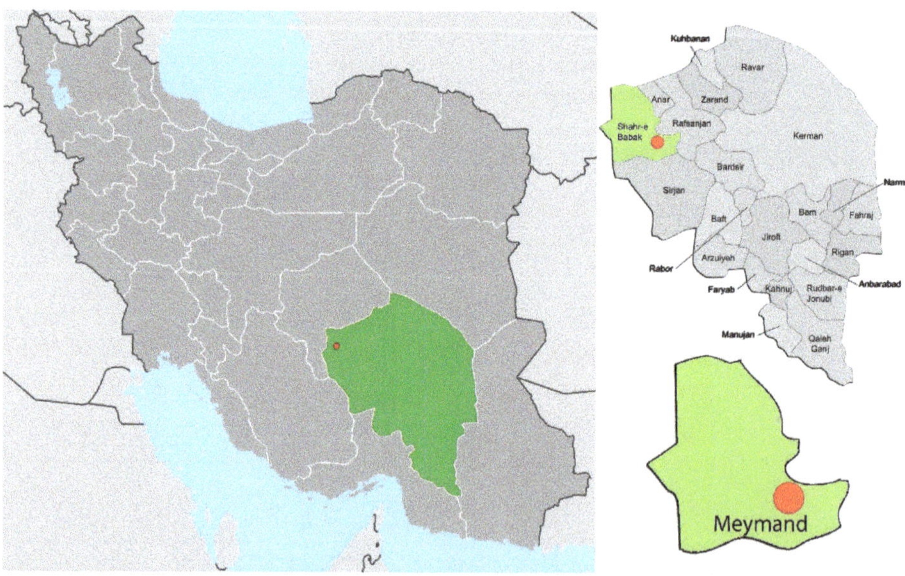

Figure 4. Location map of Meymand in Iran.

The village is located on the edges of a shallow valley less than one kilometer in length (Figure 5). The majority of its buildings serve residential purposes, and only a few buildings were used as school, mosque, hussainiya (congregation hall for Shia commemoration ceremonies), bath and fire temple. Intra-village communications are carried out through a central floodway as well as narrow paths that discharge surface waters and reduce the risk of flooding when it rains. The fabric was formed in several stages that can be distinguished from each other according to the architectural characteristics of each era [19].

Figure 5. Eastern view of Meymand village.

In this World Heritage Site, a historic bath, a school, a mosque, a Hussainiya, a fire temple and the remaining 360 buildings were used for residential purposes. Each of the residential buildings of this complex has their own architectural aspects.

The required water for this village was supplied from a spring at the bottom of the central river in the village, and these buildings did not have direct access to water. To heat these buildings in winter, burning coal and wood were used, and the resulting smoke was discharged by a convective flow through the wooden doors of the rooms.

People of Meymand used to live in the village only in the cold months of the year and were busy farming and stockbreeding in the neighboring plains and valleys in the rest of the year [27]. Because of the cold climate of the valley, their houses have short and small openings that are proportional to their dwellers' height and minimize heat loss (ibid.). Moreover, the blackened and soot-covered rooms have thick walls that reduce heat loss to a minimum. They are cold in summer and quite warm in winter. They reach the comfort zone with minimum energy. Rural density and cohesion, as well as the multistory structure of Meymand village, have made its spatial communications dense and close (Figure 6). The village has distinct neighborhood divisions on the basis of the families who lived there [28].

Limited tools were used to dig these buildings, and the 3 main common tools of all diggers in all historical periods are a carving pen, hammer and carving pickaxe [22]. In Meymand, through interviewing local people and observing the common tools used for stonework, it was determined that the three mentioned tools were the main drilling tools of this building in the past (Figure 7).

Figure 6. Aerial image of Meymand village.

Figure 7. Sample of equipment used for excavation.

4. Results

4.1. Excavation Techniques

Three distinct excavation techniques were employed in the rock-cut architecture of Meymand village. They vary according to the area in which they were used. In simpler words, in areas where delicacy in construction was necessary, light tools and fine techniques such as chiseling were employed. On the contrary, more sophisticated techniques were used when the volume of excavation was considerable. In sum, the excavation techniques are included as follows:

4.1.1. Excavation Method

The excavation technique was chosen by the architect and the excavator. However, three distinct techniques were employed to excavate rock-cut houses. Apart from edges, niches and sills, which demanded delicacy in excavation, the main space inside the rooms was excavated steadily in small pieces, or huge pieces of rock were removed. The context material plays a fundamental role in the excavation technique. The said techniques were applied only in buildings whose context was made of pyroclastic rocks such as tuff and pumice. These buildings are found in Meymand, Hilevar, Kandovan and Abazar (Figure 8).

Figure 8. Pumice stones found in Meymand, Kandovan and Abazar and their difference with the dense sediment found in Bafran.

In buildings whose context was made of dense clay sediments, a steady excavation technique was used. Examples of this technique are found in Kouhpayeh village of Kerma and Bafran near Naein (Table 1). The difference in context material was accompanied by difference in structural system, and an arched system was used in contexts such as clay, which have less strength (Figures 9 and 10). The arrangement of selected layers is also important in excavation, and the rock-cut architects always sought to choose a context that was either a solid mass free of layers or had horizontal layers (Figure 11). Therefore, the context may have better borne the compressive forces and would not slide under the pressure of gravity.

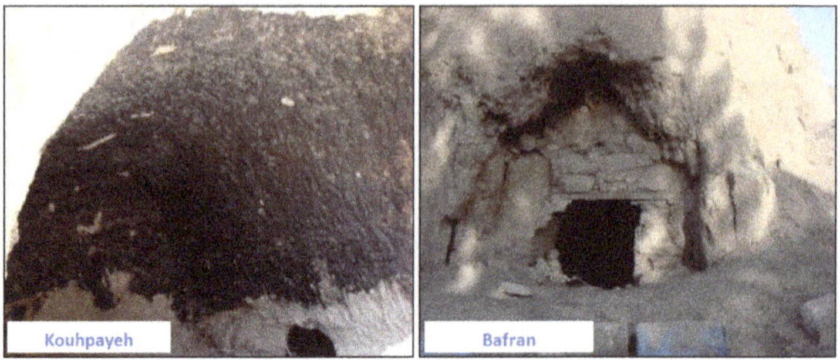

Figure 9. Arched excavation in the form of barrel arch in Koupayeh of Kerman and Bafran of Naein.

Figure 10. Flat roof in Maymand buildings.

Figure 11. Horizontal arrangement of layers in pyroclastic context of Qebleh Daghi of Azarshahr and sedimentary context of Kouhpayeh.

4.1.2. Rough out Excavation

In this technique, heavy pickaxes are used to excavate large and deep grooves around a huge mass. Then, the surface of the mass between the grooves is divided into distinct spots against which a wedge is placed and hammered to the point when the mass crumbles into small pieces [29]. This technique requires considerable skill because failure to control the blows and locate the early grooves leads to the improper demolition of the rock, and thus an irregular room space (Figure 12), yet it accelerates the excavation process.

Figure 12. Mass excavation technique in Meymand houses.

4.1.3. Fine Shaping and Detailing Excavation

This technique has limited application. In this technique, light tools such as light pickaxes are used to excavate doorways, carve façade decorations, polish the corners, level out the floor and ceiling and excavate lock holes (Figures 13–15). In the fine excavation technique, light yet continuous blows remove small particles from the surface; therefore, the rough surface will turn into a level one, and the desired form takes shape. Simple and limited tools are among the characteristics of this architecture. Heavy and light pickaxes, small and large chisels, heavy mallets and light hammers, metal wedges and wheelbarrows are the said tools. This is interesting because each stage of the architecture of the construction demands the use of several tools whose application sometimes adds to the complicacy of the task. Therefore, it is an advantage of rock-cut architecture that it uses a limited range of simple tools. Moreover, it is possible to make the tools available in the region with simple industrial facilities such as traditional blacksmithing.

4.1.4. Rough Shaping Excavation

In this simple technique, heavy pickaxes are used to strike consecutive blows up and down at an almost 45 degree angle to remove small pieces of stone, and the process continues steadily until the whole mass breaks up. This technique enables the excavator to control the volume of excavation and reduce the roughness of the surface. In general, the three techniques were all applied in rock-cut buildings. However, the steady excavation technique was more prevalent thanks to executive balance (Table 1). The walls should also be varnished to reach the desired internal form. This also demanded the application of the steady excavation technique.

Apart from space creation and excavation techniques (Figure 16), the quality of excavation in Meymand was dependent on the size of the particles constituting the settlement context. In other words, the finer the grains, the finer the excavated area. The context masses are mainly made of tuff, which is a fine-grained rock and thus gives the excavator a free hand to excavate the corners, niches and openings. On the contrary, spaces excavated in agglomerate and breccia masses are organic and have no corners and angles. Tuff and

Pumice particles are round, while Agglomerate and Breccia particles are large and angular. This difference also changed the quality of excavation.

Figure 13. Fine excavation technique used in doorways, niches and interior space of rock-cut rooms in Meymand.

Figure 14. Fine excavation technique used in interior surface of rooms.

Figure 15. Difference in the size of the particles led to difference in the quality of excavation in Ansaroud Grand Mosque and Pish Avesta Temple.

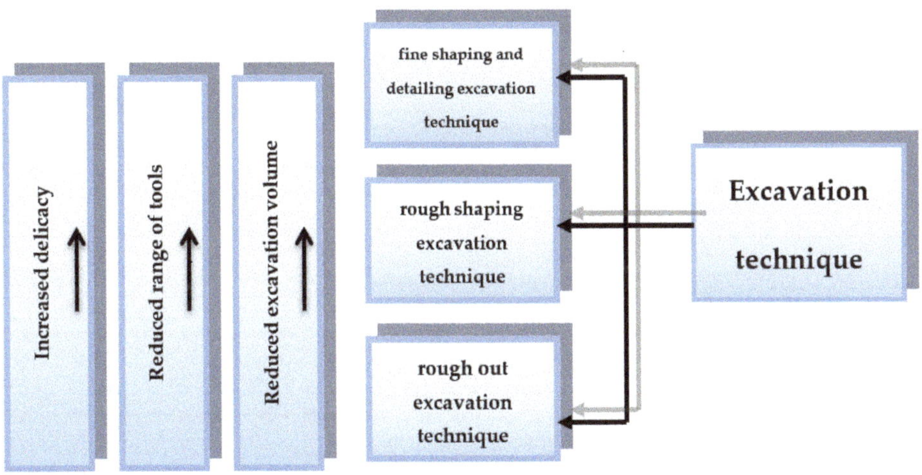

Figure 16. The impact of excavation technique on volume, delicacy and tools.

5. Discussion

5.1. Excavation Design

Apart from appropriate excavation techniques and tools, a predetermined design was needed to excavate the buildings. In other words, observing the excavation prerequisites and conducting the excavation process required a detailed plan. The process covered the following stages: excavation of the trench to reach the upright surface needed to excavate the building; excavation to reach a semi-open space in front of the entrance; fine excavation of the entrance; and simultaneous and parallel excavation of rooms. Each of the above stages required certain techniques and tools (Table 2).

Table 2. The relationship between architectural elements and excavation tools used in Meymand.

Architectural Element	Excavation Technique	Excavation Tool
Kicheh (entrance trench)	Rough out	Mallet and wedge
Soffeh (semi-open space next to kicheh)	Rough shaping and fine shaping and detailing	Heavy and light pickaxe
Dargah (Doorway)	Fine shaping and detailing	Light pickaxe
Room	Rough out and rough shaping	Heavy pickaxe, wedge and mallet

5.2. Excavation Stages

The buildings of Meymand were generally excavated in the following order:

5.2.1. Stage 1: Excavation of Kicheh (Entrance Trench)

The aim of this stage is to turn the steep surface into an upright one in order to excavate the interior spaces of the building, a surface in front of which the excavator can easily stand and perform the excavation. The excavators first located the excavation path and determined its length, and then applied the mass excavation technique with heavy tools to excavate the trench. This architectural element is similar to an inverted triangle. Its total length depends on the steepness of the context. The less steep the context, the longer the trench. For instance, if a house is built on the surface of a hill that is 30 degrees steep, the trench will be longer than six meters. The excavation was performed roofless until an upright surface with a reasonable height proportional to the excavator's height (2.5 to 3 m) took shape. The upright surface acted as the proper and standard context for building the interior semi-open and closed space of the building (Figure 17). The difference between the highest level of the valley, which naturally has upright surface for excavation, and the lower levels, which have steep surfaces, lies in this stage. In fact, the houses built at the highest level did not need this stage of excavation. Therefore, their volume of excavation was less than that of the houses built in the lower levels. The main difference made by the absence of this architectural element is that houses built in the lower levels are more compatible with the climate conditions because they have an extra element (i.e., the trench) which makes balance between the inside and outside temperature. The dwellers of the houses which are equipped with trenches acknowledge that they are more comfortable.

5.2.2. Stage 2: Excavation of Soffeh (Semi-Open Space)

In this stage, the roofed spaces of the building were excavated mainly using the steady excavation technique, for the general shape of the spaces was important and appropriate upright surfaces should be reached to excavate the room entrances. As soon as this element took general shape, the surface of the walls was excavated using the fine excavation technique. At this stage, semi-open spaces took shape, which acted as intermediaries between the outside space and the closed rooms inside the house. In fact, they acted as distribution space. In addition, the dwellers of the house performed most of their daily activities in this part of the building. The excavators performed the excavation from above when necessary and then shaped the floor of soffeh to reach the desired height level. The bottom level was usually excavated 20 to 30 cm deeper than the entrance trench's (Figure 18). The rock layer of the ceiling in the thinnest part where the soffeh and kicheh meet is above 50 cm thick to lend the building enough strength. As we moved inside the building, the thickness increased due to the steep surface of the settlement context.

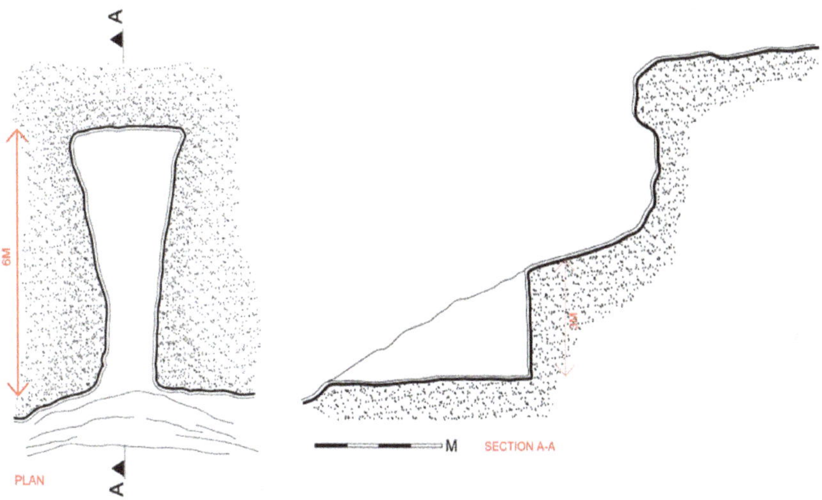

Figure 17. The first stage of excavation (excavation of entrance trench, i.e., Kicheh) in cross-section and plan view.

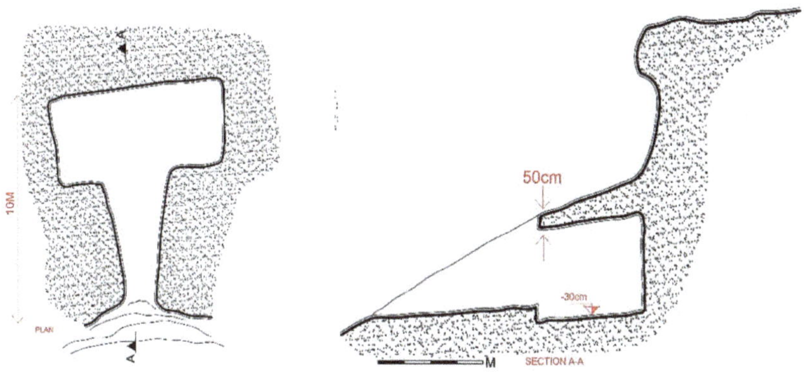

Figure 18. The second stage of excavation (excavation of semi—open space, i.e., Soffeh) in cross-section and plan view.

5.2.3. Stage 3: Excavation of Dargah (Doorway)

This is technically the most sensitive stage of the excavation. It was performed using the fine excavation technique and light tools. In this stage, the surface was lightly scraped. Usually, three sides of the soffeh were ready to excavate the rooms; therefore, the doorways were first located and then excavated (Figure 19). The average thickness and depth of the doorways were approximately 50 cm. This stage is really sensitive because the excavators are able to correct their mistakes only in this stage. In fact, it is not possible to correct the mistakes in later stages since there is no surface to rebuild the doorways (Figure 20). On the other side, the need to build doors in sizes compatible with the cold climate of the region adds to the sensitivity of the stage because the measures of the entrance should be selected in a manner that minimizes the thermal exchange in the hot and cold seasons of the year (Figure 21). Furthermore, they should be excavated in shapes that make it possible to install or change wooden doors and their locks.

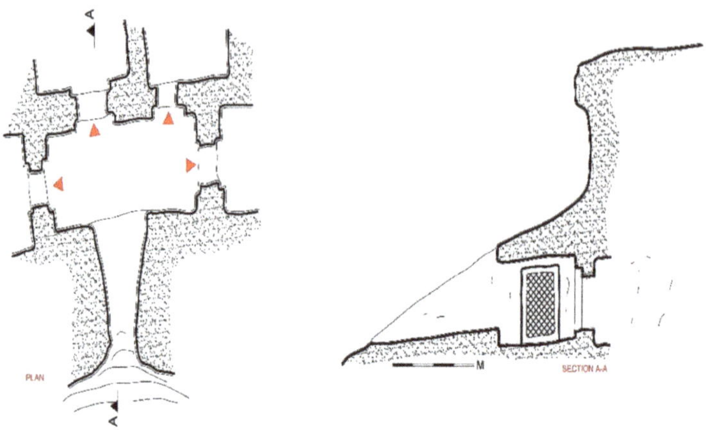

Figure 19. The third stage of excavation of doorways (Dargah) in cross-section and plan view.

Figure 20. An unfinished kicheh in Meymand shows that the doorways should be completed before moving to other elements.

Figure 21. Details of a front door from inside.

5.2.4. Stage 4: Excavation of Room

Steady excavation is the main technique used in this stage. The excavator first excavated the general shape of the room with desired measurements, size and height. Then, he employed the fine technique and used light tools to scrape the surface of the walls, floor, ceiling and corners to reach the desired shape (Figure 22). The quality of excavation is dependent on the size of the particles constituting the excavation context. The finer the grains, the smoother and more appropriate the form of the rooms. If there is no restriction in the neighborhood of this building and the bedrock has a uniform texture in terms of strength, it is possible to add space inside these buildings by drilling. Evaluation of the average penetration depth and opening width of the rooms in Meymand houses shows that the diggers tried not to penetrate more than 18 m and the width of the openings more than 6 m, because as the openings increase their strength decreases. Additionally, by increasing the penetration depth, the ventilation and the lighting of the building are also reduced (refer to the doctoral dissertation). According to archaeological evidence, it took at least 3 weeks to prepare and dig a room in this village.

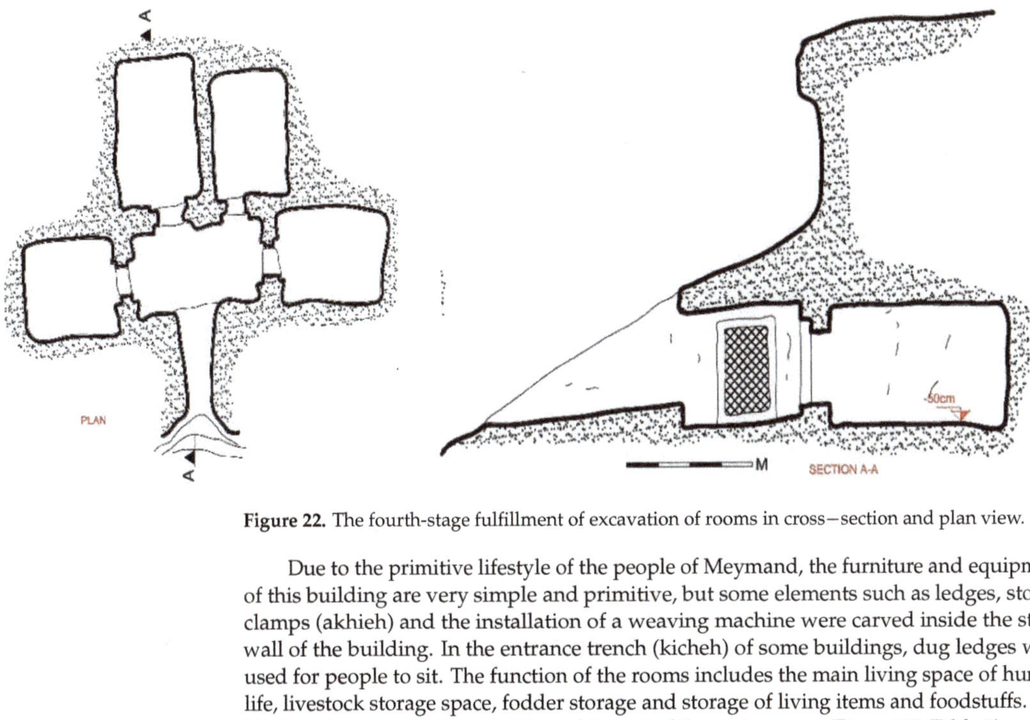

Figure 22. The fourth-stage fulfillment of excavation of rooms in cross–section and plan view.

Due to the primitive lifestyle of the people of Meymand, the furniture and equipment of this building are very simple and primitive, but some elements such as ledges, stoves, clamps (akhieh) and the installation of a weaving machine were carved inside the stone wall of the building. In the entrance trench (kicheh) of some buildings, dug ledges were used for people to sit. The function of the rooms includes the main living space of human life, livestock storage space, fodder storage and storage of living items and foodstuffs. The kitchen does not exist separately and is part of the main rooms (Figure 23, Table 3).

5.2.5. Stage 5: Construction of Other Elements

In this stage, wooden doors, canopies, stone restrooms and stone walls were constructed, and the building reached the point of completion (Figure 24).

Figure 23. The function of rooms in Maymand's house architecture.

Table 3. The description of the function of rooms in Maymand's house architecture according to Figure 23.

Number	Local Name	Name	Function	Description
1	Otaq Boni	Main room	Living and kitchen together	Room quite dug in bedrock used as main livelihood place
2	Anbar	Side room or store room	Store and bedroom	Room quite dug in bedrock used as subsidiary space
3	Soffeh	patio	Semi-open space	Restricted by 3 rock-cut walls and rocky or wooden roof and 1 side open to air by entrance trench
4	Aqol	cratch	Livestock care place	Dug in rock and located in entrance trench used for livestock
5	Kahdoon	Hay loft	Fodder warehouse	Dug in rock and located in entrance trench used for storage
6	Kicheh	Entrance trench	corridor	Entrance and connector between outside and inside and introducing a residential unit

Figure 24. Hippophae wood and wormwood shrubs form the canopy of a house in Meymand village.

In Meymand, the only opening in the building through which light and ventilation of the rooms are provided are their wooden entrance doors, the height of which has been shaved to the minimum required to reduce the surface heat exchanges. An evaluation of about 50 buildings in this study showed that the average height of the doors is 130 cm and their width is about 82 cm. As a result, the light penetration into the rooms is very low. There are no windows in any of Meymand's residential houses.

5.3. Excavation Design Measures in Meymand Buildings

5.3.1. Arrangement of Buildings in Different Levels

One of the main principles observed by the excavators was to avoid arranging two buildings on each other in two different levels (Figure 25). They usually tried to excavate the buildings in places where there was no house immediately above or below the excavation site. They took this measure to increase the thickness of the layers between the buildings that were excavated near each other in two different height levels. Therefore, the rock layer became thicker, and the strength of the buildings increased. This technique minimized the overlap of plans in buildings excavated in two levels.

Figure 25. Special arrangement of buildings (excavators avoided laying two buildings on each other in two different levels).

5.3.2. Varying Height of the Longitudinal Sections

Comparison and contrast of the longitudinal sections of the buildings reveals a fundamental similarity in alignment. This principle was carefully observed in large units. In other words, the bottom level of the units was by average 40 to 60 cm lower than that of the main entrance. This has some advantages: first, it facilitates ventilation, and thus, the smoke of cooking fire is better removed. Second, the thickness of the rock layer of the ceiling is increased thanks to this creative measure. Third, thermal exchange is reduced in the cold season of the year because the room ceilings are lower than those of the semi-open entrance trench (Figure 26).

Figure 26. Bottom level of the rooms are lower than that of the entrance trench.

5.3.3. Increasing the Area by Increasing the Depth

In parts of the village, there are natural upright walls suitable for excavation due to geomorphological conditions. Compared to the buildings excavated in steep surfaces, buildings excavated in natural upright walls are harder to be expanded radially. Therefore, the excavators neither expanded the buildings radially nor increased the span of the surfaces. Instead, they excavated as deeply as possible to increase the number of the rooms, and consequently, the general length of the building (Figure 27). Comparison between the buildings excavated in the highest level and those excavated in the lower levels reveals that the total length of the rooms in the highest level is by average 30 percent more than those excavated in the lower levels. The former have one or two rooms while the latter have by average four rooms (Figure 28). The excavators generally increased the depth of these buildings because they did not have the semi-open space. The absence of the semi-open space led to increased thermal exchange in the cold seasons of the year. The depth was increased in order to decrease the thermal exchange.

Figure 27. There are upright walls in the highest level of the village in which deep spaces are excavated.

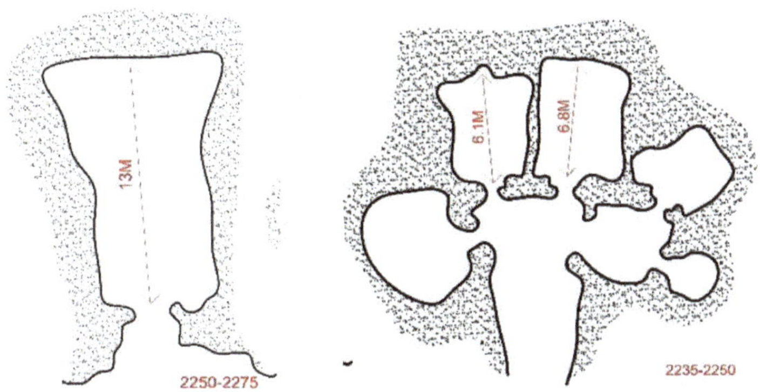

Figure 28. The difference in plan between the houses in the highest level (2250–2275) and those in the lower levels (2235–2250) from the sea level and increasing the area of rooms by increasing their depth.

5.3.4. Special Design of the Entrance Trench (Kicheh)

Comparison of the general shape of the entrance trenches in houses excavated in the steep surface of the hills reveals that the width of the entrance is less than that of the space in which the semi-open space and rooms are located. This creates a triangular shape in the plan of this part (Figure 29). Therefore, the volume of excavation is reduced, and more light is allowed into the building without increasing its contact with open air. Thermal exchange is also reduced thanks to the triangular shape of the entrance trench (Figures 30–34).

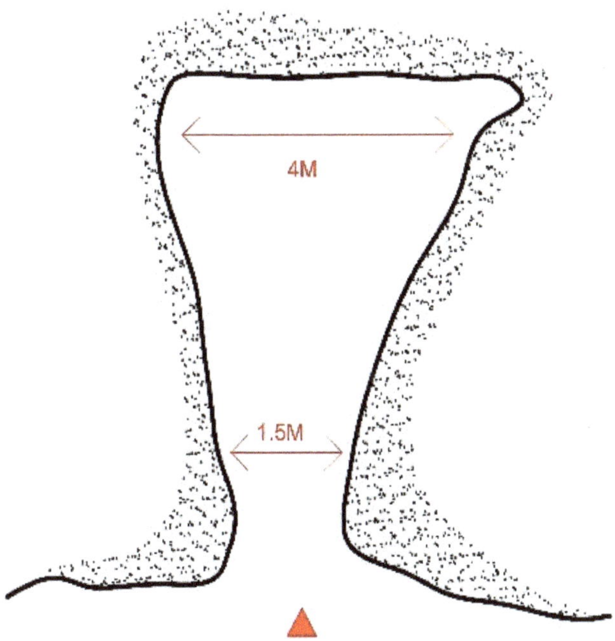

Figure 29. Triangular shape of the rock-cut architecture of Meymand.

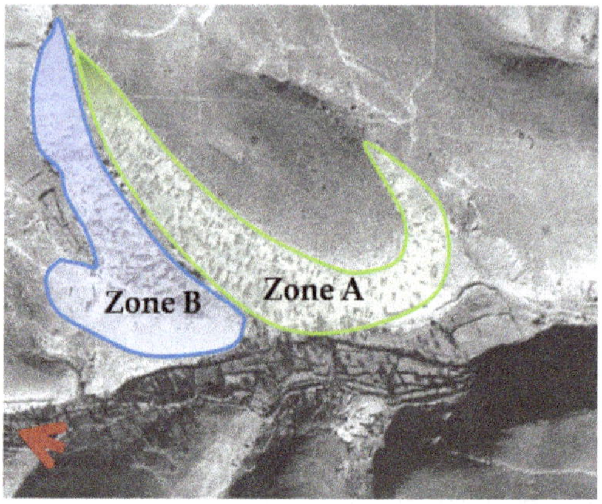

Figure 30. Zone A and Zone B located on the aerial image of Meymand village.

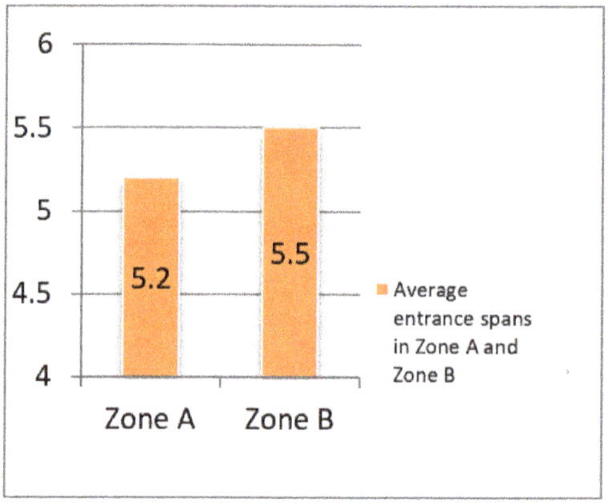

Figure 31. Average entrance spans of rooms in Zone A and Zone B.

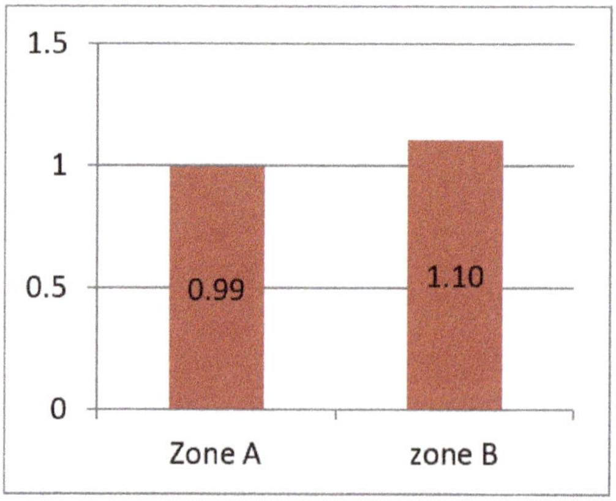

Figure 32. Average areas of front doors in Zone A and Zone B.

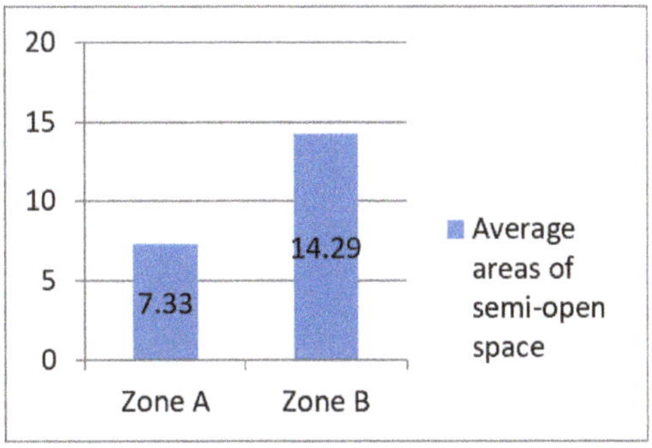

Figure 33. Average areas of semi-open spaces in Zone A and Zone B.

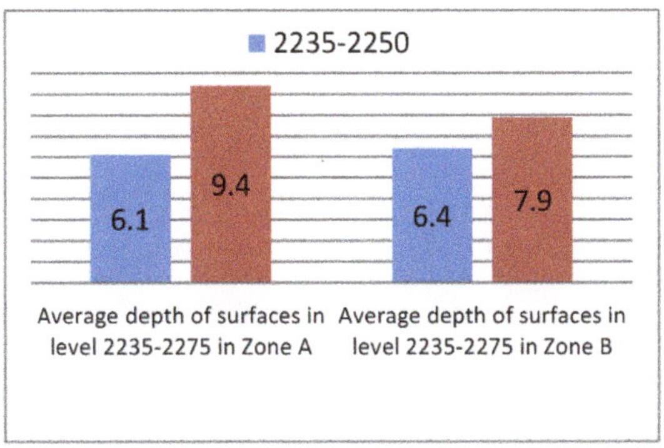

Figure 34. Average depth of the rooms in different height levels in Zone A and Zone B.

6. Conclusions

Rock-cut architecture is a special type of vernacular architecture. It uses a rock context that is excavated to create the architectural space. In fact, it is a productive type of architecture. The rock excavation is almost an irreversible act, something that the following generations can hardly restore.

In general, it follows distinct design patterns and employs certain techniques. Rock-cut architecture is dependent on the tools available to the architects, geomorphological conditions of the excavation area and the type and material of the settlement context. Most of the Iranian rock-cut residential buildings have similar design standards and spatial measures due to structural, climate and measure similarities. The standards were mainly set to guarantee climate and life comfort. Although the rock-cut buildings are dispersed throughout Iran, three main techniques were employed to excavate them. Most of them were excavated using two or three of the said techniques.

The main point that can be concluded from the study of the rock-cut architecture of Meymand is that the most important design approach was achieving a space with minimal dimensions in accordance with the basic needs of daily life and avoiding harsh environmental conditions. The fine excavation and precise digging of surfaces in the buildings of this site have not received much attention and were only the creation of shelter without any decorations with minimal energy consumption and spending time. Therefore, the excavation of surfaces is limited to removing large protrusions and leveling them. This feature can be seen in all types of buildings in this complex.

Application of the fine technique and the size of the particles constituting the settlement context are the two factors determining the quality of excavation. Iran provides the proper background for excavation and construction of rock-cut buildings in the future. Low costs of construction, reduced energy consumption and building materials, and climate compatibility are the advantages of rock-cut architecture that can serve as a model for contemporary architecture. Revival of rock-cut architecture and excavation of rock-cut buildings requires a deeper insight into the body and technology of this unknown architecture.

Author Contributions: Conceptualization, M.M.; methodology, M.M. and R.A.; software, M.M. and F.A.; validation, M.M. and F.A.; formal analysis, M.M.; investigation, M.M. and F.A.; resources, F.M.S.; data curation, F.A. and R.A.; writing—original draft preparation, M.M. and F.A.; writing—review and editing, R.A. and F.M.S.; visualization, M.M.; supervision, R.A. and F.M.S.; project administration, R.A. and F.M.S.; funding acquisition, F.A. All authors have read and agreed to the published version of the manuscript.

Funding: This research received no external funding.

Institutional Review Board Statement: Not applicable.

Informed Consent Statement: Not applicable.

Data Availability Statement: Not applicable.

Conflicts of Interest: The authors declare no conflict of interest.

References

1. Fabbri, K.; Pretelli, M. Heritage buildings and historic microclimate without HVAC technology: Malatestiana Library in Cesena, Italy, UNESCO Memory of the World. *Energy Build.* **2014**, *76*, 15–31. [CrossRef]
2. Zhu, J.; Tong, L. Experimental study on the thermal performance of underground cave dwellings with coupled Yaokang. Renew. *Energy* **2017**, *108*, 156–168. [CrossRef]
3. Martínez-Molina, A.; Tort-Ausina, I.; Cho, S.; Vivancos, J.L. Energy efficiency and thermal comfort in historic buildings: A review. *Renew. Sustain. Energy Rev.* **2016**, *61*, 70–85. [CrossRef]
4. Del, M.S.T.T.; Sadooghi, Z.; Tabrizi, S.K. Recognition of defensive factors in the architectural heritage of Iran's organic ancient shelters. *Front. Archit. Res.* **2022**, in press. [CrossRef]
5. Zhao, X.; Nie, P.; Zhu, J.; Tong, L.; Liu, Y. Evaluation of thermal environments for cliff-side cave dwellings in cold region of China. Renew. *Energy* **2020**, *158*, 154–166. [CrossRef]
6. Cozzolino, M.; Caliò, L.M.; Gentile, V.; Mauriello, P.; Di Meo, A. The Discovery of the Theater of Akragas (Valley of Temples, Agrigento, Italy): An Archaeological Confirmation of the Supposed Buried Structures from a Geophysical Survey. *Geosciences* **2020**, *10*, 161. [CrossRef]
7. Goudini, J. Defining Troglodytic Architecture and its Design Methods Based on Iranian Examples. *Tunn. Undergr. Space Eng.* **2020**, *9*, 285–304. [CrossRef]
8. Barbero-Barrera, M.M.; Gil-Crespo, I.J.; Maldonado-Ramos, L. Historical development and environment adaptation of the traditional cave-dwellings in Tajuña's valley, Madrid, Spain. *Build. Environ.* **2014**, *82*, 536–545. [CrossRef]
9. He, Y.; Chu, Y.; Song, Y.; Liu, M.; Shi, S.; Chen, X. Analysis of design strategy of energy efficient buildings based on databases by using data mining and statistical metrics approach. *Energy Build.* **2022**, *258*, 111811. [CrossRef]
10. Rostam, N.G.; Hojjati, A.; Mahdavinejad, M.; Mirlohi, M. Natural energy efficient materials for rock cut architecture in case of Kandovan, Iran. *Adv. Mater. Res.* **2014**, *935*, 202–206. [CrossRef]
11. Alkaff, S.A.; Sim, S.C.; Efzan, M.N.E. A review of underground building towards thermal energy efficiency and sustainable development. *Renew. Sustain. Energy Rev.* **2016**, *60*, 692–713. [CrossRef]
12. Hanachi, P.; Mehran, M. Politics, paterns and rehabilitation patterns in Meymand village fabric. *J. Fine Art* **2007**, *30*, 53–62.
13. Mehran, M. *Statement of Restoration and Design in Meymand's Historic Fabric*; Tehran University: Tehran, Iran, 2007.
14. Ozturk, F. Negotiating between the Independent and Groups of Courtyard Complexes in Cappadocia. In Proceedings of the 30th Annual Conference of the Society of Architectural Historians, Australia and New Zealand, Gold Coast, QLD, Australia, 2–5 July 2013.
15. Zhu, J.; Tong, L.; Li, R.; Yang, J.; Li, H. Annual thermal performance analysis of underground cave dwellings based on climate responsive design. *Renew. Energy* **2020**, *145*, 1633–1646. [CrossRef]
16. Zhu, J.; Liu, Y.; Li, R.; Chen, B.; Chen, Y.; Lu, J. Thermal Storage Performance of Underground Cave Dwellings under Kang Intermittent Heating: A Case Study of Northern China. *Processes* **2022**, *10*, 595. [CrossRef]
17. Wootton, W.; Russell, B.; Rockwell, P. Stoneworking Techniques and Processes. The Art of Making in Antiquity. Stoneworking in the Roman World. 2013, pp. 1–35. Available online: http://www.artofmaking.ac.uk/content/essays/3-stoneworking-techniques-and-processes-w-wootton-b-russell-p-rockwell/ (accessed on 7 January 2022).
18. Lamesa, A. Methods for Rock-hewn Worksite Analysis: The Church of Göreme n° 4b (Cappadocia): A Case Study. *Anatolia Antiq.* **2020**, *XXVIII*, 23–45. [CrossRef]

19. Mangeli, M. The Impact Modality of Climate and Context in Rock-Cut Architecture Genesis in Meymand Village Shahre Babak; (With Conservation Approach). Ph.D. Thesis, Isfahan Art University, Isfahan, Iran, 2018.
20. Ashrafi, M. A Study in Troglodytic Architecture Typology. *Archit. Urban. Lett.* **2012**, 25–47.
21. Kiani, M.; Hashemi, M.; Ajalloeian, R.; Benavente, D. Investigating the geological and geomechanical characteristics governing the weathering behavior of Meymand tuff. *Environ. Earth Sci.* **2022**, *81*, 45. [CrossRef]
22. Mangeli, M.; Abouei, R.; Saradj, F.M. A New Look at Unique Characteristics of Iran's Rock-Cut Architecture Settlements (Case Study: The World Heritage Site of Meymand Village, Shahre Babak). *J. Stud. Hum. Settlements Plan.* **2017**, *12*, 785–802.
23. Hoseini, Z. *Shahr-e Babak the Land of Turquoise*; Kermanology Center: Kerman, Iran, 2002.
24. Heritage, O.C. *Documenting Reports of Meymand*; Cultural Heritage Organization of Iran: Kerman, Iran, 2006.
25. Asadabadi, R. *Study and Recognition of Meymand Area*; Cultural Heritage Organization: Kerman, Iran, 1996.
26. Karimi, F. New look at the Iran's Rock art based on field and ancient study. *Iranology Q.* **2007**, *3*, 50–62.
27. Ashrafi, M. Architectural Sustainability in the Cultural Landscape of Maymand. *J. Iran. Archit. Stud.* **2022**, *9*, 97–122. [CrossRef]
28. Mangeli, M.; Sattaripour, A. A report on potentials of restoration and revitalization of Meymand historical village. In Proceedings of the 4th International Conference of Sustainable Development and Planning, Limassol, Cyprus, 13–15 May 2009. [CrossRef]
29. Bessac, J.C. Étude technique et interprettations du monument rupestre de Qadamgah (Fars). *Iran. Antiq.* **2007**, *42*, 185–206. [CrossRef]

designs

MDPI

Article

New Simulation Tool for Architectural Design in the Realm of Solar Radiative Transfer

Joseph Cabeza-Lainez [1,*], Jose-Manuel Almodóvar-Melendo [1], Paula Revenga-Dominguez [2], Inmaculada Rodríguez-Cunill [3] and Yingying Xu [4]

1 Department of Architectural Composition, University of Seville, 41012 Sevilla, Spain
2 Department of Art History, University of Cordoba, 14071 Cordoba, Spain
3 Department of Painting, University of Seville, 41002 Sevilla, Spain
4 Department of Oriental Studies, University of Seville, 41001 Sevilla, Spain
* Correspondence: crowley@us.es

Abstract: In this paper, we devise a system for architectural simulations that considers the volumetric and three-dimensional properties or the energy sources involved in the energy exchanges within or around edifices and built or urban spaces. The advances are based in optics theory evolving from the assumptions presented in the book *The Photic Field* by P. H. Moon and D. E. Spencer, with added improvements suggested by D. DiLaura. Such procedure is deftly performed by means of solving complex integral equations, which were unavailable until recently and originate in the research developed by the authors. This experimental software is called DianaX. The advantages of this new system allow for a clearer visualization of the performance of buildings in terms of radiated energy. Reductions in the amounts of used energy can be achieved precisely by means of the design process of the software, which can be considered in some respects as a Design Tool. With this tool, the analysis of heritage building paradigms is feasible as it assesses the potential of new foreseen projects taking into account new artificial lighting devices that deviate from the conventional linear or point approach in the domain. The main finding demonstrated is the feasibility and appropriateness of this method to address the problems posed. As future prospects, we would like to increase the catalogue of designs that can benefit from the conscious use of our tool for scientific design.

Keywords: lighting design; architectural design; luminaire's design; extended sources; LEDs; radiant exchange; energy exchange; simulation

Citation: Cabeza-Lainez, J.; Almodóvar-Melendo, J.-M.; Revenga-Dominguez, P.; Rodríguez-Cunill, I.; Xu, Y. New Simulation Tool for Architectural Design in the Realm of Solar Radiative Transfer. *Designs* **2022**, *6*, 72. https://doi.org/10.3390/designs6050072

Academic Editor: Farshid Aram

Received: 27 July 2022
Accepted: 23 August 2022
Published: 26 August 2022

1. Introduction

Traditionally, the energy simulation of architectural spaces is not intended for design as it does not take into account the spatial complexity of shapes [1]. As in other types of simulation, such as those for structural engineering, the acting magnitudes are customarily conducted through linear or at best bi-dimensional fields. This is due to several reasons: on the one hand, thermal fields are generally considered as scalar and, on the other, the idea cherished by engineering of a minimum inhabitable space with fixed planar floor and ceiling discourages the exploration of richer spatial configurations that can offer more visual attraction.

The traditional tools used to tackle this problem contemplate only the modelling of the sky conceived as a discretized hemisphere and mainly take into account the sky conditions which have shown too much variability and generally tend to disregard over-abundant solar radiation [2]. The design of reflective and protective means over-imposed to glazed apertures has not advanced very much due to this kind of approach Instead, we focus on the manifold geometric subtleties that may appear in the design of the interior of buildings, identifying which types of design strategies enhance the diffusion and distribution of radiation that is finally enjoyed by the users and not in achieving theoretical energy balances at the expense of creating discomfort.

In this way, we can analyse very different designs corresponding to sundry tectonic cultures and asses their applicability and adequacy from the perspective of perception and phenomenology and not only of building codes and regulations. Then, new software capable of overcoming the former constraints has been experimentally developed by Cabeza-Lainez and is called DianaX; it is available in reference [1] and also by special appointment with the department.

Technically speaking, to perform the aforementioned study, we need to solve a non-negligible mathematical complexity in this issue, which has resulted in the abandonment of the research of heat transfer when the problem is three-dimensional as it happens in architectural design. In the following, we explain briefly the physical fundamentals of our approach.

If the surfaces in Figure 1, named A_i and A_j, radiate in a diffuse mode, the problem is reduced to finding the balance of energy between them. Neglecting the potential losses of the procedure [3,4], we only need to solve the mutual interchange of radiation that is provided by the shapes involved as the two sources present the same intensity.

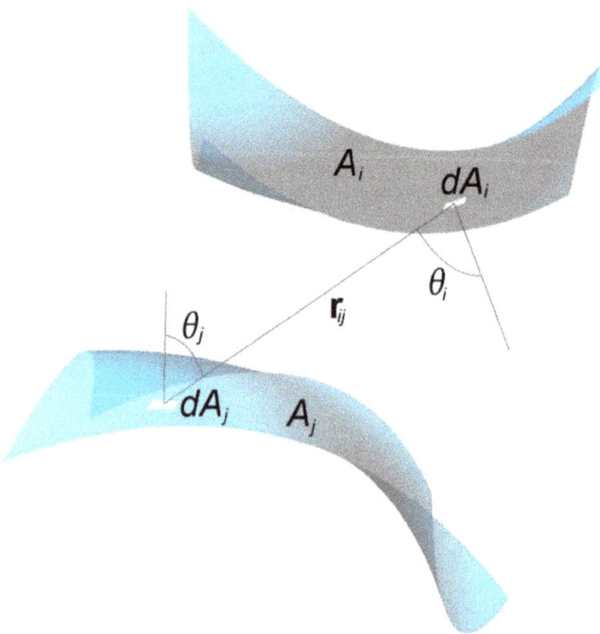

Figure 1. Arrangement of surface-source elements. Source: Salguero-Andujar.

The magnitudes θ_i and θ_j correspond to the angles enclosed by the normal lines to the unit areas dA_i and dA_j (Figure 1), where r_{ij} stands for an arbitrary distance that colligates the unit surface sources dA_i and dA_j.

Setting to facilitate calculations, a phase of negligible rebound reflections from the first source to the second one, the radiation is not redirected and the energy exchange amounts to:

$$d\phi_{ij} = (E_i - E_j)\cos\theta_i \cos\theta_j \frac{dA_i dA_j}{\pi r_{ij}^2} \tag{1}$$

where E_i and E_j correspond to the level of energy (in W/m^2) expelled by the respective sources i and j.

In the following, we designate a non-dimensional entity called, *configuration factor* (sometimes *form factor*), F_{ij} that provides the ratio of radiation that is expelled from source i

caught by source j. Thus, in accordance, the energy that departs from source i and reaches source j can be defined as $E_iA_iF_{ij}$ and, correspondingly $E_jA_jF_{ji}$ amounts to the energy that passes from source j to source i.

In this case, Equation (2) can be expressed in the form:

$$d\phi = E_iA_iF_{ij} - E_jA_jF_{ji}. \quad (2)$$

When we focus on a temporarily constant stage of the problem, where we cannot identify variations in the energy exchange and the estimated fugues of flux that can be attributed to the wayof emission become nominal, we can say that $d\phi = 0$, which means that $E_iA_iF_{ij} = E_jA_jF_{ji}$, an expression known as Lambert's reciprocity theorem [3].

In 1764, the Swiss polymath Lambert published his singular treatise *Photometria*, written in Germanized Latin. In it, he explained his celebrated sixteenth theorem (XVI), also known as the reciprocity principle that states that: "If two surfaces are equally luminous and face each other in some manner, the flux that reaches from each one of them onto the other must be the same" [5].

Such theorem, in turn, allows us to formulate the question in terms of symbolic calculus and, thus, we arrive to the canonical equation that rules over every radiative exchange.

The former generates an asymmetric algebra based in two principles [6]. Cabeza-Lainez obtained without integration the exact solution to numerous shapes applicable in design. This is the basis for the new experimental simulation tool that we present in this paper and that is called DianaX.

2. Methodology

In order to explain how the proposed method works, we will take one of the simplest possible volumes that is enclosed by two surfaces, namely, the spherical cap. In it, by geometric operations deriving from the algebra mentioned previously and knowing the involved areas, the radiant exchange between the cap and itself, F_{11}, is always h/D [6], where h is the height of the cap perpendicular to the base of diameter D = 2 R, with R being the radius of the whole sphere. This expression, which coincides with the ratio of areas of the cap and the whole sphere (see Figure 2), is a particular case of Cabeza-Lainez's first principle, simple enough to facilitate all the ensuing calculations and development of the software.

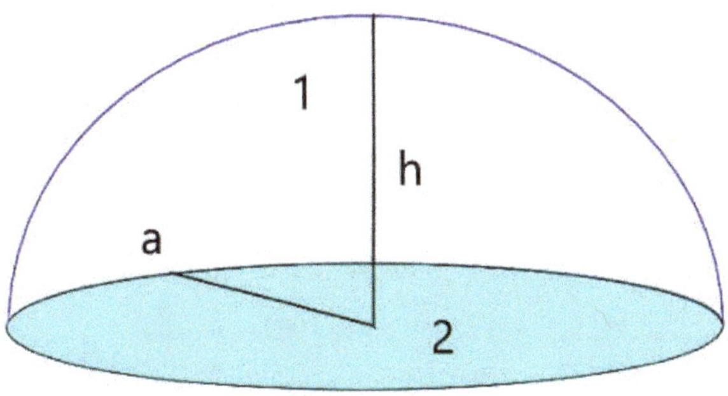

Figure 2. A spheric fragment (surf. 1) with vertical elevation h and sphere radius R, over a circular base (surf. 2) with radius a, different from R.

It is known that the relations between the areas of the cap and any limiting circle are:

$$A_1 = 2\pi Rh \quad (3)$$

$$A_2 = \pi a^2 \quad (4)$$

Generally speaking, a $\neq$ R:

$$A_{tot} = 4\pi R^2 \quad (5)$$

$$F_{11} = A_1/A_{tot} = h/(2R) = h/D \quad (6)$$

A_1 is obtained from Equation (3) as function of R and h. For the whole sphere, whose total area is A_{tot} (5), F_{11} = h/D = 1 and, for the hemisphere, the only case in which R = a, F_{11} = h/D = 1/2 and so forth.

From here stem the first four principles of Cabeza-Lainez [6], which allow us to solve all algebraic operations, including addition and the scalar product of configuration factors that may appear in three-dimensional energy exchanges. For example, a spherical cap is the simplest section that we can apply to a sphere but there are more possibilities detailed in the references [7].

By means of the procedures detailed in the bibliography [8,9], it is possible to assemble computer tools of high accuracy to know in detail the distribution of radiant energy, be it heat, light or sound, in any kind of space or precinct, taking into account their constructive features.

This tool is called method of form factors and has been developed by Cabeza-Lainez. It is based in finding the exchange ratios by virtue of the factors previously defined, which can be of application in each particular case. The tool implies an important advance to know the distribution radiation in all its forms, but it also entails the accurate design of the elements involved in these exchanges, contributing to the mitigation of climate change, for instance, and adding significant psychologic and physiologic bonuses due to the introduction of direct solar radiation in the spaces [10] and the perceptive advantages that it entails.

In the following section, we analyse several designs to find the behaviour of radiative energy in complex spaces, mainly emitted by fenestration or luminaries. We divide this into three sections, buildings from the past, especially temples or churches, contemporary buildings and projects yet to be realized.

3. Results of Utilizing the Simulation Tools

In this section, we present several architectural paradigms as well as projects yet to be realized in which the authors had some intervention. They have been simulated with the experimental software DianaX, developed by Cabeza-Lainez [1], and the figures and quantities presented correspond to the output of this tool. The main input in all cases is external solar radiation in the location considered as well as the characteristics of the building materials, such as glazing and opaque walls; interior veneers can also be important in some cases.

3.1. Simulations of Radiative Transfer in Buildings of the Past

3.1.1. The Roman Pantheon

This is a well-known architectural paradigm in which we identify two sources of radiation, the oculus and a rotating solar patch depicted in the inside of the magnificent dome (Figure 3). Sectional simulation contributes to the understanding of this singular space.

3.1.2. The Church of St. Louis of France (Seville)

It is an 18th-century baroque church in the historic center of Seville, possessing only eight clerestory windows and a lantern over the dome. Radiation here serves to enhance and extensive iconographic program for illustration of the Jesuit novices [8,11].

After the careful calculations performed by virtue of our software, we obtained the graphs presented in Figures 4–8 and the values of illuminance as output, on which we comment briefly in the following paragraphs.

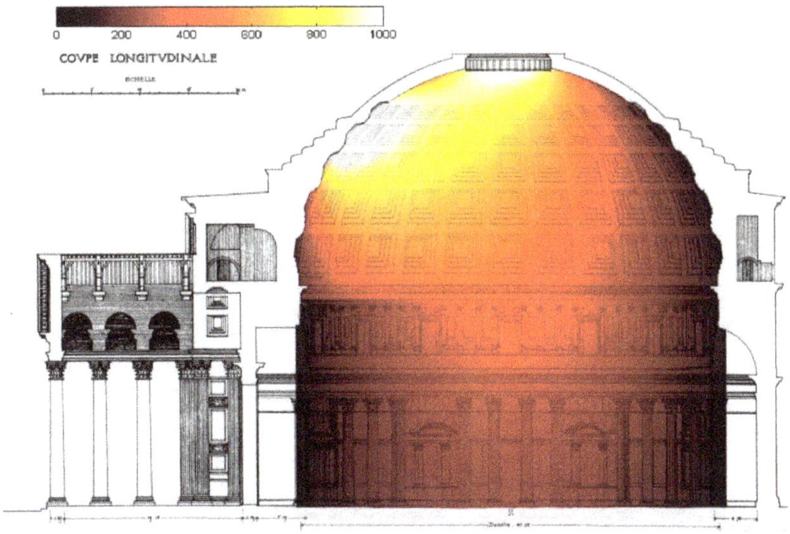

Figure 3. The Roman Pantheon's lighting simulation values in Lux. Source: Cabeza-Lainez.

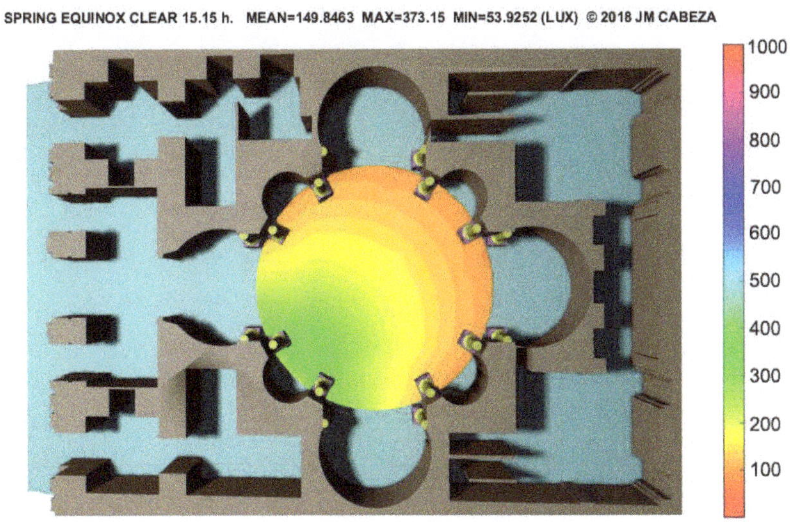

Figure 4. Simulations for the typical spring sunny day values in lux. This case occurs in 74% of the total time (76% during harvest). The distribution is reversed in the south–north axis at 8.45 h. Source: Authors.

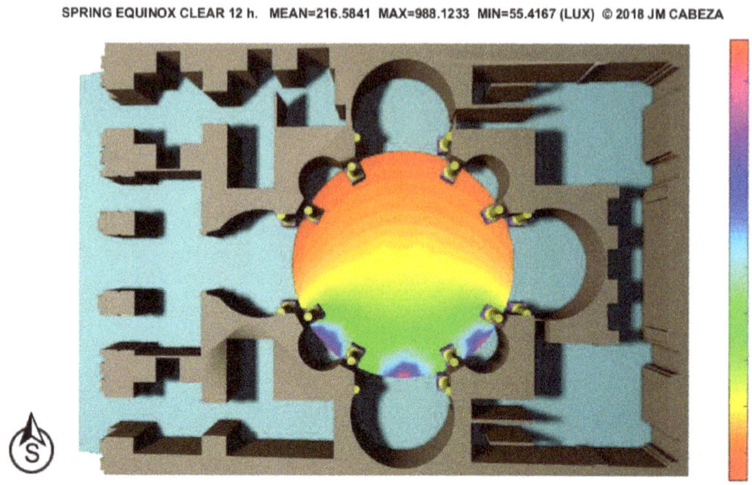

Figure 5. Typical sunny day at noon in Spring. All the values are in unit of lux. Some points may reach illuminances around 1000 lux. Source: Authors.

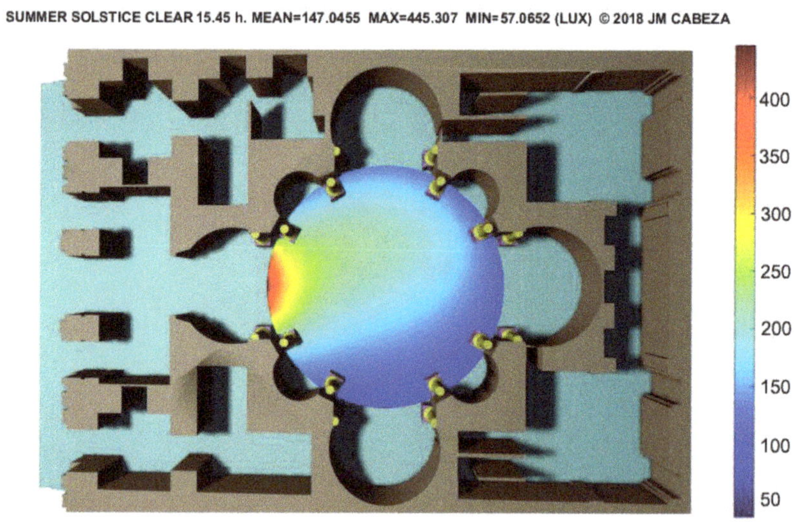

Figure 6. Summer standard sunny day. Reversed in the vertical axes at 8.45 h. Source: Authors.

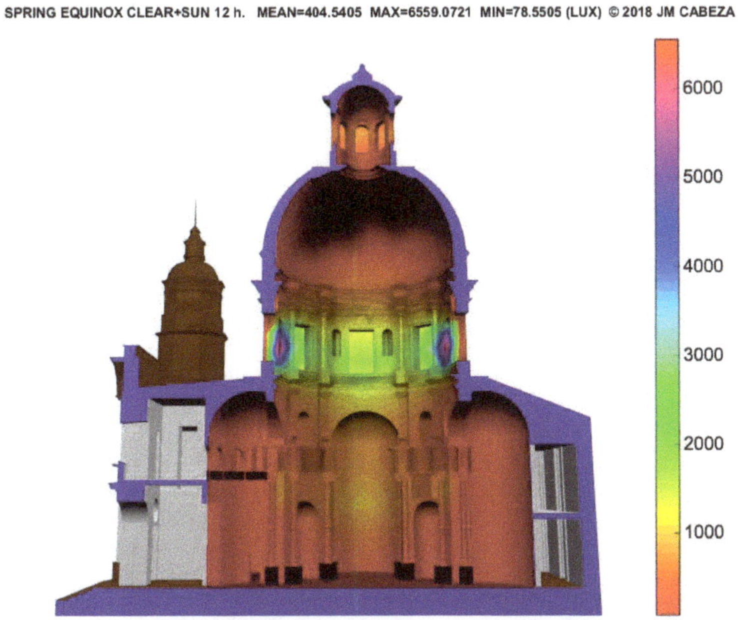

Figure 7. Daylight section on the 21st of March, on a sunny day at noon. Values near 300 lux at the lower points. Source: Cabeza-Lainez.

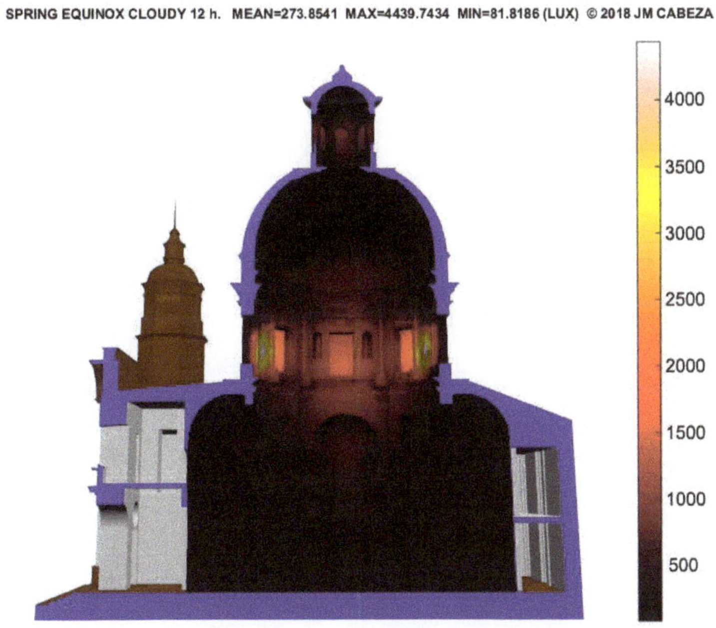

Figure 8. Daylight section on the 21st of September in a cloudy day. Maximum values of 200 lux appear in the lower levels of the space. Source: Cabeza-Lainez.

In spring, the situation of luminous radiation is favourable; during the entire daylighting time, the average values are in the vicinity of 150 lux and values ranging from 200 to 300 lux are normally found (Figure 4). At noon (Figure 5), the mean illuminance is about 200 lux and a maximum of some 1000 lux can be obtained at most parts of the usable floor plan. The distribution of radiation is variegated but does not lead to excessive contrast, as it moves in crescendo from the threshold of 200 lux up to 500 in the altarpiece of the northern arm of the church, where sunrays fall around the beginning of spring.

During the warm season around late June, the illuminance values may reach 150 lux throughout the day (Figure 6), but at midday, some figures over 300 lux are registered.

The simulations reveal the adequate performance of light under the typical scenario in Seville, which is characterised by sunny skies (Figure 6; Figure 7). Following the Spanish Meteorological Agency for the standard values in a year at the selected location, we completed the input parameters with the tool Energy Plus and the registers of the local weather institution. Outside winter, the occurrence of sunny skies is near 90% for the city of Seville.

The illumination's vertical components are not too flat or steep and maintain an adequate angle that allows for a compatible perception of the displays that cover the surfaces of the church as was the baroque norm, but it is certainly not the case of the Pantheon (Figure 3) and the buildings that were directly derived from it or had to be accommodated in ancient Roman structures. We present the simulations of the section of the church for spring clear sky (Figure 7) and cloudy sky (Figure 8) respectively.

The dynamic characteristics of lighting are less acute for the rare situation of cloudy sky, but can still be noticed and have been validated through sensor measurements. The sensors were placed at regular intervals in the lower levels of the spaced and acquired registers for alternate periods of six months.

To summarize, the results presented in Figures 4–8 show that the daylighting field is adequate, starting at a minimum of 100 lux, in plan for an overcast situation (less than 10% of the time). In other conditions, values of over 300 lux can be obtained and averages of 200 lux are safely reached in the core of the church for more than half of the daylighting hours.

3.1.3. The Church of St. Andrew on the Quirinal (Rome)

This church of elliptic plan is an important referent of Baroque architecture, which was constructed by G. L. Bernini also for the Society of Jesus [12]. Due to its singular shape, so different to a cuboid, it is difficult to simulate, but our tool is capable of analyzing such unusual curved geometries with advantage and precision (Figure 9).

In the sectional distribution of radiation, we can appreciate subtleties such as the role played by the lantern and the chapel of the saint with windows embedded in the wall and hidden from view in the outside (Figure 10).

The distribution in the plan is interesting not only because of the high intensities achieved, but also due to the subsidiary role played by the eight chapels that remain mostly in the dark, in contrast with the chapel of the saint in the minor axis of the ellipse and that is illuminated in a clearer way (Figure 11).

In the next subsection, we discuss in brief detail two paradigms of contemporary modern architecture: the Rautatalo building in central Helsinki by Alvar Aalto [13] (Figure 12) and the Glass House by Philip Johnson [14] (Figure 13).

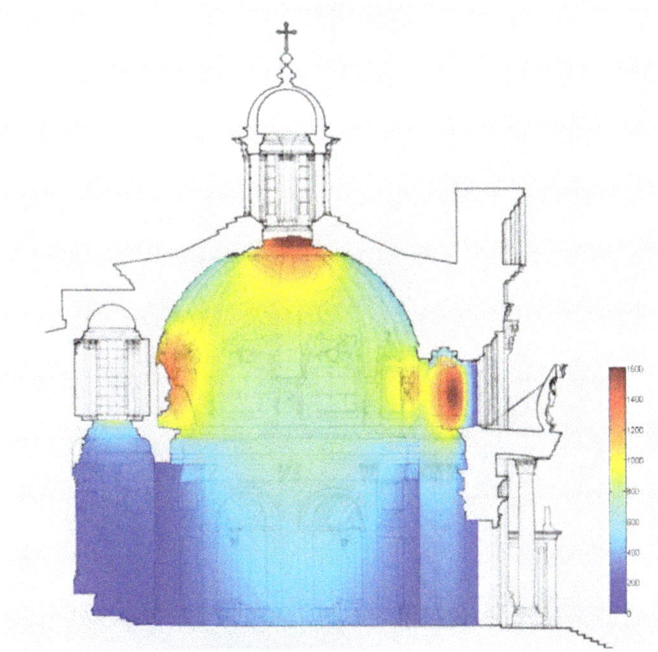

Figure 9. Daylight Section of the Sant'Andrea Church, showing the subtle behaviour of the niche chapel.

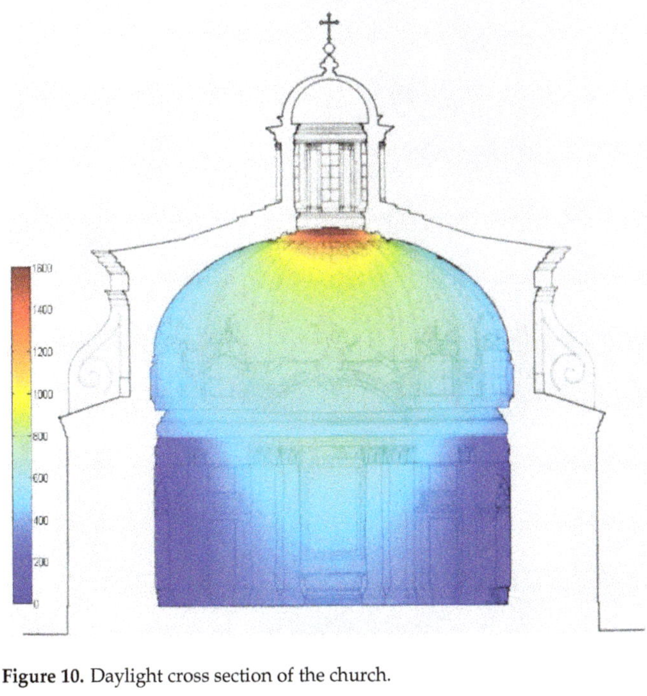

Figure 10. Daylight cross section of the church.

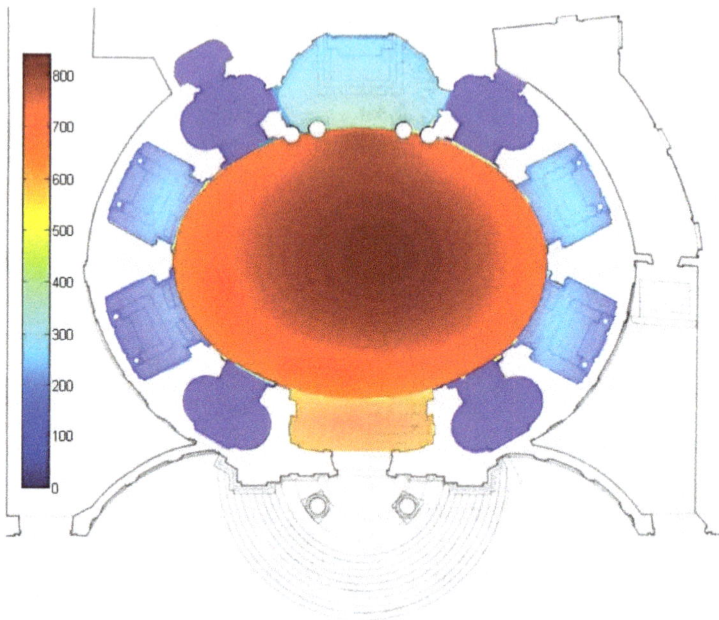

Figure 11. Distribution of light in the plan, showing the contrast between the central space and the chapels.

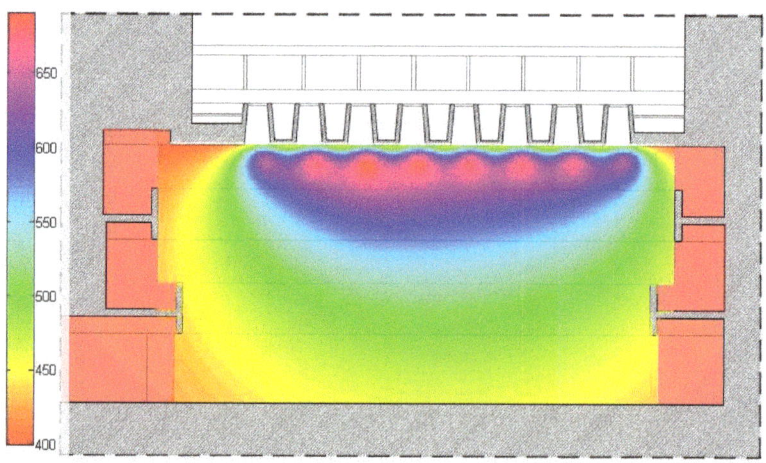

Figure 12. Rautatalo, Helsinki. Sectional values in summer.

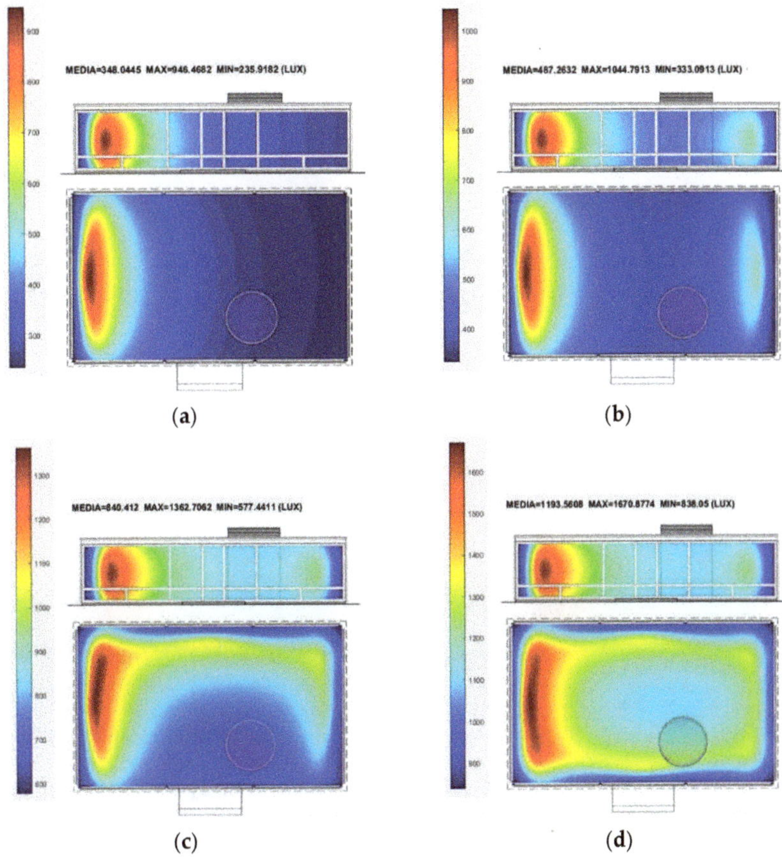

Figure 13. Lighting breakdown following the four façades of the building. (**a**) Glazing only on the left side. (**b**) Left and right sides. (**c**) Both sides and glazing on the upper side. (**d**) Glazing on the four sides.

3.2. Simulations of Contemporary Buildings by A. Aalto and P. Johnson

The Rautatalo atrium features 40 conical lightwells, revealed as circles in the ceiling that add a distinct type of illumination to this singular space conceived as a department store (now the venue for Nordea bank). Lighting levels are much reinforced in the plan and sections of the atrium, but the ceiling remains in stark contrast with the brightly illuminated circles and the rest of the surface [15].

Regarding the Glass house [14], a bolder simulation experiment was conducted since we studied different cases as if only one façade was glazed and, then, successively adding the other three façades step by step (Figure 13a–d). Thus, the contribution of each one of the four façades can be better appreciated. The first one is the southern façade (Figure 13a) and then north (Figure 13b), west (Figure 13c) and east (Figure 13d).

The second proposal that we consider interesting is glazing on the four faces but comparison with a window of half the height of the original (upper part) (Figure 14).

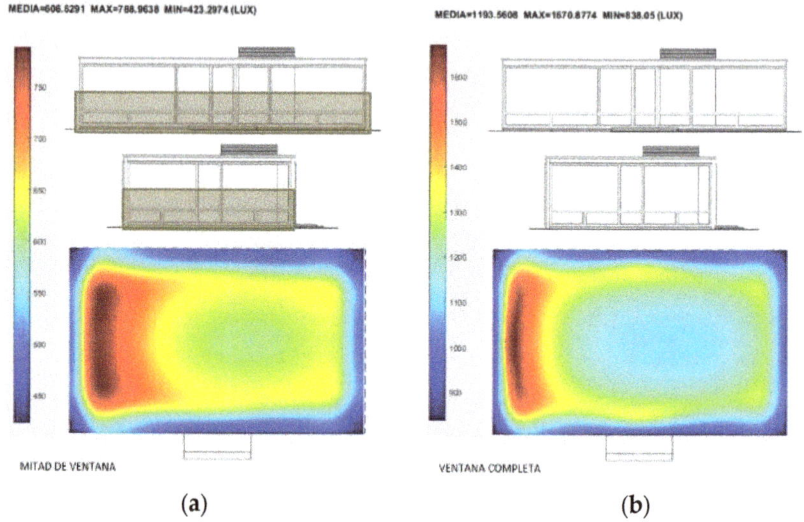

Figure 14. Glass House results in plan. (**a**) Half height windowsill at 1.8 m. (**b**) Full window.

The Glass House by P. Johnson has a rectangular plan of 10 × 17 m, and its height is of 3.65 m; external intensities of radiation considered are the average in a clear spring day in New Canaan, Connecticut, USA, that is, 5000 lux to the South, 3000 lux to the East and West and finally 2000 lux in the north. The resulting illuminance is obtained at the floor level.

The reflection coefficients considered were set as: brick floor 0.2, grey ceiling 0.5 and glazed envelope at 0.3 since the transmittance of the glass was considered at 0.7. It is important to stress that, in order to calculate daylighting radiation, we did not consider external obstructions, annexed buildings or vegetation nor externally reflected sunrays. In this example, direct radiation was not taken into account.

The levels found in the case of daylight show that the highest values are in the vicinity of the glass wall but are significantly reduced towards the inner part of the home [16]. If there was one single glass wall, the distribution of light would be rather uneven, but as we add more glazed walls to the equation, luminous uniformity rises and visual comfort is improved.

It is noticeable that the amount of light available is rather high, reaching 1193 lux on average. In this sense, we produced the simulation of Figure 14, in which we reduce the height of the window to a half as if we had an opaque window sill of 1.8 m.

By reducing the size of the openings, we can see how the illuminance is reduced to a half, from 1193 lux with full windows to 606 lux as average, in Figure 14a. Obviously, a totally glazed façade reinforces the connection with the environment and the landscape, but it is still interesting that, with half the size of the glass, there are significant levels that offer optimum visual comfort and save energy as well. The conclusion is that, if not for subjective design considerations, the excess glazing would be redundant from a radiation point of view [17–20].

In the last subsection of this section, we present three unbuilt projects, namely, a cultural centre in San Sebastian (Donosti) by Arch. Pablo Rico Pérez, the railway hub at Barcelona's airport that was the first entry at a competition in 2010 by Archs. Cesar Portela and Antonio Barrionuevo with assistance from J. Cabeza-Lainez, and finally a project of artificial lighting at a model school by Arch. Lorenzo Muro.

3.3. *Simulations of Architectural Projects*

3.3.1. Conversion Project of an Old Tobacco Factory in Donosti (San Sebastian) into a New Cultural Center Called Tabakalera

This project was a competition entry in which all the refurbishments proposed were strictly based on simulations with DianaX. In this case, the simulation included temperature levels to be achieved by the implementation of diverse strategies (Figures 15–21).

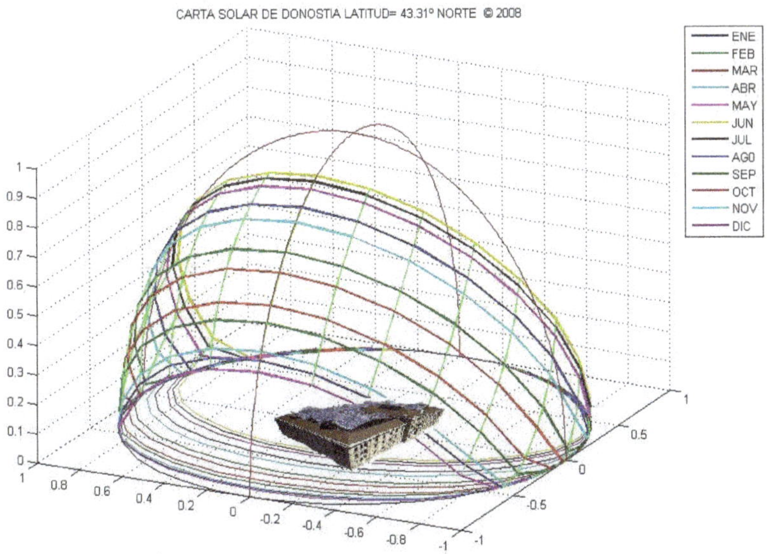

Figure 15. Solar chart for the new cultural centre.

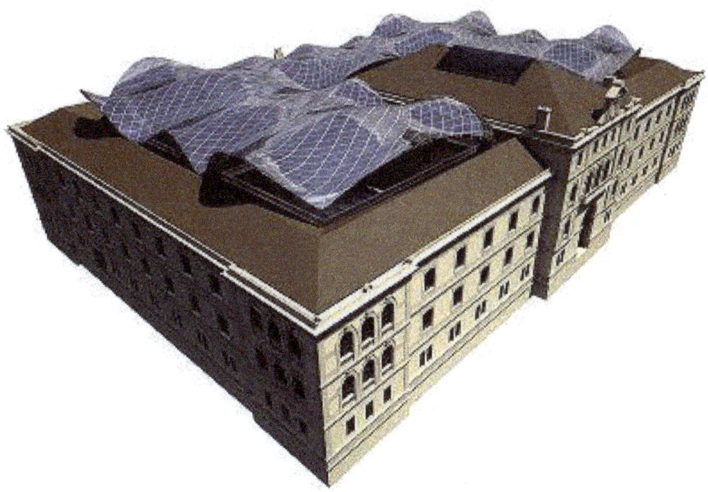

Figure 16. View of the retrofitted main building, a project by archs. Pablo Rico Perez and J. M Cabeza-Lainez.

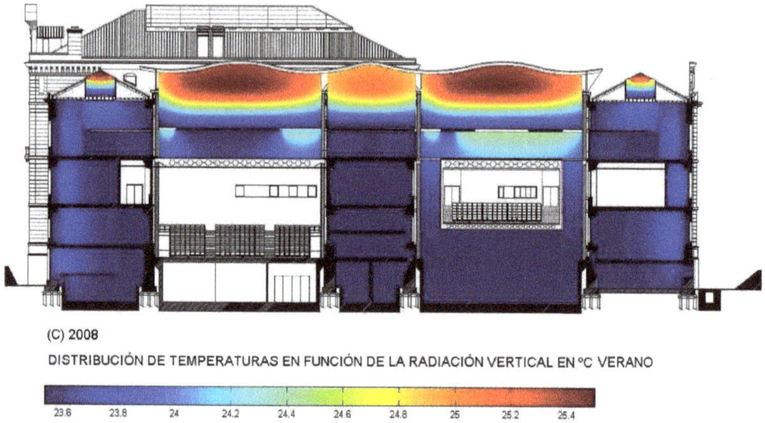

Figure 17. Vertical distribution of temperatures in deg. C, during summer.

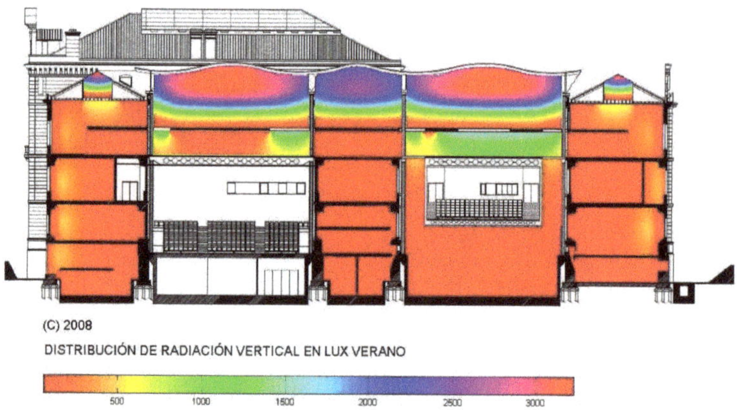

Figure 18. Radiation distribution in lux, in summer.

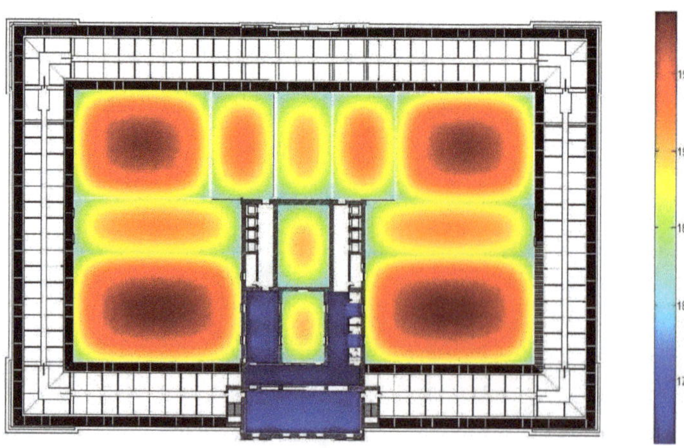

Figure 19. Temperature distribution in deg. C, during winter, inside the retrofitted spaces.

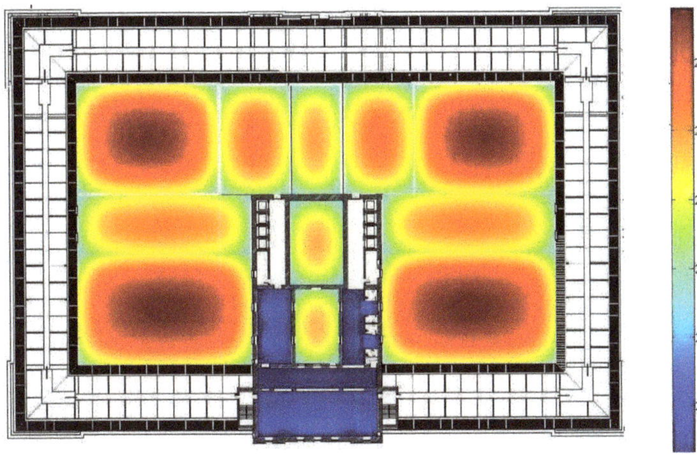

Figure 20. Temperature distribution in deg. C, during summer.

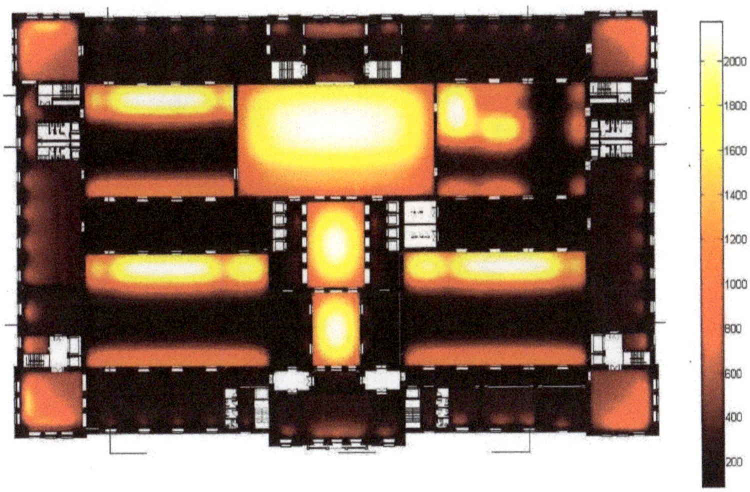

Figure 21. Radiation distribution in winter on the second floor.

The project is organized through former courtyards that are now partly covered and reused, and include PV systems on the ceiling and new types of filters for radiation [21–23]. It follows the inner distributions of temperature and radiation both in the section and plan Figures 17–21.

3.3.2. Project for a Transport Hub at Barcelona Airport

This project included the railway station as well as the subway and buses, and it was meant to be a hub for all transport systems in the airport. It was named Campo dei Miracoli (Italian for field of miracles) [24–27]. Its main feature was a covered elliptic atrium with special blind systems to control the light in the lobby of the central station [28–30] (Figures 22–24).

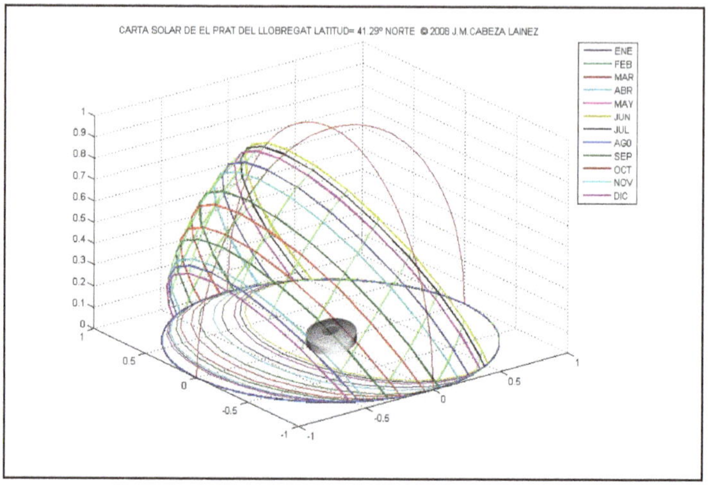

Figure 22. Solar chart for the railway hub.

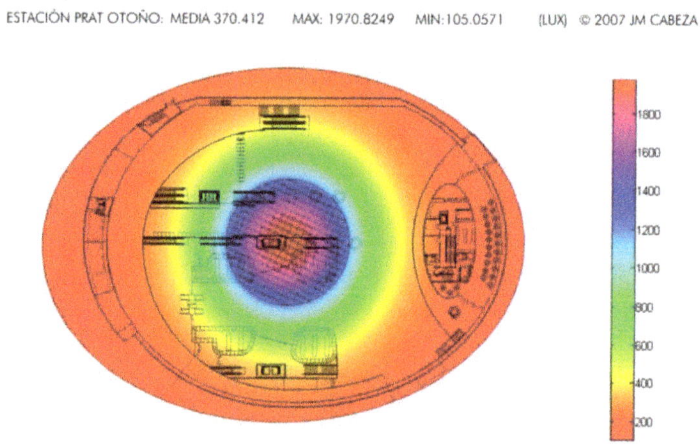

Figure 23. Distribution of radiation in autumn in plan. Values in Lux.

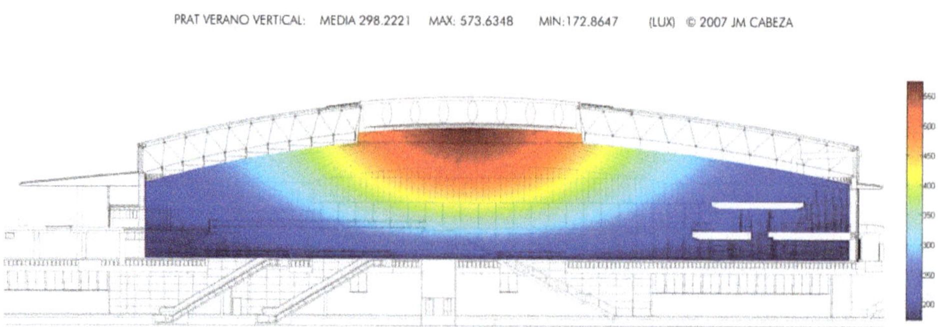

Figure 24. Distribution of radiation in summer in section. Values in Lux.

3.3.3. Artificial Lighting of School Comparing Two Types of Luminaires and Our SOFTWARE against Lightscape 3.2

In this example, we can assess how interior lighting is viable with surface source instead of the customary linear sources.

Firstly, we introduce a model classroom, which has specific regulations in Spain, including directions related to its dimensions and to the lighting systems employed. Two different conditions were simulated: a continuous lighting ceiling (floodlight) and three lines of luminaires distributed in the ceiling [31]. Both configurations have in origin the same total luminous flux (total).

The classroom is defined as an open space with a rectangular plan of 6.90 m. by 8.40 m. and a height of 3.0 m. The walls have a smooth finish with the usual reflection coefficients and no daylight was considered in the example. Such coefficients are: ceramic floor 0.2, walls 0.5 and ceiling 0.7. The luminaires are composed of LED panels with a diffuser. The working plan is set on the floor (Figure 25)

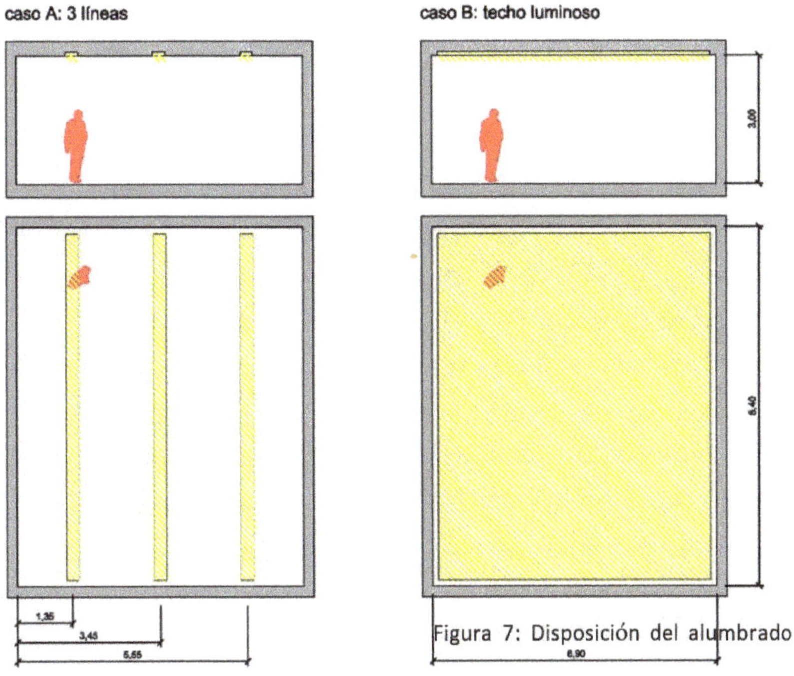

Figure 25. Linear lighting (**right**) versus surface lighting (**left**).

The illuminance level to be achieved in the classroom is of 500 lux (School regulations and UNE 12464.1-European norm for interior lighting).

1 Linear lighting. Three line fixtures of luminaires are set on the ceiling and orientated longitudinally and parallel to the wider side of the room. The dimensions of such lines are fixed at 0.30 m. by 8.40 m and 12,600 lumens. In the classroom, there are three lines 2.1 m apart from axis to axis and centred with respect to the Y-axis (Figure 25).

2 Surface source. A luminaire covering the whole breadth of the ceiling is proposed with dimensions 6.9 m by 8.40 m and to a height of 3 m. This luminaire has 37,800 lm.

As can be observed in the results output (Figure 26), the luminous ceiling can be an option with similar levels on the floor as compared with the linear source, but the distribution is much more homogeneous, especially in the upper section of the walls

(Table 1; Table 2 and Figure 26). Such quality is particularly important in classrooms because it is in this area where the blackboards are usually located [32].

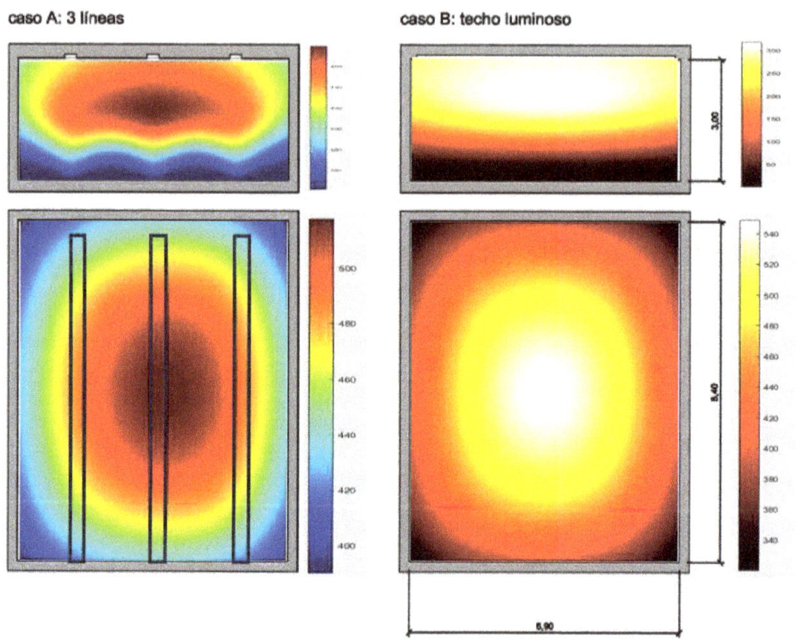

Figure 26. Superficial lighting vs. linear lighting. Calculations with DianaX by Lorenzo Muro and Joseph Cabeza-Lainez.

Table 1. Surface lighting vs. linear lighting calculation with DianaX. Prepared by the authors.

3 lines (0.30 × 8.40 m)	surface (6.90 × 8.40 m)
12600 lm each line	37800 lm surface
5000 lx each line	652.17 lux surface
Reflection 0.70 0.50 0.20	Reflection 0.30 0.50 0.20 (reflection from the ceiling = translucent glazing 30%)

Table 2. Superficial lighting vs. linear lighting. Calculations results with DianaX by Lorenzo Muro and Joseph Cabeza-Lainez.

Em [lx] 466	Em [lx] 454
Emin [lx] 390	Emin [lx] 320
Emax [lx] 518	Emax [lx] 550
Emin/Em 0.837	Emin/Em 0.706

We do not possess evidence of the classroom provided with luminous ceiling in Spain, but it is a system used in similar environments, such as offices, and the architectural lighting pioneer P. H. Moon often commented on them. A great deal of buildings representative of the modern movement of authors such as Mies van der Rohe, Eero Saarinen or SOM [33] have offices illuminated by continuous luminous ceiling. In most of them, the lighting designer is Richard Kelly (1910–1977), a forerunner of architectural lighting. There are instead thousands of classrooms illuminated by linear sources since the regulations demand it. However, starting from our simulations, the rules can be reformulated in search for a better visual environment (Figure 27 and Table 3).

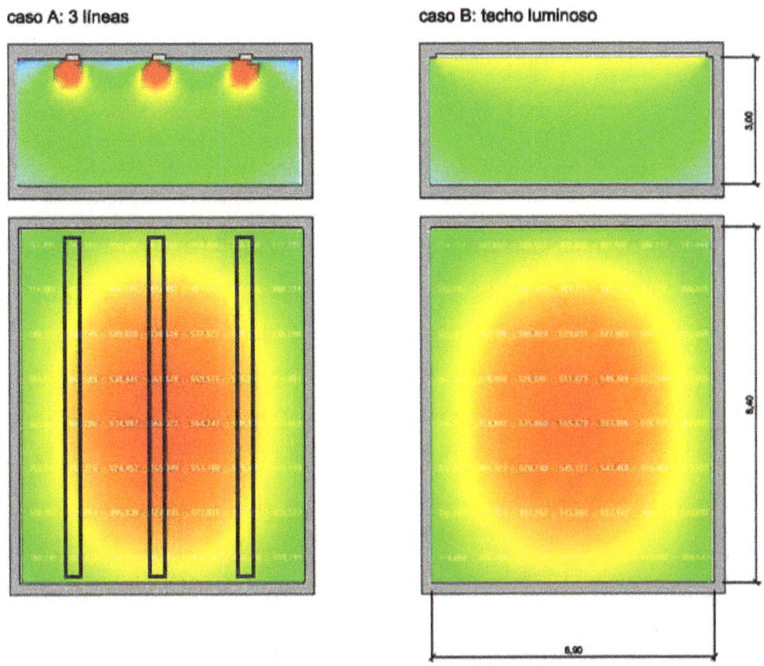

Figure 27. Surface lighting vs. linear by Lorenzo Muro.

Table 3. Surface lighting vs. linear lighting. Calculations with Lightscape 3.2. by Lorenzo Muro.

Em [lx] 452	Em [lx] 452
Emin [lx] 221	Emin [lx] 230
Emax [lx] 571	Emax [lx] 559
Emin/Em 0.489	Emin/Em 0.509

To confirm the data obtained with other programs frequently used by designers, the same calculations were analysed with the software Lightscape 3.2, obtaining very similar values in both cases and thus we considered the DianaX tool as a valid tool for radiative simulation.

4. Discussion

In this article, we showed that several approaches of simulation, involving environmental properties, can be employed satisfactorily to improve the design of spatial configurations without disregarding the intangible values of the buildings considered. Such kind of holistic approach is novel in the realm of Architecture, Heritage and especially Project Design [34]. This is the result of a series of experiences that we have launched on a new field that we have called scientific design in the sense that the decision-making process is informed by objective determinations founded on science [1,5].

In this paper, we demonstrated that our simulation tool can be applied in many different situations that appear in the design of architectural spaces, be it from the past, the present or even future projects [35,36]. Several branches of building activities, for example, can be enhanced in the provision of supplementary means for energy or other aspects.

Other procedures and tools have failed in this task because they do not possess versatility to include geometries other than the parallelepiped and need to assign any possible form to it. Additionally, they do not deal with direct sunlight referring only to the sky condition, which makes calculations rather clumsy and variable [37].

What is dominant, on the contrary, in our method is the thorough knowledge of the geometries involved in the problem and the careful insight into the mathematical definitions in which these forms are enclosed. Some especially crafted forms could prove more appropriate than others to convey the forces that allow for the habitability and amenity [38,39] of the designs that we are bound to be created when facing any new development within urban or landscape boundaries.

Author Contributions: Conceptualization, J.C.-L. and I.R.-C.; methodology, J.C.-L.; software, J.C.-L.; validation, J.-M.A.-M. and P.R.-D.; formal analysis, P.R.-D.; investigation, Y.X.; resources, P.R.-D.; writing—review and editing, I.R.-C.; visualization, Y.X.; supervision, J.C.-L.; All authors have read and agreed to the published version of the manuscript.

Funding: This research received no external funding.

Institutional Review Board Statement: Not applicable.

Informed Consent Statement: Not applicable.

Data Availability Statement: Not applicable.

Acknowledgments: J.C.-L. would like to acknowledge the support demonstrated by Guohui Liu, Jialei Wu, Juan Francisco Ojeda and Carmen Barrio de Alarcon. We are grateful to Lorenzo Muro for his kindness and help.

Conflicts of Interest: The authors declare no conflict of interest.

References

1. Cabeza-Lainez, J.M. *Fundamentos de Transferencia Radiante Luminosa o La Verdadera Naturaleza del Factor de Forma y sus Modelos de Cálculo*; Netbiblio: Seville, Spain, 2010.
2. Ashdown, I. Radiative Transfer Networks Revisited. *J. Illum. Eng. Soc.* **2001**, *1*, 38–51. [CrossRef]
3. Lambert, J.H. *Photometri: Sive de Mensura et Gradibus Luminis, Colorum et Umbrae*; DiLaura, D., Ed.; IESNA: New York, NY, USA, 2001.
4. Holman, J.P. *Heat Transfer*; Mac Graw-Hill: New York, NY, USA, 1997.
5. Sasaki, K.; Sznajder, M. Analytical view factor solutions of a spherical cap from an infinitesimal surface. *Int. J. Heat Mass Transf.* **2020**, *163*, 120477. [CrossRef]
6. Cabeza-Lainez, J.M.; Pulido-Arcas, J.A. New configuration factors for curved surfaces. *J. Quant. Spectrosc. Radiat. Transf.* **2013**, *111*, 71–80. [CrossRef]
7. Cabeza-Lainez, J. Architectural Characteristics of Different Configurations Based on New Geometric Determinations for the Conoid. *Buildings* **2022**, *12*, 10. [CrossRef]
8. Almodovar-Melendo, J.-M.; Cabeza-Lainez, J.-M.; Rodriguez-Cunill, I. Lighting Features in Historical Buildings: Scientific Analysis of the Church of Saint Louis of the Frenchmen in Sevilla. *Sustainability* **2018**, *10*, 3352. [CrossRef]
9. DiLaura, D.L. *New Procedures for Calculating Diffuse and Non-Diffuse Radiative Exchange Form Factors*; ASME: New York, NY, USA, 1999.
10. Cabeza-Lainez, J.M. The Quest for Light in Indian architectural heritage. *J. Asian Archit. Build. Eng.* **2008**, *7*, 39–46. [CrossRef]
11. Almodóvar-Melendo, J.M.; Cabeza-Lainez, J.M. Nineteen thirties architecture for tropical countries: Le Corbusier's brise-soleil at the Ministry of Education in Rio de Janeiro. *J. Asian Archit. Build. Eng.* **2008**, *7*, 9–14. [CrossRef]
12. Revenga Domínguez, P. *Barroco*; Grupo Cultural Ediciones: Madrid, Spain, 2008; ISBN 978-84-8369-099-4.
13. Revenga Domínguez, P. Un alboroto magnífico. In *Palas y las Musas. Diálogos entre la Ciencia y el Arte*; México, D.F., Ed.; Siglo XXI: Mexico DF, Mexico, 2016; Volume 2, ISBN 978-607-03-0782-9.
14. Cabeza-Lainez, J.M.; Almodóvar-Melendo, J.M. Daylight, Shape, and Cross-Cultural Influences through the Routes of Discoveries: The Case of Baroque Temples. *Space Cult.* **2018**, *21*, 340–357. [CrossRef]
15. Cabeza-Lainez, J.; Almodovar-Melendo, J.-M.; Dominguez, I. Daylight and Architectural Simulation of the Egebjerg School (Denmark): Sustainable Features of a New Type of Skylight. *Sustainability* **2019**, *11*, 5878. [CrossRef]
16. Salguero-Andújar, F.; Cabeza-Lainez, J.-M. New Computational Geometry Methods Applied to Solve Complex Problems of Radiative Transfer. *Mathematics* **2020**, *8*, 2176. [CrossRef]
17. Cabeza-Lainez, J.M.; Salguero-Andújar, F.; Rodríguez-Cunill, I. Prevention of Hazards Induced by a Radiation Fireball through Computational Geometry and Parametric Design. *Mathematics* **2022**, *10*, 387. [CrossRef]

18. Salguero-Andújar, F.; Prat-Hurtado, F.; Rodriguez-Cunill, I.; Cabeza-Lainez, J. Architectural Significance of the Seokguram Buddhist Grotto in Gyeongju (Korea). *Buildings* **2022**, *12*, 3. [CrossRef]
19. Rubio-Bellido, C.; Pulido-Arcas, J.A.; Cabeza-Lainez, J.M. Understanding climatic traditions: A quantitative and qualitative analysis of historic dwellings of Cadiz. *Indoor Built Environ.* **2018**, *27*, 665–681. [CrossRef]
20. Moore, F. *Concepts and Practice of Architectural Daylighting*; Van Nostrand Reinhold: New York, NY, USA, 1991.
21. Modest, M.F. View Factors. In *Radiative Heat Transfer*, 3rd ed.; Academic Press: Boston, MA, USA, 2013; pp. 129–159.
22. Petty, M. "The Edge of Danger": Artificial lighting and the dialectics of domestic occupation in Philip Johnson's Glass and Guest Houses. In Proceedings of the Occupation: Negotiations with Constructed Space, Brighton, UK, 2–4 July 2009.
23. Moon, P.H.; Spencer, D.E. *The Photic Field*; The MIT Press: Cambridge, MA, USA, 1981.
24. Robbins, C.L. Daylighting. In *Design and Analysis*; Van Nostrand Reinhold: New York, NY, USA, 1986.
25. Charles, P.P.; Thomas, C.R. Building performance simulation in undergraduate multidisciplinary education: Learning from an architecture and engineering collaboration. In Proceedings of the Building Simulation, Glasgow, Scotland, 27–30 July 2009.
26. William, M.C.L. *Perception and Lighting as Formgivers for Architecture*; McGraw-Hill: New York, NY, USA, 1977.
27. Cabeza Lainez, J.M. *Solar Radiation in Buildings, Transfer and Simulation Procedures*; Babatunde, E.B., Ed.; InTech: Rijeka, Croatia, 2012; ISBN 978-953-51-0384-4.
28. Ashdown, I. *Radiosity: A Programmer's Perspective*; John Wiley & Sons Inc.: New York, NY, USA, 1994. Available online: http://www.helios32.com (accessed on 22 July 2021).
29. MacAllister, A.S. Graphical Solutions of Problems Involving Plane-Surface Lighting Sources. *Lighting World* **1910**, *56*, 17–24.
30. Ne'eman, E. Visual Aspects of Sunlight in Buildings. *Lighting Res. Technol.* **1974**, *6*, 159–164. [CrossRef]
31. Yamauchi, J. Theory of Field of Illumination. *Res. Electro-Tech. Lab.* **1932**, *2*, 3–16.
32. Hopkinson, R.G.; Petherbridge, P.; Longmore, J. *Daylighting*; Heinemann: London, UK, 1966.
33. Moon, P.H. *The Scientific Basis of Illuminating Engineering*; Dover Publications: New York, NY, USA, 1962.
34. Yamauchi, J. The Light Flux Distribution of a System of Inter-Reflecting Surfaces. *Res. Electro-Tech. Lab.* **1927**, *3*, 11–18. (In Japanese)
35. Almodóvar-Melendo, J.-M.; Quesada-García, S.; Valero-Flores, P.; Cabeza-Lainez, J. Solar Radiation in Architectural Projects as a Key Design Factor for the Well-Being of Persons with Alzheimer's Disease. *Buildings* **2022**, *12*, 603. [CrossRef]
36. Almodovar-Melendo, J.M.; La Roche, P. Roof ponds combined with a water-to-air heat exchanger as a passive cooling system: Experimental comparison of two system variants. *Renew. Energy* **2019**, *141*, 195–208. [CrossRef]
37. Berque, A. *La Rizière et la Banquise*; Colonisation et Changement Culturel a Hokkaido; Publications Orientalistes de France: Paris, France, 1980.
38. Berque, A. *Poetique de la Terre-Histoire Naturelle*; Belin: Paris, France, 2014.
39. Yamauchi, J. *The Amount of Flux Incident to Rectangular Floor through Rectangular Windows*; Researches of the Electro-Technical Laboratory: Tokyo, Japan, 1929; No. 250.

designs

MDPI

Article

Strength of Composite Columns Consists of Welded Double CF Sigma-Sections Filled with Concrete—An Experimental Study

Mohamed A. Reda [1,2], Ahmed M. Ebid [2,*], Sherif M. Ibrahim [1] and Mohamed A. El-Aghoury [1]

1 Structural Engineering Department, Faculty of Engineering, Ain Shams University, Cairo 11535, Egypt
2 Structural Engineering & Construction Management Department, Faculty of Engineering & Technology, Future University in Egypt, Cairo 11865, Egypt
* Correspondence: ahmed.abdelkhaleq@fue.edu.eg

Abstract: In-filled tubes section is a very successful configuration for axially loaded members such as columns and struts. Steel shell tube filled with concrete has many advantages, such as eliminating the need for shuttering, reinforcement bars or ties besides increasing both flexural and axial capacities and enhancing the ductility. The main disadvantage of in-filled tubes is the need for a shell thick enough to prevent the local buckling and hence the local decomposition. Previous studies tried to solve this problem using intermediate stiffeners or shear connectors. This research presents another approach to solve this problem using double cold-formed sigma-sections (face to face) as steel shell tubes. Sixteen specimens with different lengths, cross section dimensions and shell thicknesses were tested under both concentric and eccentric compression loads. Ultimate capacities, lateral deformations and normal strains were recorded. The theoretical capacities were calculated using AISC-LRDF-94, EN-1994-04 and CSI-COL software considering full composite action, and the deviations from the experimental results were 24%, 24% and 13%, respectively.

Keywords: cold formed; composite columns; sigma sections; in-filled column

Citation: Reda, M.A.; Ebid, A.M.; Ibrahim, S.M.; El-Aghoury, M.A. Strength of Composite Columns Consists of Welded Double CF Sigma-Sections Filled with Concrete—An Experimental Study. *Designs* **2022**, *6*, 82. https://doi.org/10.3390/designs6050082

Academic Editor: Farshid Aram

Received: 23 August 2022
Accepted: 13 September 2022
Published: 21 September 2022

1. Introduction

Composite steel and concrete members, such as Concrete-filled Steel Tubes (CFST), have been widely used since the early decades of the 20th century in constructing both buildings and bridges [1]. Combining the high tensile strength of steel with the high compressive strength of concrete gives the (CTSF) members the advantages of low cost, high ductility and hence high performance under seismic loads [2,3]. Several research studies were carried out to study the behavior (CFST) columns with different cross sections (circular, square and rectangular).

Lin et al., 2005 [4] carried out an experimental study using 50 (CFST) specimens to investigate the effect of cross-sectional shape (circular or square), dimensions, yield strength and local slenderness ratio (D/t) on the axial capacity of the column. They developed an equation to predict the axial capacity considering the studied factors. The results of this research were valid for short columns only. Zhang and Guo, 2007, 2011 [5,6] tested 26 rectangular section (CFST) specimens with different slenderness ratios (L/r), aspect ratios (b/d), steel to concrete ratios (As/Ac) and relative eccentricity ratios (e/d) to figure out the effect of each parameter on the capacity of the column. They compared the test results with international codes. However, the study considered only rectangular sections. Bahrami et al., 2011 [7] presented a theoretical study using non-linear finite element analysis models to investigate the behavior and capacity of (CFST) columns under axial compressive loading. Different cross sections, number of stiffeners and thicknesses of steel sheets were tested to investigate their effects on both behavior and capacity. The study was concerned with square columns only. Ren et al., 2014 [8] tested 44 axially loaded short (CFST) columns with special cross sections (triangular, fan-shaped, D-shaped, 1/4 circular and semi-circular).

They reported the failure mode and ultimate capacity for each column, but their conclusions are valued for stub (short) columns only.

Double-tube composite sections were investigated by Wang et al., 2017 [9], Ding et al., 2019 [10] and Ci et al., 2020 [11]. They presented experimental studies to investigate the capacity of concrete-filled double-tube circular columns under axial compression loads considering the effect of concrete strength, yield strength, inner and outer tube thickness, and the ratio between the diameters of inner and outer tubes. The tested samples were simulated using 3D-FEM models, and the experimental results were compared with both numerical models and international design codes. The tested samples were only short circular and rectangular columns.

Cold-formed short, square composite columns were studied by Tao et al., 2009 [12]. They suggested a design procedure for stiffened concrete-filled thin-walled steel tubular columns based on experimental study results. They found that adding fibers to concrete can effectively increase ductility.

Stainless-steel shells were used instead of traditional steel shells in (CFST) members to increase the corrosion resistance and add some architectural effects. Young and Ellobody, 2005 [13] tested a number of rectangular cold-formed stainless-steel tubes filled with concrete under axial compression and suggested design recommendations for such short columns. Dai et al., 2020 [14] tested and analyzed the compressive behavior of 18 specimens of stainless-steel short and circular columns filled with concrete. He et al., 2020 [15] studied the buckling effect on slender circular (CFST) with stainless-steel shell and high-strength concrete using 12 experimental tests. They identified different failure modes and presented a set of load deformation curves. In the same context, Kazemzadeh et al., 2020 [16] presented a series of experimental and numerical investigations on the local and post local buckling of stainless-steel composite columns to investigate the axial slenderness limit of different column cross sections (box, circular and I-section).

El-Aghoury et al., 2016 [17,18] studied both the ultimate capacity and buckling behavior of axially loaded columns consisting of combined CFS sigma pure steel sections (empty shells). They used 3D numerical modeling to investigate both the global and local buckling behavior of the double sigma section.

This research presented a novel composite cross section that has never been tested before; this novel section consists of double cold-formed sigma-sections (face to face) filled with concrete. The aim of this study is to experimentally investigate the ultimate capacity of the previously described composite column considering the effect of the global slenderness ratio of the column (L/r), aspect ratio of the cross section (b/d), local slenderness ratio (c/t) and relative eccentricity ratio (e/d). Besides that, the experimental capacities were compared with theoretical capacities calculated using AISC-LRFD, EN-1994 and CSI-COL software, considering full composite action behavior.

2. Materials and Methods

2.1. Concrete

Table 1 presents the mix proportions of the used concrete, where the fineness modulus of the used sand was 2.30 and the measured slump was 37 mm. Figure 1 summarizes the grain size distribution and the physical properties of both coarse and fine aggregates. The design strength of the concrete mix was 20 Mpa. The actual compressive cube strength and elastic modulus of concrete were measured by testing 150 mm size cubes and 150 × 150 × 300 mm size prisms, respectively. The average cube compressive strengths were 17.2 and 21.1 Mpa after 7 and 28 days, respectively. The measured concrete elastic modulus (Ec) after 28 days was 21.3 Gpa.

Table 1. The concrete mix proportions (by weight/m^3).

Item	OPC Grade R42.5	Coarse Aggregate (Crushed Stone)	Fine Aggregate (Natural Sand)	Potable Water
Unit	kg	kg	kg	Liter
Value	250	1300	650	125

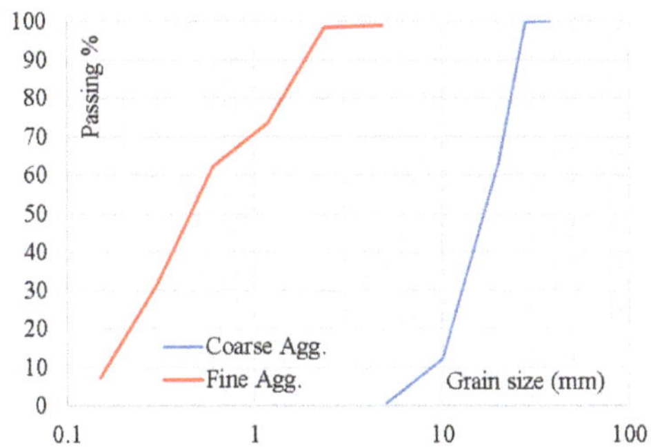

Figure 1. Grain size distribution of fine and coarse aggregates.

2.2. Steel

All sigma sections were fabricated using 1.5, 2.0 and 2.5 mm thick steel sheets. Three coupon samples for each thickness were tested according to ASTM E8 [19], using a displacement-controlled servo-hydraulic tensile testing machine. The loading rate was 0.8 mm/min to measure yield stress (Fy) and ultimate strength (Fu). Table 2 summarizes the test results.

Table 2. Steel properties.

Sample	Thickness (mm)	Fy (Mpa)	Fu (Mpa)
1	1.5	336	445
2	2.0	329	438
3	2.5	322	431

2.3. Test Specimens

The experimental program was designed to study the effect of global slenderness ratio (λ), local slenderness ratio (c/t), aspect ratio (d/b) and load eccentricity ratio (e/d), where © is the longest straight portion of the steel shell. Accordingly, 16 specimens were fabricated and tested in the labs of the Housing and Building National Research Center (HBRC), Giza, Egypt. Each specimen consists of two cold-formed sigma sections welded face to face using a single bevel butt weld and filled with concrete. The specimen's dimensions were selected within the shown ranges in Table 3.

Table 3. Considered values for studied variables.

Variable	Considered Values
Column height (L)	1000, 2000 and 2500 mm
Cross section depth (d)	100, 150 and 200 mm
Load eccentricity ratio (e/d)	0.00, 0.125 and 0.25
Steel shell thickness (t)	1.5, 2.0 and 2.5 mm

All specimens have the same cross section width (b) of 100 mm. Table 4 summarizes the configurations of each specimen, while Figure 2 shows the specimens manufacturing process and all fabricated steel shells tube.

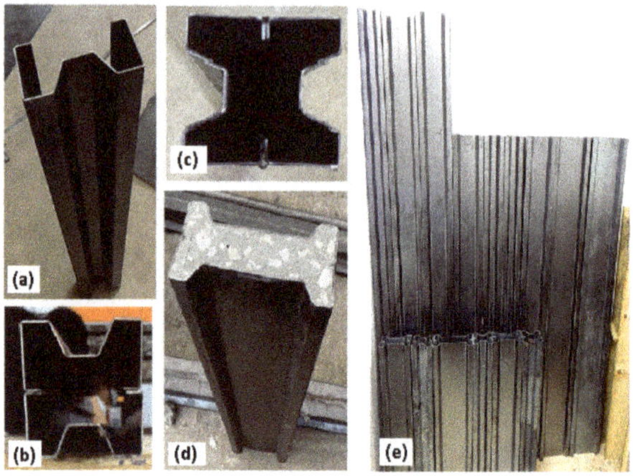

Figure 2. Specimens manufacturing process (**a**) Forming sigma sections, (**b**) assembling column section, (**c**) welding the two stigmas, (**d**) casting concrete and (**e**) all fabricated steel shells.

Table 4. Summary of tested specimens' configurations.

Spec.	L	b	d	t	c	e	r_{min}	λ_{max}	d/b	c/t	e/d
ID	(mm)	(mm)	(mm)	(mm)	(mm)	(mm)	(mm)				
1	2500	100	150	1.5	75	37.5	29	86	1.5	50	0.25
2	2500	100	100	2.0	50	12.5	32	78	1.0	25	0.13
3	2500	100	200	2.5	125	0.0	29	86	2.0	50	0.00
4	2500	100	150	2.5	75	0.0	31	82	1.5	30	0.00
5	2000	100	150	1.5	75	37.5	29	69	1.5	50	0.25
6	2000	100	200	1.5	125	25.0	28	72	2.0	83	0.13
7	2000	100	100	2.0	50	0.0	32	63	1.0	25	0.00
8	2000	100	150	2.5	75	37.5	31	66	1.5	30	0.25
9	2000	100	100	2.5	50	12.5	32	62	1.0	20	0.13
10	2000	100	200	2.5	125	0.0	29	69	2.0	50	0.00
11	1000	100	100	2.0	50	25.0	32	31	1.0	25	0.25
12	1000	100	150	1.5	75	20.0	29	34	1.5	50	0.13
13	1000	100	100	1.5	50	0.0	31	32	1.0	33	0.00
14	1000	100	150	2.0	75	0.0	30	33	1.5	38	0.00
15	1000	100	100	2.5	50	12.5	32	31	1.0	20	0.13
16	1000	100	200	2.5	125	25.0	29	34	2.0	50	0.13

2.4. Test Setup and Instrumentation

Each specimen was equipped with two strain gauges at mid height of the column on the 100 mm width faces measuring the maximum and minimum axial strains due to concentric and eccentric loads. In addition, two LVTDs were used to measure the lateral deformations at mid height of the column in both major and minor directions of the cross section. Finally, the testing frame is equipped with a load cell to measure the applied load. Figure 3 illustrates test setup and instrumentation.

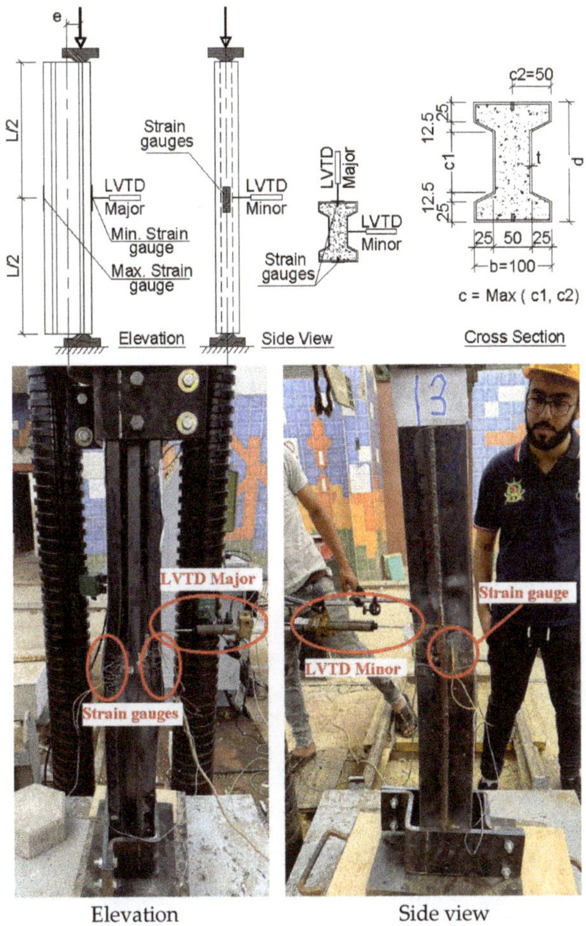

Figure 3. Testing setup, instrumentation and typical dimensions of specimen cross section.

2.5. Test Procedure

All specimens were tested under axial compression (concentric or eccentric) load up to failure using an AMSLER compression testing machine with 5000 kN capacity. To ensure full contact between the column head and loading plate, the top and bottom 20 mm was cut from each specimen using a concrete saw to obtain a smooth and leveled loading surface.

Two-ball seats were welded to the upper and lower steel loading heads at the required eccentricity to achieve the hinged support. The lower ball seating was fixed to the lower machine head to prevent any displacement of the specimens during testing. The upper plate is allowed to rotate around a fixed sphere. This ensures that the applied load is always passing through the desired location (concentric or eccentric) and perpendicular to the specimen cross section. The specimens were loaded up to failure with a loading rate of 50 kN per minute.

3. Experimental Results

For each specimen, failure load, failure mode, maximum lateral deformations and axial strains were recorded as summarized in Table 5, where LVTD minor and major readings are the lateral movements of the mid-height specimen in minor and major cross section axis, respectively. Negative strain and positive strain values correspond to compressive stresses and tensile stresses, respectively. Figure 4 shows the failure modes of tested specimens.

Table 5. Summary of experimental program results.

Spec. ID	P Exp	Δ Minor Measured	Δ Minor Calc.	Δ Major Measured	Δ Major Calc.	Strain-Min	Strain-Max	Mode of Failure
	(KN)	(mm)	(mm)	(mm)	(mm)	(µε)	(µε)	
1	226	**18**	**25**	5	20	−1062	10	GB
2	233	**40**	**21**	**53**	**6**	−935	−329	GB, F
3	394	**29**	**25**	7	0	−731	−661	GB
4	464	**28**	**23**	2	0	−968	−891	GB
5	243	3	16	**13**	**14**	−952	−293	F
6	353	3	18	7	12	−1080	−373	LB
7	285	**31**	**13**	15	0	−807	−788	GB, F
8	383	**29**	**14**	**19**	**15**	**−1709**	−2	GB, F and C
9	442	1	13	**29**	**6**	**−1497**	−518	F, C
10	580	**79**	**16**	68	0	−937	−922	GB, F
11	278	2	3	**9**	**3**	**−1611**	16	F, C
12	350	1	4	**2**	**3**	−1260	−371	F
13	324	**4**	**3**	1	0	−935	−827	GB
14	405	**3**	**3**	1	0	−913	−891	GB
15	451	2	3	**7**	**2**	**−1749**	−519	F, C
16	586	**8**	**4**	**6**	**4**	−1028	−881	GB, F

GB: global buckling; LB: local buckling; C: compression failure; F: flexural failure.

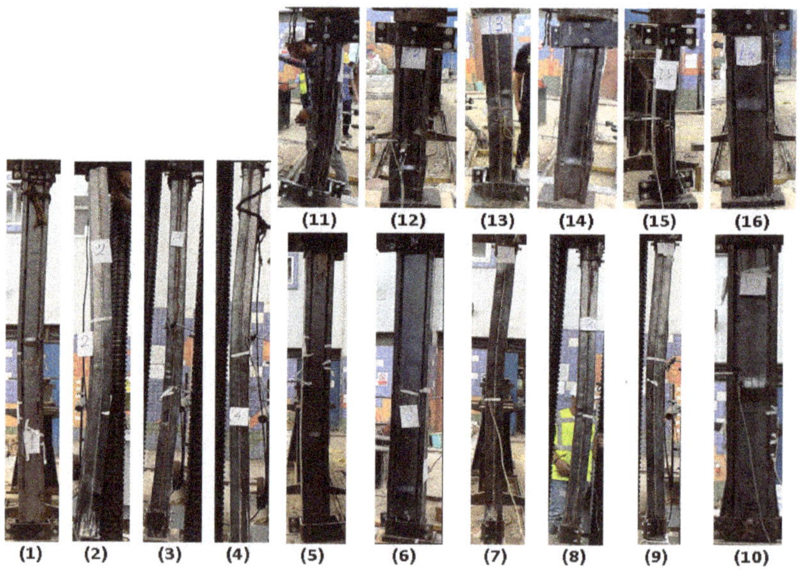

Figure 4. Failure modes of tested specimens.

The failure mode of each specimen was investigated based on the measured axial strains and lateral deformations as follows:

- The lateral deformation due to eccentricity (in major direction) is estimated for each specimen using Equation (1):

$$\Delta_{\text{major}} = \frac{P_{\text{exp}}\ e\ L^2}{8\ EI} \tag{1}$$

where (EI) is the fully bonded composite flexural stiffness; hence, if the measured Δ-major was close to or exceeded this value, that indicated flexural failure.

- The lateral deformation due to global buckling (in a minor direction) is estimated for each specimen using Equation (2):

$$\Delta_{minor} = \frac{\lambda^2\, b}{30{,}000} \tag{2}$$

where (λ) is the fully bonded composite global slenderness ratio. If the measured Δ-minor was near or more than this value, that indicated global buckling failure.

- Yield strain equals (Fy/Es) which is (0.33/210 = 1570 μ-strain). If the minimum measured strain was within this range or more, this value indicated overstressing failure.
- Finally, if none of the above conditions occurred, this indicated a local buckling failure.

The marked numbers in Table 5 illustrate the critical values used in classifying the failure mode.

4. Theoretical Analysis

The ultimate capacities of the 16 tested columns were estimated according to the AISC-LRFD-94 code [20] and EN-1994-04 code [21] using the well-known CSI-COL. The capacities from the three theoretical methods were compared to experimental capacities. Table 6 summarizes these results.

Table 6. Summary of theoretical analysis results.

Spec.	EXP.	AISC	Error	EN-1994	Error	CSI-COL.	Error
ID	(kN)	(kN)	(%)	(kN)	(%)	(kN)	(%)
1	226.0	172.0	−24%	225.0	0%	246.0	9%
2	233.0	233.0	0%	284.0	22%	253.0	9%
3	394.0	393.0	0%	489.0	24%	422.0	7%
4	464.0	365.0	−21%	423.0	−9%	404.0	−13%
5	243.0	268.0	10%	300.0	23%	263.0	8%
6	353.0	310.0	−12%	354.0	0%	392.0	11%
7	285.0	329.0	8%	347.0	13%	328.0	7%
8	383.0	351.0	−8%	324.0	−15%	379.0	−1%
9	442.0	336.0	−24%	350.0	−21%	431.0	−2%
10	580.0	539.0	−7%	570.0	−2%	516.0	−11%
11	278.0	301.0	8%	250.0	−10%	269.0	−3%
12	350.0	343.0	−2%	332.0	−5%	343.0	−2%
13	324.0	328.0	1%	350.0	8%	332.0	3%
14	405.0	494.0	9%	517.0	14%	496.0	9%
15	451.0	406.0	−10%	385.0	−15%	400.0	−11%
16	586.0	690.0	18%	714.0	22%	655.0	12%

4.1. AISC-LRFD-94 Code

The design philosophy of this code depends on estimating the equivalent strength of the composite column (F_{ym}) and equivalent elastic modulus (E_m), as shown in Equations (3) and (4), while the equivalent radius of gyration is the same as an empty steel shell.

$$F_{ym} = F_y + c_1\, f_{yr}\, (A_r/A_s) + c_2\, F_c\, (A_c/A_s) \tag{3}$$

$$E_m = E_s + c_3\, E_c\, (A_c/A_s) \tag{4}$$

where

- f_s and A_s are yield strength and cross sectional area of steel section
- F_c and A_c are compressive strength and cross sectional areas of concrete
- f_{yr} and A_r are yield strength and cross sectional area of reinforcement bars
- c_1 = 1.00, c_2 = 0.68 and c_3 = 0.40 for in-filled composite columns.

For a small eccentric load, where the vertical load is more than 20% of the axial capacity of the section, the well-known interaction formula shown in Equation (5) is used.

$$(P/P_n) + 0.89\,(M_x/M_{xn}) + 0.89\,(M_y/M_{yn}) = 1.0 \tag{5}$$

where

- P, M_x and M_y are the actual vertical load and bending moments in X and Y directions
- P_n is the axial capacity of a section considering buckling, as shown in Equations (6) and (7)
- M_{xn} and M_{yn} are the flexural capacity of column section in X and Y directions without axial load and considering plastic stress distribution

$$P_n = F_{cr}\,A_s \tag{6}$$

$$\begin{aligned} F_{cr^-} &= (1 - 0.348\,\lambda_m{}^2)\,F_{ym} \text{ for } \lambda_m \le 1.1 \\ &= 0.648\,F_{ym}/\lambda_m{}^2 \text{ for } \lambda_m > 1.1 \end{aligned} \tag{7}$$

where

- λ_m is the normalized slenderness ratio = $\frac{L\sqrt{(Fy/Em)}}{\pi r}$.

4.2. EN-1994-04 Code

Unlike the AICS-LRFD code, EN-1994 depends on drawing an interaction diagram for the composite section with vertical load, considering the buckling effect on *Y*-axis and the bending moment on *X*-axis, as shown in Figure 5a.

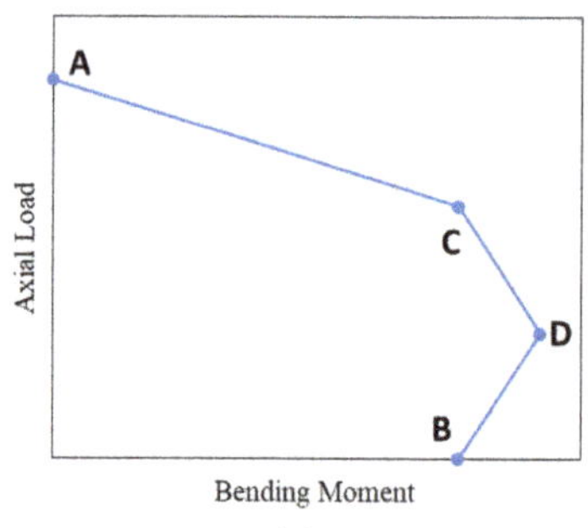

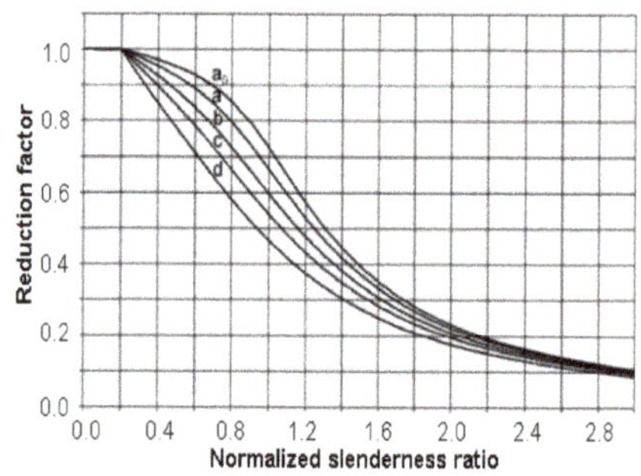

Figure 5. (**a**) Typical interaction diagram of EN-1994. (**b**) Buckling reduction factor of EN-1994.

Point (A) presents the pure axial capacity of the composite section considering the buckling effect as shown in Equation (8)

$$P_n = X\,(0.85\,f_{c'}\,A_c + F_{yr}\,A_r + F_y\,A_s) \tag{8}$$

where X is the buckling reduction factor from curve "b" (for in-filled hollow sections) in Figure 5b.

Point (B) presents the pure plastic flexural capacity of the composite section without axial load. Point (C) has the same moment value as (B), and axial load equals the axial capacity of only the concrete section. Finally, point (D) has half the axial capacity of (C) and the plastic flexural capacity of the composite section, considering the effect of the corresponding axial load.

Once the interaction diagram is generated, the capacity of the section at a certain eccentricity could be determined by drawing a straight line from the origin with a slope equals to (N/M = 1/e) and finding its intersection with the interaction diagram.

4.3. Using CSI-COL Software (Computers & Structures INC., California, USA)

A 3D-interaction diagram (PMM-interaction diagram) was generated for each specimen using the well-known CSI-COL Ver. 6.2 software considering the built-in Mander model for unconfined concrete behavior and the built-in structural steel model. The stress–strain curve of the Mander model [22] is automatically generated for certain (fc′ and Ec) values, while the structural steel model requires (Fy and Fu) to generate the stress–strain curve. Figure 6 presents both the generated and the experimental stress–strain curves for steel and concrete and a typical 3D-interaction diagram.

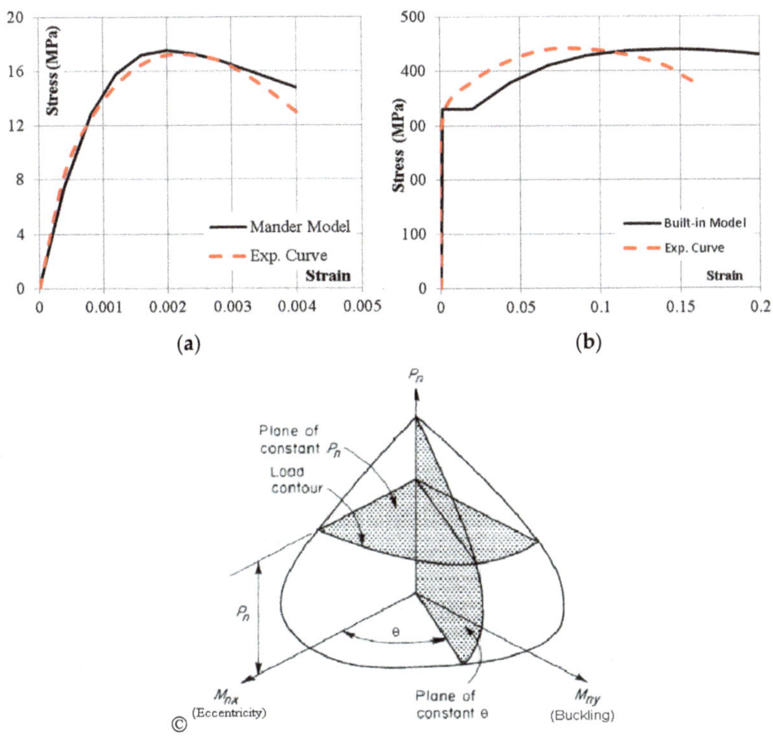

(c) Typical 3D-inetraction diagram

Figure 6. Used material modes (**a**) for concrete, (**b**) for steel and (**c**) typical 3D interaction diagram.

In order to present both experimental and theoretical capacities graphically as shown in Figure 7, a certain section in the 3D interaction diagram is constructed. The angle of this section (θ) depends on the value of eccentricity moment ($M_{major} = P \times e$) and buckling moment ($M_{minor} = P \times \Delta_{minor}$), as shown in Equation (9):

$$\theta = \tan^{-1}(M_{minor}/M_{major}) \tag{9}$$

In the constructed 2D interaction diagram, the presented bending moment is the resultant moment as shown in Equation (10):

$$M_{result} = (M_{minor}{}^2 + M_{major}{}^2)^{0.5} \tag{10}$$

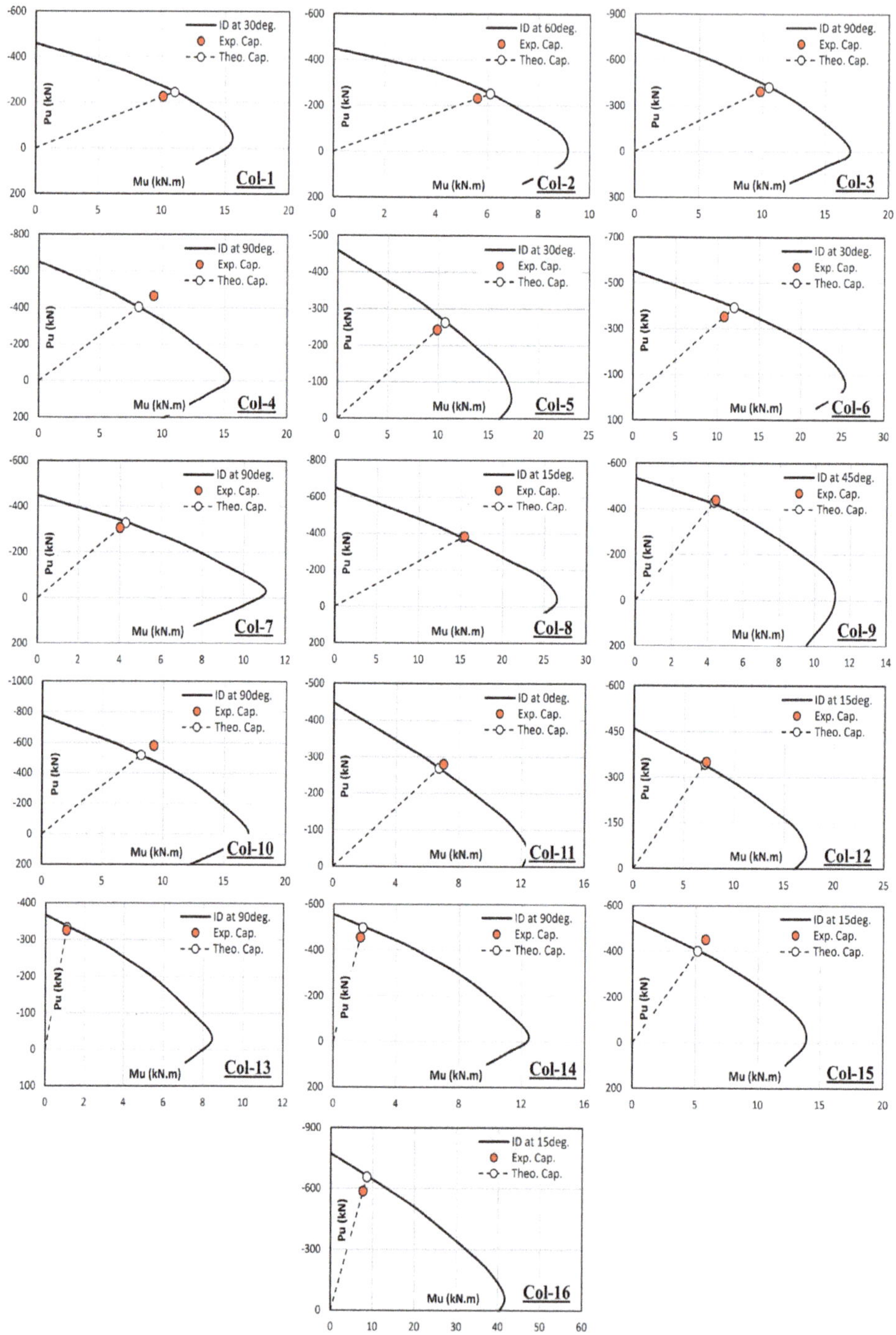

Figure 7. CSI-COL interaction diagrams for the tested specimens.

5. Discussion

The summarized experimental results in Section 3 included the measured failure loads, maximum lateral deformations, extreme axial strains and the investigated models of failure. However, in order to study the impact of each considered parameter on the axial capacity and failure mode, the reduction factor (P_{exp}/P_{max}) was calculated for each specimen, where P_{max} is the maximum axial capacity of the composite section without considering the buckling effect as per Equation (11).

$$P_{max} = 0.85\, f_{c'}\, A_c + F_y\, A_s \tag{11}$$

Figure 8a presents the relation between the buckling reduction factor and the global normalized slenderness ratio for different relative eccentricities. The best fitting curves indicated the following:

- The reduction factor (P_{exp}/P_{max}) decreases with increasing the global normalized slenderness ratio regardless of the relative eccentricity.
- For centric loading (e/d = 0) the reduction factor ranged between 90% and 60% for global normalized slenderness ratio of 0.4–1.1, respectively, which matches (EN-1994) curve "b" in Figure 5b.
- For small eccentricity (e/d = 0.13), where the axial force acts inside the core of the section, the reduction factor is slightly less than the centric case, which indicates that plastic failure due to vertical load dominates the behavior.

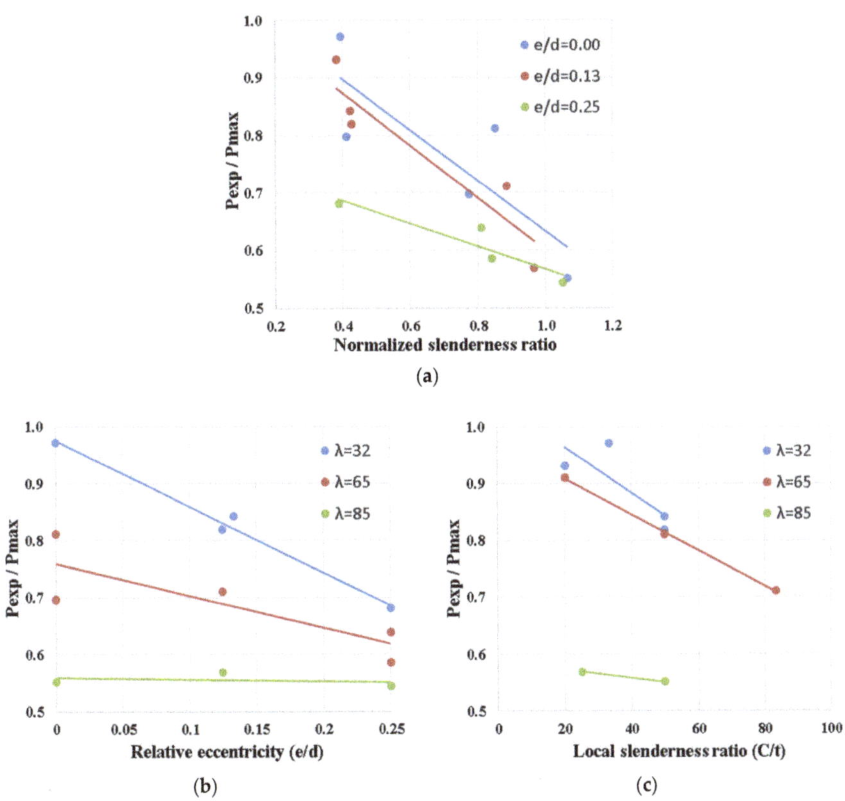

Figure 8. Relation between (P_{exp}/P_{max}) and (**a**) normalized slenderness ratio, (**b**) relative eccentricity and (**c**) local slenderness ratio.

- For critical eccentricity (e/d = 0.25), where the axial force acts on the core edge of the section, the reduction factor is significantly reduced, which indicates that flexural plastic failure dominates the behavior.
- Finally, all fitting lines almost intersected at a normalized slenderness ratio of 1.1, which matches the elastic buckling limit of AISC that showed in Equation (7).

Figure 8b illustrates the relationship between the buckling reduction factor and relative eccentricities for different global slenderness ratios. The best fitting curves indicated the following:

- The rate of reduction factor decreasing is decreased with increasing the slenderness ratio. The maximum decreasing rate was observed for the lowest slenderness ratio (λ = 32 or λ = 0.4), while the reduction factor was almost constant for the highest slenderness ratio (λ = 85 or λ = 1.05).
- The slenderness limit that separates plastic failure from elastic buckling is about (λ = 90 or λ = 1.10), which matches the recommendations of the AISC code.

Figure 8c shows the relation between buckling reduction factor and local slenderness ratio for different global slenderness ratios. It could be noted that the impact of the local slenderness ratio (C/t) is clearly observed in the case of low and medium global slenderness ratios (λ = 32, 65 or λ = 0.4, 0.8), where the failure occurred due to overstressing. On the other hand, for high global slenderness ratios (λ = 85 or λ = 1.05), the effect of local buckling was very minor as the columns failed due to elastic buckling.

The comparison between both experimental results from Section 3 and theoretical analysis from Section 4 indicates the following points:

- Both experimental and analytical results showed that the core of the considered composite section lies at e/d = 0.25. At this relative eccentricity, the minimum normal stress is almost zero (as shown in Table 5, specimens 1, 8 and 11).
- The summarized results in Table 6 and Figure 9 show wed deviation of 24% between the experimental capacities and the calculated ones using AISC and EN-1994 codes, while the deviation was about 13% for the CSI-COL capacities.
- The enhanced accuracy of CSI-COL may be justified as follows:
    - CSI-COL considered the Mander model for concrete behavior, which is so close to the experimental one (as shown in Figure 6a). On the other hand, most design codes (including AISC and EN-1994) considered a simplified equivalent block distribution to simulated concrete behavior.
    - CSI-COL considered a non-linear stress–strain relation for steel sections, while design codes used simplified bilinear elastic–perfect plastic relation.
    - The formulas in design codes were developed based on regular cross sections (rectangular or circular in-filled tubes), while CSI-COL considered the actual non-regular cross section of double sigma face-to-face.
    - The software used generates a 3D-interaction diagram, while AISC uses an interaction formula and EN-1994 uses a simplified polygon 2D interaction diagram.
- Although CSI-COL showed better accuracy than design codes, it still has a significant deviation from the experimental results. This deviation could be justified as follows:
    - The built-in stress–strain curve for steel sections in CSI-COL has a significant deviation from the actual one, as shown in Figure 6b.
    - CSI-COL does not consider the effect of local buckling of the thin steel shell.
    - There must be some imperfections and random errors in manufacturing and testing the samples, which cannot be considered in any theoretical analysis.

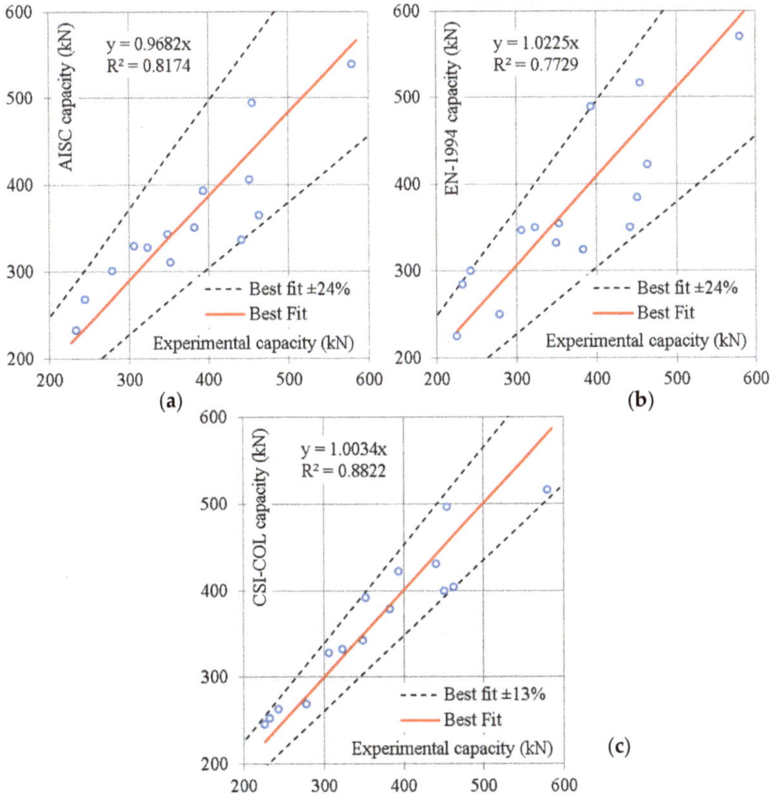

Figure 9. Comparison between both experimental and theoretical capacities. (**a**) Using AISC-LRFD, (**b**) Using EN-1994, (**c**) Using CSI-COL software.

6. Conclusions

This paper studied both experimental and theoretical behaviors of composite columns consisting of two cold-formed sigma sections welded face to face and filled with concrete. The study considered axial and eccentric compression loading for different slenderness ratios. The results of this study could be summarized in the following points.

- Experimental tests, design codes and CSI-COL software results indicated that over-stressing plastic failure is distinguished from elastic buckling failure at a normalized slenderness ratio of about 1.1 ($\lambda \approx 90$).
- Both theoretical calculations and strain measurements showed that the core edge of this section is located at relative eccentricity (e/d) equals 0.25.
- The calculated capacities using CSI-COL software, AISC-LRFD-94 and EN-1994-04 had deviations of (13%, 24% and 24%) from the experimental capacities, respectively.
- Although the local buckling failure was experimentally observed for local slenderness ratio (c/t = 80), none of the three used theoretical methods was able to capture this behavior. An advanced non-linear 3D FEM modeling may be needed to simulate this phonon.
- The good matching between experimental and CSI-COL capacities indicated that the studied section could achieve full composite behavior without using any shear connectors.
- The results of this study are limited by the size of the tested sample; more full-scale samples should be tested to verify the accuracy of the concluded results.

- Further studies may be carried out using more advanced 3D-FEM modeling to investigate the local buckling behavior of this section.

Author Contributions: Conceptualization, M.A.E.-A.; methodology and editing, A.M.E.; formal analysis and investigation, M.A.R.; supervision, S.M.I. All authors have read and agreed to the published version of the manuscript.

Funding: This research received no external funding.

Data Availability Statement: Not applicable.

Conflicts of Interest: The authors declare no conflict of interest.

References

1. Alimohammadi, H.; Hesaminejad, A.; Yaghin, M.L. Effects of different parameters on inelastic buckling behavior of composite concrete-filled steel tubes. *Int. Res. J. Eng. Technol.* **2019**, *6*, 603–609.
2. Liu, D.; Gho, W.; Yuan, J. Ultimate capacity of high-strength rectangular. *J. Constr. Steel Res.* **2003**, *59*, 1499–1515. [CrossRef]
3. Zhu, L.; Ma, L.; Bai, Y.; Li, S.; Song, Q. Large diameter concrete-filled high strength steel tubular stub. *Thin-Walled Struct.* **2016**, *108*, 12–19. [CrossRef]
4. Han, L.; Yao, G.; Zhao, X. Tests and calculations for hollow structural steel (HSS) stub columns filled with self-consolidating. *J. Constr. Steel Res.* **2005**, *61*, 1241–1269. [CrossRef]
5. Zhang, S.; Guo, L. Behavior of High Strength Concrete-Filled Slender RHS Steel Tubes. *Adv. Struct. Eng.* **2007**, *10*, 337–351. [CrossRef]
6. Guo, L.; Zhang, S.; Xu, Z. Behavior of Filled Rectangular Steel HSS Composite Columns under Bi-Axial Bending. *Adv. Struct. Eng.* **2011**, *14*, 295–306. [CrossRef]
7. Bahrami, A.; Badaruzzaman, W.H.W.; Osman, S.A. Finite Element Analysis of Ultimate Load Capacity of Slender Concrete-Filled Steel Composite Columns. In Proceedings of the International Conference on Advanced Science, Engineering and Information Technology 2011, Bangi-Putrajaya, Malaysia, 14–15 January 2011. [CrossRef]
8. Ren, Q.X.; Han, L.H.; Lam, D.; Hou, C. Experiments on special-shaped CFST stub columns under axial compression. *J. Constr. Steel Res.* **2014**, *98*, 123–133. [CrossRef]
9. Wang, Z.B.; Tao, Z.; Yu, Q. Axial compressive behavior of concrete-filled double-tube stub columns with stiffeners. *Thin-Walled Struct.* **2017**, *120*, 91–104. [CrossRef]
10. Ding, F.X.; Wang, W.J.; Lu, D.R.; Liu, X.M. Study on the behavior of concrete-filled square double-skin steel tubular stub columns under axial loading. In *Structures*; Elsevier: Amsterdam, The Netherlands, 2020; pp. 665–676. [CrossRef]
11. Ci, J.; Jia, H.; Chen, S.; Yan, W.; Song, T.; Kim, K.S. Performance analysis and bearing capacity calculation on circular concrete filled double steel tubular stub columns under axial compression. In *Structures*; Elsevier: Amsterdam, The Netherlands, 2020; pp. 554–565. [CrossRef]
12. Tao, Z.; Uy, B.; Han, L.H.; Wang, Z.B. Analysis and design of concrete-filled stiffened thin-walled steel tubular columns under axial compression. *Thin-Walled Struct.* **2009**, *47*, 1544–1556. [CrossRef]
13. Young, B.; Ellobody, E. Experimental investigation of concrete-filled cold-formed high strength stainless steel tube columns. *J. Constr. Steel Res.* **2005**, *62*, 484–492. [CrossRef]
14. Dai, P.; Yang, L.; Wang, J.; Zhou, Y. Compressive strength of concrete-filled stainless steel tube stub columns. *Eng. Struct.* **2020**, *205*, 110106. [CrossRef]
15. He, A.; Liang, Y.; Zhao, O. Flexural buckling behaviour and resistances of circular high strength concrete-filled stainless steel tube columns. *Eng. Struct.* **2020**, *219*, 110893. [CrossRef]
16. Azad, S.K.; Li, D.; Uy, B. Axial slenderness limits for austenitic stainless steel-concrete composite columns. *J. Constr. Steel Res.* **2020**, *166*, 105856. [CrossRef]
17. El Aghoury, M.A.; Hana, M.T.; Amoush, E.A. Axial stability of columns composed of combined sigma CFS. In Proceedings of the Annual Conference of Structural Stability Research Council, Orlando, FL, USA, 12–15 April 2016.
18. el Aghoury, M.A.; Hana, M.T.; Amoush, E.A. Strength of combined sigma cold formed section columns. In Proceedings of the EUROSTEEL 2017, Copenhagen, Denmark, 13–15 September 2017.
19. *ASTM E8/E8M−13a*; Standard Test Methods for Tension Testing of Metallic Materials. ASTM International: West Conshohocken, PA, USA, 2022. [CrossRef]
20. American Institute of Steel Construction (AISC). *Manual of Steel Construction-Load & Resistance Factor Design-Volume I-Structural Members, Specifications, & Codes*; American Institute of Steel Construction: Chicago, IL, USA, 1994; ISBN 1-56424-042-8.
21. *EN 1994-1-1:2004*; Eurocode 4: Design of Composite Steel and Concrete Structures–Part 1-1: General Rules and Rules for Buildings. European Committee for Standardization (CEN): Brussels, Belgium, 2004.
22. Mander, J.B.; Priestley, M.J.N.; Park, R. Theoretical stress-strain model of confined concrete. *J. Struct. Eng.* **1988**, *114*, 1804–1826. [CrossRef]

designs

MDPI

Article

Estimating the Buckling Load of Steel Plates with Center Cut-Outs by ANN, GEP and EPR Techniques

Jagan Jayabalan [1], Manju Dominic [2], Ahmed M. Ebid [3,*], Atefeh Soleymani [4], Kennedy C. Onyelowe [5] and Hashem Jahangir [6]

1 Department of Civil Engineering, Galgotias University, Greater Noida 203201, India
2 Department of Civil Engineering, Rajadhani Institute of Engineering and Technology, Thiruvananthapuram 695102, India
3 Department of Structural Engineering, Future University in Egypt, New Cairo 11865, Egypt
4 Department of Structural Engineering, Shahid Bahonar University of Kerman, Kerman 76169-14111, Iran
5 Department of Civil Engineering, Michael Okpara University of Agriculture, Umudike 440101, Nigeria
6 Department of Civil Engineering, University of Birjand, Birjand 97174-31349, Iran
* Correspondence: ahmed.abdelkhaleq@fue.edu.eg

Abstract: Steel plates are used in the construction of various structures in civil engineering, aerospace, and shipbuilding. One of the main failure modes of plate members is buckling. Openings are provided in plates to accommodate various additional facilities and make the structure more serviceable. The present study examined the critical buckling load of rectangular steel plates with centrally placed circular openings and different support conditions. Various datasets were compiled from the literature and integrated into artificial intelligence techniques like Gene Expression Programming (GEP), Artificial Neural Network (ANN) and Evolutionary Polynomial Regression (EPR) to predict the critical buckling loads of the steel plates. The comparison of the developed models was conducted by determining various statistical parameters. The assessment revealed that the ANN model, with an R^2 of 98.6% with an average error of 10.4%, outperformed the other two models showing its superiority in terms of better precision and less error. Thus, artificial intelligence techniques can be adopted as a successful technique for the prediction of the buckling load, and it is a sustainable method that can be used to solve practical problems encountered in the field of civil engineering, especially in steel structures.

Keywords: cut-outs; buckling load; artificial intelligence; steel plates; axial loading

Citation: Jayabalan, J.; Dominic, M.; Ebid, A.M.; Soleymani, A.; Onyelowe, K.C.; Jahangir, H. Estimating the Buckling Load of Steel Plates with Center Cut-Outs by ANN, GEP and EPR Techniques. *Designs* **2022**, *6*, 84. https://doi.org/10.3390/designs6050084

Academic Editor: Farshid Aram

Received: 26 August 2022
Accepted: 19 September 2022
Published: 22 September 2022

1. Introduction

1.1. Background

Plate buckling analysis has a wide range of applications in a variety of engineering domains, such as civil and structural engineering, mechanical engineering, marine, aerospace engineering, etc. [1], particularly when a lightweight design is the main objective [2]. Plate buckling is a type of destabilization that occurs when a sudden deflection occurs under compressive load. Plates buckle when exposed to compressive stresses greater than the critical limit [3,4]. A critical level of stress develops whenever the plates are subjected to compressive load, causing this to occur. The origin of plate buckling is governed by partial differential equations, which makes the interpretation exceedingly difficult [5,6]. A plate with cut-outs has a more sophisticated buckling analysis than a plate without cut-outs [7]. Cut-outs in box girders and certain load-bearing spars save material, can be used as windows or doors, or simply improve the design aesthetics. They also provide ventilation, accessibility for maintenance, installation, and damage assessment. As a result, an in-depth understanding of perforated and non-perforated plate buckling is necessary [8–13]. Perforated plates have lower structural strength than plates without holes, and buckling behavior is one of the most significant failures to be considered in these systems' safe and reliable operation.

1.2. Literature Review

The existence of cut-outs of varied sizes results in the formation of free edges in steel plates, which causes high stresses, resulting in plate stiffness degradation and early fracture [14–16]. Therefore, for effective design, it is critical for the design engineer to comprehend the stability, overall strength, and failure parameters of steel plates with cut-outs of varied shapes and sizes. A full grasp of buckling stresses and corresponding mode shapes is the cornerstone of dependable structural designs and constructing a straightforward technique can give an acceptable method. Because of the issue's difficulty, most academics investigated the buckling performance of rectangular sheets with perforations and isotropic support using the FEM or Rayleigh-Ritz techniques [17–20].

Numerical, experimental, and analytical methods and their combinations have been used in several important investigations. The impact of a rectangular cut-out on a plate was studied by Suneel Kumar et al. [21]. Several cut-out dimensions, slenderness ratios, and area ratios were employed to get the required result. The impact of changing different factors on the ultimate strength of a plate subjected to axial compression was determined. Ansys simulation was used to verify the findings. The consequences of perforations on the buckling behavior of planes in a linear manner were investigated by Maiorana et al. [22]. Under compressive stresses, the boundary condition completely supported all edges, and circular and square holes were studied. The findings were reported on graphs for circular and square perforations with variations in the position of the cut-outs. Sandeep Singh et al. [23] investigated the effect of partial edge compression, modification in aspect ratio, and the effects of cut-outs on buckling load, discovering that partial edge compression had a larger effect on buckling strength than uniform edge compressive load, and panels without holes had a greater critical load than panels with perforation. Although the effect of partial edge compression on buckling load was less clear. Dadrasi [24] investigated the buckling performance of punctured steel plates with rectangular shapes when subjected to uniaxial compressive stress. Cut-outs in a circle or square were utilized in a variety of loading bands, numeral, and empirical findings. For finite element analysis, ABAQUS software was employed, and for an experimental study, a group of servo-hydraulic INSTRON8802 was utilized. The findings for plates with or without cut-outs were inspected. They found that the critical buckling load ascended as the loading band broadened and, furthermore, that the buckling load for plates with a circular cut-out was higher than for plates with square cut-outs. Djelosevic et al. [25] investigated the elastic stability of panels with varied perforation geometry. For circular, square, and rectangular apertures, several variables' influence on the plate's elastic stability, such as opening form, size, and direction, was investigated, and a sensitivity factor was constructed. The presence of holes, they found, decreased the deformation energy. Whenever thin rectangular plates with cut-outs in the shape of a circle were uniaxially loaded, instability arose far before the yield point, and it occurred prior to the yield point of the panel material when the plates were thick, according to Mauro et al. [26]. The effect of change in plate thickness with a hole regarding plate buckling was investigated by Mohamazadeh and Noh [27]. The buckling coefficient and buckling stresses were determined using the Gerard and Becker formulas. They compared plates with cut-outs against plates without perforations and discovered that if there was an increase in the thickness of the plate, the buckling load and stress also increased. They [28] also examined the effect of perforations on the buckling of thin plates. ABAQUS was used to perform the simulations, and critical values for buckling load and corresponding stress were computed for variations in hole diameter and plate thickness. Shariati et al. [29] studied the buckling behavior of a panel in the shape of a rectangle with a circular hole at various points on the panel under various boundary circumstances, and found that the boundary circumstances had a substantial effect on the buckling behavior. For different aspect ratios, Jana [30] investigated the buckling performance of a rectangular sheet composed of a simple support border for uniform axial compressive loads with a circular cut-out. For changes in perforation size, aspect ratio, and thickness of the plate, eigenvalue buckling analysis was performed. Using Ansys and MATLAB, he discovered that the ideal perforation location for the highest buckling load is in the middle of the axis on which the stress is applied. Giulio Lorenzini et al. [31] investigated the impact of several types of cut-outs on the buckling of the plate and discovered that by selecting the right cut-out method, a high performance might be achieved, i.e., performance could be improved by eliminating a similar quantity of material. Using the Ritz energy equation, Adah et al. [32] develop MATLAB software to determine the critical buckling load for a rectangular plate with an axial compression force. Researchers determined critical buckling coefficients for a variety of boundary conditions and then compared their results to existing literature. By examining the increase in buckling stress caused by cut-outs, Blesa, Gracia, and Rammerstorfer [33] determined that cut-outs can enhance plate buckling strength while decreasing weight. Caio César Cardoso da Silva and colleagues [34] looked into the impact of hexagonal opening geometric arrangements on buckling mechanical characteristics and found that a longitudinally hexagonal cut-out outperforms an oblique hexagonal cut-out. The basic buckling of annular and circular plates with guided edges [12,35] and elastic edges [36,37] was investigated by Rao and Rao. They did not, however, consider the perforation in their research. Using the finite element approach, Sinha et al. [5] performed a buckling study on stainless-steel plates either with or without perforation. The influence of plate length, thickness, and diamond-shape hole-size on buckling load values was investigated. The purpose of Gore and Lokavaraput's research [1] was to explore how material characteristics and geometrical modifications influenced the buckling load bearing capacity of rectangular flat sheets joined on both sides. A comparison was made between a solid plate and a plate with a perforation. Variations in the orientation of the elliptical perforation and the greatest buckling load that could be attained were explored. Fu and Wang [38] devised a new phenomenological galaxy formation model for the critical buckling load of perforated plates with characteristic equations based on the Timoshenko shear beam theory. The suggested model's results were compared to those produced using FEM and found to be in excellent agreement. Using FEM, Hosseinpour et al. [39] investigated the behavior of steel plates with a central circular cut-out when exposed to compressive axial force. As a consequence, an ANN-based formula for determining perforated steel plate's ultimate strength was developed, and its reliability was contrasted to that of previous research formulas.

As stated above, the buckling examination of plates with varied perforation shapes, sizes, orientations, and locations has been the subject of various research. Their findings showed that plate buckling is influenced by perforations' form, size, and direction. Yet, there are several opportunities to investigate the influence of these factors in various combinations [5]. Plates are integral structural members with wide applications in civil engineering, aerospace, and shipbuilding. They are used in construction as it reduces the weight of structures considerably. They are used as the main load-carrying structural member in all these applications. Columns are one-dimensional load-carrying members and plates are two-dimensional load-carrying members used in buildings, bridges, airplanes, and ships. Buckling is the major failure phenomenon observed in plates. Buckling happens when plates are subjected to axial compressive loads. It is important to ensure that plates fail by yielding rather than by buckling. This can be done by keeping the critical buckling stress above yield stress. The support conditions of the plates affect the critical buckling load. The width-thickness ratio of the plates can be adjusted to ensure that buckling stress is above yield stress for different support conditions. Analytical formulas are available for calculating the critical buckling load in plates with different support conditions [40]. Openings or cut-outs are provided in plate elements for increasing the serviceability of the plate structural elements used in structures. The presence of openings always changes the load-carrying capacity and structural stability of plate structures. Openings create a redistribution of stresses and change the buckling behavior. It develops stress concentration around the openings. Openings reduce the mass of the plates, and it has been found that the presence of openings increases the critical buckling load capacity of plates [41,42].

Analysis of plates with openings is very complex, and due to this complexity, finite element methods (FEM) are usually sought after for such analyses, which are validated using experimental methods. Many FEM and experimental studies have been conducted to study the behavior of plates with openings [43,44]. While experimental studies are destructive and costly, FEM studies are less costly, non-destructive, and less time-consuming. Using the data sets from experimental and FEM studies, the critical buckling load of plates with openings has been calculated using the buckling coefficient method [45] and predicted using various statistical methods [46]. Meanwhile, Wu et al. [47] deployed the differential quadrature method (DQM) to analyze the isotropic and composite laminated plates and shells with particular consideration to the hierarchical finite element method (HFEM). This research utilized the layer-wise theory with linear expansion in each layer to develop a p-version curved laminated composite. It was found that this method used fewer degrees of freedom and less input data to model a complicated case based on interpolation on arc length coordinates. The Bezier method was also used by Kabir et al. [48] for nonlinear vibration and post-buckling of random checkerboard composites reinforced with graphene nano-platelets. This robust Bezier-based solution recommended a probabilistic model to determine a matrix modulus of the graphene nano-platelets reinforced composite.

Artificial intelligence (AI) has many applications to the civil engineering field, specifically in predicting the buckling load of stiffened panels, imperfect reticulates shells, and thin cylindrical shells [49–52]. It can reduce the complexity of practical civil engineering problems to a large extent by making use of already existing experimental studies and results. Civil engineering design practices have moved from their infancy to a state of maturity through the development of design codes. These design codes have been developed based on years of exhaustive experimental studies conducted in relevant areas. AI is one step ahead of these methods due to its capacity to handle large data sets [53,54]. This will bring more accuracy to the results predicted. In this research paper, the critical buckling load of rectangular steel plates with circular cut outs loaded with uniaxial compressive load was predicted with different AI techniques.

2. Materials and Methods

2.1. Steel Setup and Experimental Data Collection

In the present study, a critical buckling load of rectangular steel plates with circular cut-outs loaded with uniaxial compressive load as depicted in Figure 1 was studied.

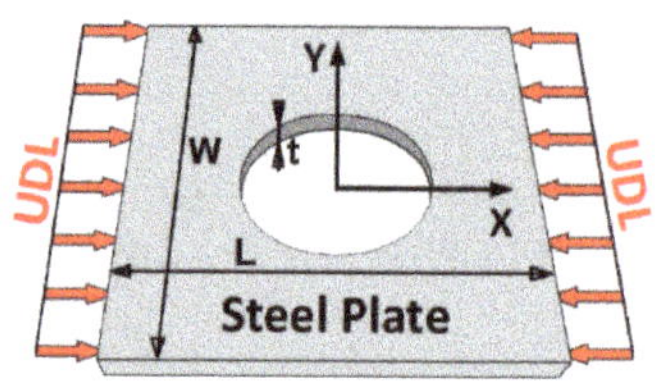

Figure 1. Uniaxial loaded steel plate with a centrally placed circular opening under uniformly distributed loading (UDL).

The support conditions considered are simply supported on all four sides (SSSS), clamped free clamped free (CFCF) and simply supported clamped simply supported clamped (SCSC). Figure 2 depicts the various support conditions.

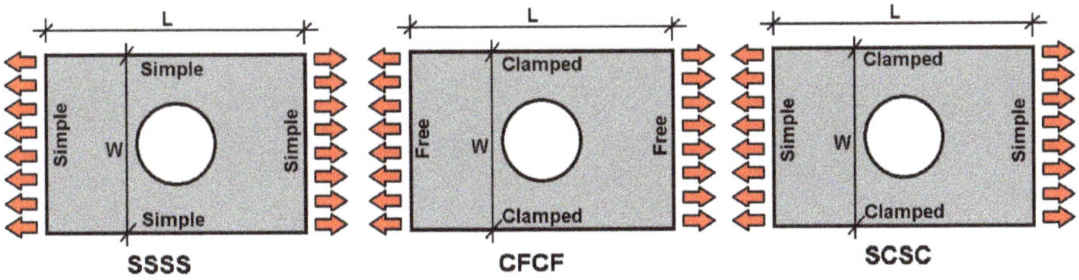

Figure 2. Support conditions of the steel plate.

The independent variables considered are the aspect ratio (length/breadth ratio) of the plate, the thickness of the plate, and the radius of the circular opening, whereas the dependent variable is the buckling load. AI techniques like genetic programming, artificial neural network, and evolutionary polynomial regression have been used to predict the critical buckling loads of steel plates.

2.2. Collected Database and Statistical Analysis

At the end of the loading exercise, 103 experimental test results were collected for rectangular steel plates with centrally placed circular holes with different configurations as presented in the Appendix A to determine their buckling load. Each record contains the following data:

- Aspect ratio (length/width) (L/W),
- Slenderness ratio (width/thickness) (W/t),
- Loss ratio (hole diameter/width) (D/W),
- Boundary conditions (buckling coefficient in width dir. x buckling coef. in length dir.) (Kx.Ky), where K = 2.00 for clamp-free, 1.00 for simple-simple, 0.75 for simple-clamp, and 0.50 for clamp-clamp,
- Relative buckling stress (buckling stress/yielding stress) (Fb/Fy), where buckling stress = buckling load/net area = $\frac{Pb}{(L-D)t}$

The collected records were divided into training (73 records) and validation sets (30 records). The validation set (testing set) was randomly selected and it was the hold-out of the training process and used to test the trained model. Tables 1 and 2 summarized their statistical characteristics and the Pearson correlation matrix. Finally, Figure 3 shows the histograms for both inputs and outputs.

Table 1. Statistical analysis of collected database.

	(L/W)	(W/t)	(D/W)	(Kx.Ky)	(Fb/Fy)
		Training set			
Min.	1.00	41.67	0.08	0.50	0.01
Max.	10.00	700.00	0.90	4.00	1.02
Avg	1.80	92.10	0.21	2.51	0.36
SD	1.51	113.76	0.17	1.60	0.32
Var	0.83	1.24	0.82	0.64	0.88
		Validation set			
Min.	1.0	20.0	0.1	0.5	0.0
Max.	10.0	500.0	0.6	4.0	1.0
Avg	2.1	90.0	0.2	2.0	0.3
SD	1.8	98.5	0.2	1.6	0.3
Var	0.82	1.09	0.83	0.79	0.91

Table 2. Pearson correlation matrix.

	L/W	W/t	D/W	Kx.Ky	Fb/Fy
L/W	1.00				
W/t	0.04	1.00			
D/W	−0.14	−0.21	1.00		
Kx.Ky	−0.28	−0.44	0.28	1.00	
Fb/Fy	−0.31	−0.27	−0.11	−0.23	1.00

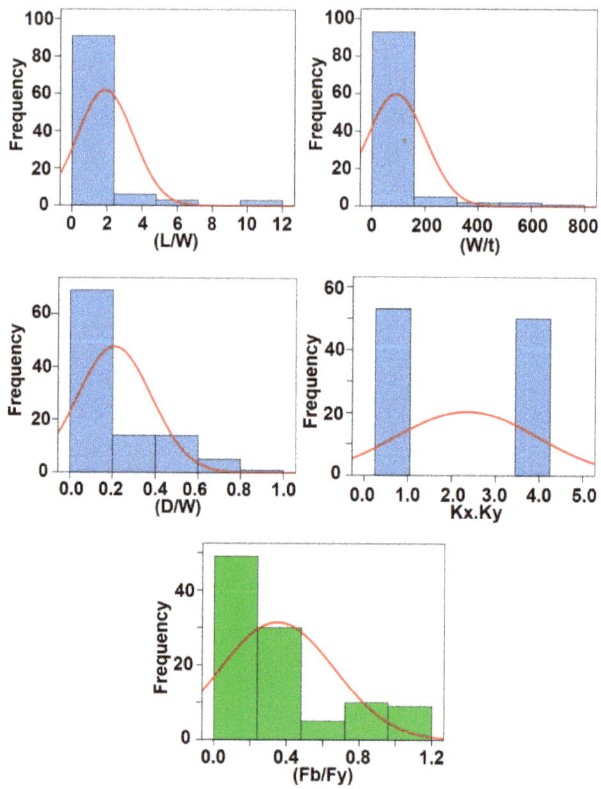

Figure 3. Distribution histograms for inputs (in blue) and outputs (in green).

2.3. Research Program

Three different artificial intelligence (AI) techniques were used to predict the buckling stress of perforated plates or plates with cut outs. These techniques are gene expression programming (GEP), artificial neural network (ANN) and polynomial linear regression optimized using a genetic algorithm which is known as evolutionary polynomial regression (EPR). All three developed models were used to predict the values of relative buckling stress (Fb/Fy) using an aspect ratio (L/W), slenderness ratio (W/t), loss ratio (D/W), and boundary conditions (Kx.Ky). Each model of the three developed models was based on a different approach (evolutionary approach for GEP, mimicking biological neurons for ANN, and optimized mathematical regression technique for EPR). However, for all developed models, prediction accuracy was evaluated in terms of the sum of squared errors (SSE).

The following section discusses the results of each model. The performance accuracies of developed models were evaluated by comparing the (SSE) between predicted and calculated (Fb/Fy) values.

3. Results and Discussion

3.1. Prediction of Relative Buckling Stress (Fb/Fy)

3.1.1. Model (1)—Using GEP Technique

The developed GP model started with 50 gene/chromosomes and settled at 250 gene/chromosomes. The population size, survivor size, and number of generations were 1000, 300, and 200, respectively. Equation (1) presents the output formulas for Fb/Fy,. The average error percentage of the total set is 22.7%, while the R^2 value is 0.932.

$$\frac{Fb}{Fy} = 0.9\left(\frac{W}{L}\right)^{\exp\left(\frac{W}{C.L}\right)} + \left(\frac{\sqrt{W/D}}{Kx.Ky} + \frac{W}{5.6D}\right)\left(\frac{t}{W}\right)\left(1 - \frac{D}{W}\right)$$
$$\text{where } C = 0.422\left(\frac{\sqrt{D/W}}{(W/D) - \log(0.42\ Kx.Ky)} + 1\right) \quad (1)$$

3.1.2. Model (2)—Using ANN Technique

A GRG-trained ANN with one hidden layer and a HyperTanh activation function was used to predict the same Fb/Fy values. The used network layout of a 4-4-1 ANN model and its connation weights are illustrated in Figure 4 and Table 3. The developed ANN was created and trained using SPSS software. It was sequentially trained with a learning rate of 0.05, the model stopped training when the reduction in errors between two successive epochs was less than 1%. Since ANN has a nonlinear activation function, it cannot be converted into an equivalent equation. The average error percentage of the total dataset for this network is 10.4% and the R^2 value is 0.986. The relative importance values for each input parameter are illustrated in Figure 5, which indicated that the aspect ratio (L/W) was the most important factor, followed by the sereneness ratio (W/t), while other factors have less influence, which agrees with a previous work [55].

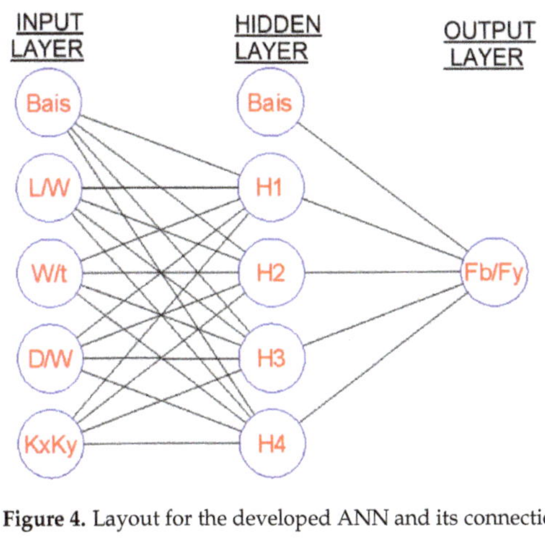

Figure 4. Layout for the developed ANN and its connection weights.

Table 3. Weight matrix for the developed ANN model.

	H1	**H2**	**H3**	**H4**	
(Bias)	13.70	1.37	9.75	8.58	
L/W	10.88	−6.93	2.93	13.16	
W/t	−3.84	3.82	21.04	−3.67	
D/W	−0.63	−0.49	−1.57	0.85	
Kx.Ky	−5.89	7.93	−3.19	−3.07	
	H1	H2	H3	H4	(Bias)
Fb/Fy	−5.23	−10.61	−27.02	−10.32	−22.31

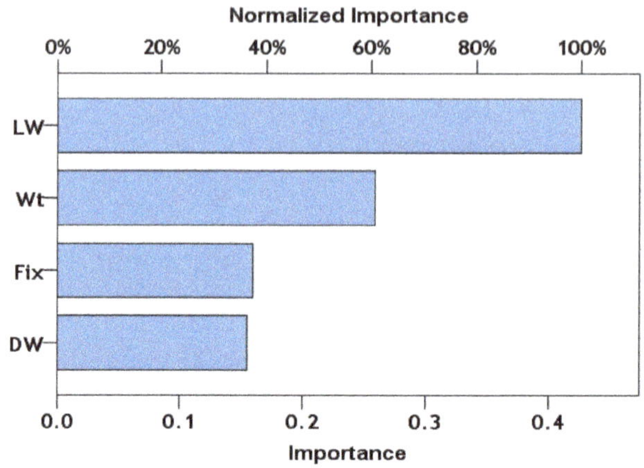

Figure 5. Relative importance of input parameters.

3.1.3. Model (3)—Using EPR Technique

Finally, the developed EPR model was limited to quadrilateral level, for 4 inputs; there are 70 possible terms (35 + 20 + 10 + 4 + 1 = 210) as follows:

$$\sum_{i=1}^{i=4}\sum_{j=1}^{j=4}\sum_{k=1}^{k=4}\sum_{l=1}^{l=4} X_i.X_j.X_k.X_l + \sum_{i=1}^{i=4}\sum_{j=1}^{j=4}\sum_{k=1}^{k=4} X_i.X_j.X_k + \sum_{i=1}^{i=4}\sum_{j=1}^{j=4} X_i.X_j + \sum_{i=1}^{i=4} X_i + C$$

The GA technique was applied to these 70 terms to select the most effective 10 terms to predict the values of Fb/Fy. The outputs are illustrated in Equation (2). The average error percentage and R^2 values were 15.3% & 0.970 for the total datasets. The results of all developed models are summarized in Table 4.

$$\begin{aligned}\frac{Fb}{Fy} = \frac{Kx.Ky.L^2}{4165\ W.D} &+ \frac{Kx.Ky.W.D + 53540\ t^2}{605\ L.t} + \frac{1.7W(2.1Kx.Ky - 1)}{L(Kx.Ky)^3} \\ &- \left(\frac{W}{L}\right)^2\left(\frac{30760\ t + W}{41565\ t}\right) + \frac{28.9\ t\left[1 - (2\ Kx.Ky)^2\right]}{W(Kx.Ky)^3} - 0.25\end{aligned} \quad (2)$$

Table 4. Accuracies of developed models.

Technique	Developed Eq.	SSE	Error %	R^2
GEP	Equation (1)	0.65	22.7	0.932
ANN	Figure 2	0.14	10.4	0.986
EPR	Equation (2)	0.30	15.3	0.970

The relations between calculated and predicted values for all developed models are shown in Figure 6.

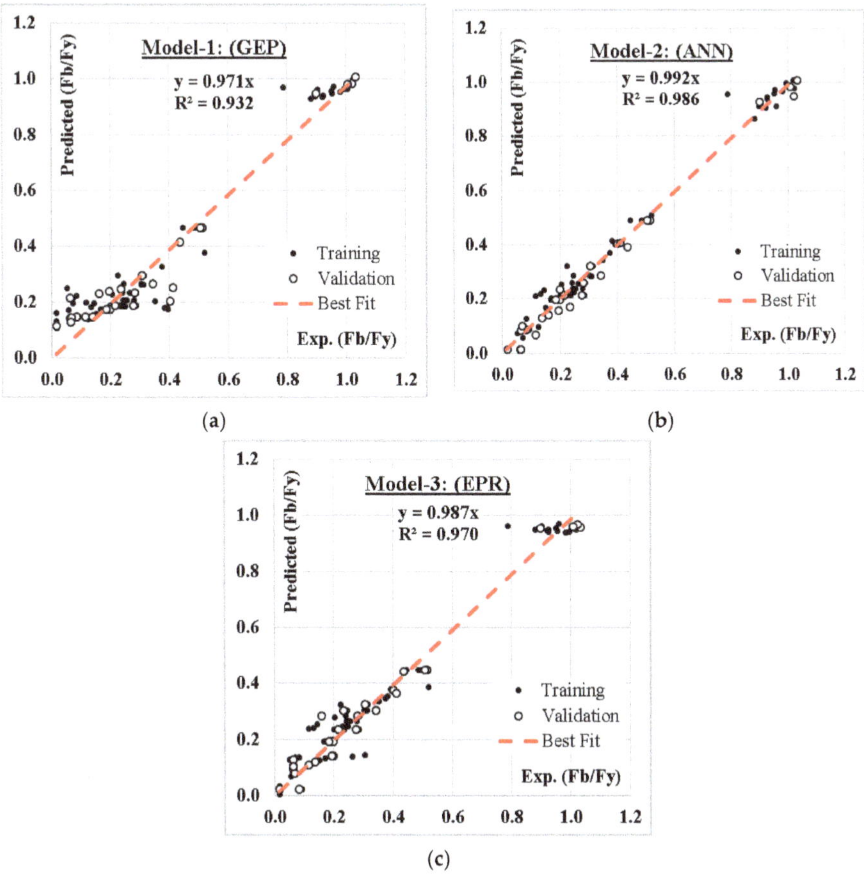

Figure 6. Relation between predicted and calculated Fb/Fy values using the developed models. (**a**) using GEP, (**b**) using ANN & (**c**) using EPR.

4. Conclusions

This research presented three models using three AI techniques (GEP, ANN and EPR) to predict the values of relative buckling stress (Fb/Fy) using an aspect ratio (L/W), slenderness ratio (W/t), loss ratio (D/W), and boundary conditions (Kx.Ky). The results of comparing the accuracies of the developed models can be concluded in the following points:

- Both ANN and EPR have the most similar prediction accuracy, 89.6% and 84.7%, respectively, while the GEP model has the lowest prediction accuracy (77.3%).
- Although, the error percentage of the ANN and EPR models were so close, the output of the EPR model was closed form equations which could be manually used or as software unlike the ANN output which cannot be manually used.
- The summation of the absolute weights of each neuron in the input layer of the developed (ANN) model indicated that aspect ratio (L/W) had major influences on the relative buckling stress rather than the slenderness ratio (W/t), while the loss ratio (D/W) and boundary conditions (Kx.Ky) had less impact on (Fb/Fy).
- The GA technique successfully reduced the 70 terms of conventional polynomial regression quadrilateral formula to only 10 terms without significant impact on its accuracy.

- Like any other regression technique, the generated formulas were valid within the considered range of parameter values, beyond this range the prediction accuracy should be verified.

Author Contributions: Conceptualization, K.C.O.; methodology, A.M.E.; formal analysis, A.S.; investigation, M.D.; data curation, J.J.; supervision, H.J. All authors have read and agreed to the published version of the manuscript.

Funding: This research received no external funding.

Institutional Review Board Statement: Not applicable.

Informed Consent Statement: Not applicable.

Data Availability Statement: Utilized data are available in the Appendix A.

Conflicts of Interest: The authors declare no conflict of interest.

Appendix A. Utilized Database

L/W.	W/t.	D/W.	Kx.Ky	Fb/Fy.	L/W.	W/t.	D/W.	Kx.Ky	Fb/Fy.
1.5.	48.	0.20.	4.0.	0.25.	1.5.	48.	0.35.	1.0.	0.08.
1.5.	48.	0.50.	4.0.	0.25.	1.5.	48.	0.20.	4.0.	0.25.
1.5.	48.	0.40.	4.0.	0.24.	5.0.	200.	0.10.	0.5.	0.02.
1.0.	63.	0.08.	1.0.	1.02.	1.5.	48.	0.10.	4.0.	0.23.
1.5.	48.	0.10.	4.0.	0.29.	1.7.	600.	0.10.	0.5.	0.02.
1.5.	48.	0.40.	4.0.	0.26.	1.5.	48.	0.30.	4.0.	0.28.
1.5.	48.	0.25.	1.0.	0.07.	1.5.	48.	0.50.	4.0.	0.30.
1.0.	83.	0.11.	1.0.	0.95.	1.5.	48.	0.70.	4.0.	0.37.
1.5.	48.	0.50.	4.0.	0.30.	1.5.	48.	0.10.	4.0.	0.24.
1.0.	50.	0.16.	1.0.	0.90.	1.0.	42.	0.16.	1.0.	0.96.
1.6.	48.	0.10.	4.0.	0.17.	2.5.	400.	0.10.	0.5.	0.02.
1.0.	63.	0.14.	1.0.	0.95.	2.0.	100.	0.50.	1.0.	0.17.
1.1.	48.	0.10.	4.0.	0.45.	10.0.	50.	0.10.	0.5.	0.07.
1.5.	48.	0.30.	4.0.	0.24.	1.5.	48.	0.30.	4.0.	0.25.
1.0.	63.	0.11.	1.0.	1.01.	1.5.	48.	0.15.	4.0.	0.13.
2.5.	80.	0.10.	0.5.	0.39.	1.5.	48.	0.08.	1.0.	0.05.
1.5.	48.	0.30.	4.0.	0.26.	3.3.	300.	0.10.	0.5.	0.02.
1.5.	48.	0.50.	4.0.	0.31.	1.5.	48.	0.10.	4.0.	0.20.
1.0.	50.	0.14.	1.0.	0.79.	1.4.	700.	0.10.	0.5.	0.01.
2.0.	100.	0.10.	0.5.	0.39.	10.0.	100.	0.10.	0.5.	0.02.
1.0.	83.	0.14.	1.0.	0.93.	1.5.	48.	0.60.	4.0.	0.22.
1.1.	48.	0.10.	4.0.	0.49.	2.0.	100.	0.60.	1.0.	0.20.
1.0.	100.	0.08.	1.0.	0.98.	1.1.	48.	0.10.	4.0.	0.51.
1.7.	120.	0.10.	0.5.	0.38.	2.0.	100.	0.30.	1.0.	0.14.
2.0.	100.	0.40.	1.0.	0.15.	1.5.	48.	0.10.	4.0.	0.28.
1.5.	48.	0.10.	4.0.	0.23.	1.5.	48.	0.50.	4.0.	0.34.
1.5.	48.	0.20.	4.0.	0.23.	2.0.	250.	0.10.	0.5.	0.06.
1.0.	83.	0.08.	1.0.	0.99.	1.1.	48.	0.10.	4.0.	0.51.
2.1.	48.	0.10.	4.0.	0.09.	2.0.	100.	0.10.	1.0.	0.11.
2.1.	48.	0.10.	4.0.	0.09.	1.5.	48.	0.40.	4.0.	0.28.
2.1.	48.	0.10.	4.0.	0.09.	3.3.	60.	0.10.	0.5.	0.40.
1.0.	83.	0.16.	1.0.	0.90.	1.5.	48.	0.40.	1.0.	0.16.
1.5.	48.	0.08.	4.0.	0.12.	2.1.	48.	0.10.	4.0.	0.08.
1.4.	350.	0.10.	0.5.	0.06.	1.0.	50.	0.08.	1.0.	1.03.
3.3.	150.	0.10.	0.5.	0.07.	2.0.	500.	0.10.	0.5.	0.02.
1.5.	48.	0.40.	4.0.	0.29.	10.0.	20.	0.10.	0.5.	0.44.
1.5.	48.	0.60.	4.0.	0.31.	1.6.	48.	0.10.	4.0.	0.19.
1.0.	100.	0.11.	1.0.	0.93.	1.7.	300.	0.10.	0.5.	0.06.
2.1.	48.	0.10.	4.0.	0.09.	1.0.	63.	0.16.	1.0.	0.90.

L/W.	W/t.	D/W.	Kx.Ky	Fb/Fy.	L/W.	W/t.	D/W.	Kx.Ky	Fb/Fy.
1.4.	140.	0.10.	0.5.	0.35.	1.6.	48.	0.10.	4.0.	0.20.
1.0.	42.	0.11.	1.0.	1.02.	1.5.	48.	0.15.	1.0.	0.06.
1.5.	48.	0.10.	4.0.	0.22.	1.5.	48.	0.60.	4.0.	0.31.
1.6.	48.	0.10.	4.0.	0.17.	1.5.	48.	0.45.	1.0.	0.20.
1.5.	48.	0.20.	4.0.	0.25.	1.5.	48.	0.50.	1.0.	0.23.
2.0.	100.	0.20.	1.0.	0.12.	1.5.	48.	0.10.	4.0.	0.22.
1.5.	48.	0.25.	4.0.	0.14.	5.0.	100.	0.10.	0.5.	0.07.
1.5.	48.	0.90.	4.0.	0.52.	1.0.	42.	0.14.	1.0.	1.02.
1.5.	48.	0.10.	4.0.	0.21.	2.5.	200.	0.10.	0.5.	0.06.
1.0.	100.	0.14.	1.0.	0.92.	1.0.	50.	0.11.	1.0.	1.01.
1.1.	48.	0.10.	4.0.	0.49.	1.6.	48.	0.10.	4.0.	0.19.
1.5.	48.	0.38.	4.0.	0.21.	5.0.	40.	0.10.	0.5.	0.41.
1.0.	100.	0.16.	1.0.	0.88.	.	.	.	.	.

References

1. Gore, R.; Lokavarapu, B.R. Effect of Elliptical Cutout on Buckling Load for Isotropic Thin Plate. In *Innovations in Mechanical Engineering*; Narasimham, G.S.V.L., Babu, A.V., Reddy, S.S., Dhanasekaran, R., Eds.; Springer: Singapore, 2022; pp. 51–69.
2. Al Qablan, H. Applicable Formulas for Shear and Thermal Buckling of Perforated Rectangular Panels. *Adv. Civ. Eng.* **2022**, *2022*, 3790462. [CrossRef]
3. Dehadray, P.M.; Alampally, S.; Lokavarapu, B.R. Buckling Analysis of Thin Isotropic Square Plate with Rectangular Cut-Out. In *Innovations in Mechanical Engineering*; Narasimham, G.S.V.L., Babu, A.V., Reddy, S.S., Dhanasekaran, R., Eds.; Springer: Singapore, 2022; pp. 71–86.
4. Taherian, I.; Ghalehnovi, M.; Jahangir, H. Analytical Study on Composite Steel Plate Walls Using a Modified Strip Model. In Proceedings of the 10th International Congress on Civil Engineering, Abriz, Iran, 5–7 May 2015.
5. Sinha, V.; Patel, R.; Ghetiya, K.; Nair, M.; Trivedi, T.; Rao, L.B. Impact of Diamond-Shaped Cut-Out on Buckling Nature of Isotropic Stainless-Steel Plate. In *Innovations in Mechanical Engineering*; Narasimham, G.S.V.L., Babu, A.V., Reddy, S.S., Dhanasekaran, R., Eds.; Springer: Singapore, 2022; pp. 119–146.
6. Soleymani, A.; Reza Esfahani, m. Effect of concrete strength and thickness of flat slab on preventing of progressive collapse caused by elimination of an internal column. *J. Struct. Constr. Eng.* **2019**, *6*, 24–40. [CrossRef]
7. Hosseini-Hashemi, S.; Khorshidi, K.; Amabili, M. Exact solution for linear buckling of rectangular Mindlin plates. *J. Sound Vib.* **2008**, *315*, 318–342. [CrossRef]
8. Orun, A.E.; Guler, M.A. Effect of hole reinforcement on the buckling behaviour of thin-walled beams subjected to combined loading. *Thin-Walled Struct.* **2017**, *118*, 12–22. [CrossRef]
9. Singh, T.G.; Chan, T.-M. Effect of access openings on the buckling performance of square hollow section module stub columns. *J. Constr. Steel Res.* **2021**, *177*, 106438. [CrossRef]
10. Hasrati, E.; Ansari, R.; Rouhi, H. A numerical approach to the elastic/plastic axisymmetric buckling analysis of circular and annular plates resting on elastic foundation. *Proc. Inst. Mech. Eng. Part C J. Mech. Eng. Sci.* **2019**, *233*, 7041–7061. [CrossRef]
11. Russell, M.J.; Lim, J.B.; Roy, K.; Clifton, G.C.; Ingham, J.M. Welded steel beam with novel cross-section and web openings subject to concentrated flange loading. *Structures* **2020**, *24*, 580–599. [CrossRef]
12. Rao, L.B.; Rao, C.K. Buckling of Circular Plates with an Internal Elastic Ring Support and Elastically Restrained Guided Edge Against Translation. *Mech. Based Des. Struct. Mach.* **2009**, *37*, 60–72. [CrossRef]
13. Jahangir, H.; Daneshvar Khorram, M.H.; Ghalehnovi, M. Influence of geometric parameters on perforated core buckling restrained braces behavior. *J. Struct. Constr. Eng.* **2019**, *6*, 75–94. [CrossRef]
14. Daneshvar, M.H.; Saffarian, M.; Jahangir, H.; Sarmadi, H. Damage identification of structural systems by modal strain energy and an optimization-based iterative regularization method. *Eng. Comput.* **2022**, *37*, 1–21. [CrossRef]
15. Jahangir, H.; Khatibinia, M.; Kavousi, M. Application of Contourlet Transform in Damage Localization and Severity Assessment of Prestressed Concrete Slabs. *Soft Comput. Civ. Eng.* **2021**, *5*, 39–67. [CrossRef]
16. Muddappa, P.P.Y.; Rajanna, T.; Giridhara, G. Effect of reinforced cutouts on the buckling and vibration performance of hybrid fiber metal laminates. *Mech. Based Des. Struct. Mach.* **2021**, *49*, 1–21. [CrossRef]
17. Mohammadzadeh, B.; Choi, E.; Kim, W.J. Comprehensive investigation of buckling behavior of plates consider-ing effects of holes. *Struct. Eng. Mech.* **2018**, *68*, 261–275. [CrossRef]
18. Abolghasemi, S.; Eipakchi, H.; Shariati, M. An analytical solution for buckling of plates with circular cutout subjected to non-uniform in-plane loading. *Arch. Appl. Mech. Vol.* **2019**, *89*, 2519–2543. [CrossRef]
19. Kim, J.-H.; Park, D.-H.; Kim, S.-K.; Kim, J.-D.; Lee, J.-M. Lateral Deflection Behavior of Perforated Steel Plates: Experimental and Numerical Approaches. *J. Mar. Sci. Eng.* **2021**, *9*, 498. [CrossRef]
20. Yidris, N.; Hassan, M. The effects of cut-out on thin-walled plates. In *Modelling of Damage Processes in Biocomposites, Fibre-Reinforced Composites and Hybrid Composites*; Woodhead Publishing: Philadelphia, PA, USA, 2019. [CrossRef]

21. Kumar, M.S.; Alagusundaramoorthy, P.; Sundaravadivelu, R. Ultimate strength of square plate with rectangular opening under axial compression. *J. Nav. Arch. Mar. Eng.* **2007**, *4*, 15–26. [CrossRef]
22. Maiorana, E.; Pellegrino, C.; Modena, C. Linear buckling analysis of perforated plates subjected to localised symmetrical load. *Eng. Struct.* **2008**, *30*, 3151–3158. [CrossRef]
23. Singh, S.; Kulkarni, K.; Pandey, R.; Singh, H. Buckling analysis of thin rectangular plates with cutouts subjected to partial edge compression using FEM. *J. Eng. Des. Technol.* **2012**, *10*, 128–142. [CrossRef]
24. Shariati, M.; Dadrasi, A. Numerical and Experimental Investigation of Loading Band on Buckling of Perforated Rectangular Steel Plates. *Res. J. Recent Sci.* **2012**, *1*, 63–71.
25. Djelosevic, M.; Tepic, J.; Tanackov, I.; Kostelac, M. Mathematical Identification of Influential Parameters on the Elastic Buckling of Variable Geometry Plate. *Sci. World J.* **2013**, *2013*, 268673. [CrossRef]
26. De Mauro Vasconcellos, R.; Liércio André, I.; Alexandra Pinto, D.; Daniel, H. Elastic and Elasto-Plastic Buckling Analysis of Perforated Steel Plates. Repositório Inst. 2013. Available online: https://repositorio.furg.br/handle/1/4929 (accessed on 1 May 2022).
27. Behzad, M.; Noh, H.-C. Use of Buckling Coefficient in Predicting Buckling Load of Plates with and without Holes. *J. Korean Soc. Adv. Compos. Struct.* **2014**, *5*, 1–7. [CrossRef]
28. Mohammadzadeh, B.; Noh, H.C. Investigation into the buckling coefficients of plates with holes considering variation of hole size and plate thickness. *Mechanika* **2016**, *22*, 167–175. [CrossRef]
29. Shariati, M.; Faradjian, Y.; Mehrabi, H. Numerical and Experimental Study of Buckling of Rectangular Steel Plates with a Cutout. *J. Solid Mech.* **2016**, *8*, 116–129.
30. Jana, P. Optimal design of uniaxially compressed perforated rectangular plate for maximum buckling load. *Thin-Walled Struct.* **2016**, *103*, 225–230. [CrossRef]
31. Lorenzini, G.; Helbig, D.; Silva, C.; Real, M.; Santos, E.; Rocha, L. Numerical evaluation of the effect of type and shape of perforations on the buckling of thin steel plates by means of the constructal design method. *Int. J. Heat Technol.* **2016**, *34*, S9–S20. [CrossRef]
32. Adah, D.E.; Onwuka, O.M. Ibearugbulem, Matlab Based Buckling Analysis of Thin Rectangular Flat Plates. *Am. J. Eng. Res.* **2019**, *8*, 224–228.
33. Gracia, J.B.; Rammerstorfer, F.G. Increase in buckling loads of plates by introduction of cutouts. *Acta Mech.* **2019**, *230*, 2873–2889. [CrossRef]
34. Da Silva, C.C.C.; Helbig, D.; Cunha, M.L.; dos Santos, E.D.; Rocha, L.A.O.; Real, M.D.V.; Isoldi, L.A. Numerical buckling analysis of thin steel plates with centered hexagonal perforation through constructal design method. *J. Braz. Soc. Mech. Sci. Eng.* **2019**, *41*, 309. [CrossRef]
35. Rao, L.B.; Rao, C.K. Fundamental Buckling of Annular Plates with Elastically Restrained Guided Edges Against Translation. *Mech. Based Des. Struct. Mach.* **2011**, *39*, 409–419. [CrossRef]
36. Rao, L.B.; Rao, C.K. Buckling of circular plate with foundation and elastic edge. *Int. J. Mech. Mater. Des.* **2015**, *11*, 149–156. [CrossRef]
37. Rao, L.B.; Rao, C.K. Buckling of Annular Plates with Elastically Restrained External and Internal Edges. *Mech. Based Des. Struct. Mach.* **2013**, *41*, 222–235. [CrossRef]
38. Fu, W.; Wang, B. A semi-analytical model on the critical buckling load of perforated plates with opposite free edges. *Proc. Inst. Mech. Eng. Part C J. Mech. Eng. Sci.* **2022**, *236*, 4885–4894. [CrossRef]
39. Hosseinpour, P.; Hosseinpour, M.; Sharifi, Y. Artificial neural networks for predicting ultimate strength of steel plates with a single circular opening under axial compression. *Ships Offshore Struct.* **2022**, *17*, 1–16. [CrossRef]
40. Timoshenko, S.P.; Gere, J.M. *Theory of Elastic Stability*; Tata McGraw-Hill Education Pvt. Ltd.: New Delhi, India, 2010.
41. Ebid, A.M.; El-Aghoury, M.A.; Onyelowe, K.C. Estimating the Optimum Weight for Latticed Power-Transmission Towers Using Different (AI) Techniques. *Designs* **2022**, *6*, 62. [CrossRef]
42. Reda, M.A.; Ebid, A.M.; Ibrahim, S.M.; El-Aghoury, M.A. Strength of Composite Columns Consists of Welded Double CF Sigma-Sections Filled with Concrete—An Experimental Study. *Designs* **2022**, *6*, 82. [CrossRef]
43. Nemeth, M.P. Buckling behavior of compression-loaded symmetrically laminated angle-ply plates with holes. *AIAA J.* **1988**, *26*, 330–336. [CrossRef]
44. Faradjian Mohtaram, Y.; Taheri Kahnamouei, J.; Shariati, M.; Behjat, B. Experimental and numerical investigation of buckling in rectangular steel plates with groove-shaped cutouts. *J. Zhejiang Univ. Sci. A* **2012**, *13*, 469–480. [CrossRef]
45. Kelpša, Š.; Peltonen, S. Local Buckling Coefficient Calculation Method of Thin Plates with Round Holes. *ce/Papers* **2019**, *3*, 841–846. [CrossRef]
46. Shi, J.; Guo, L.; Gao, S. Study on the buckling behavior of steel plate composite walls with diamond ar-ranged studs under axial compression. *J. Build. Eng.* **2022**, *50*, 2352–7102.
47. Wu, Y.; Xing, Y.; Liu, B. Analysis of isotropic and composite laminated plates and shells using a differential quad-rature hierarchical finite element method. *Compos. Struct.* **2018**, *205*, 11–25. [CrossRef]
48. Kabir, H.; Aghdam, M. A robust Bézier based solution for nonlinear vibration and post-buckling of random checkerboard graphene nano-platelets reinforced composite beams. *Compos. Struct.* **2019**, *212*, 184–198. [CrossRef]

49. Sun, Z.; Lei, Z.; Bai, R.; Jiang, H.; Zou, J.; Ma, Y.; Yan, C. Prediction of compression buckling load and buckling mode of hat-stiffened panels using artificial neural network. *Eng. Struct.* **2021**, *242*, 112275. [CrossRef]
50. Zhu, S.; Ohsaki, M.; Guo, X. Prediction of non-linear buckling load of imperfect reticulated shell using modified consistent imperfection and machine learning. *Eng. Struct.* **2020**, *226*, 111374. [CrossRef]
51. Badarloo, B.; Jafari, F. A Numerical Study on the Effect of Position and Number of Openings on the Performance of Composite Steel Shear Walls. *Buildings* **2018**, *8*, 121. [CrossRef]
52. Tahir, Z.U.R.; Mandal, P.; Adil, M.T.; Naz, F. Application of artificial neural network to predict buckling load of thin cylindrical shells under axial compression. *Eng. Struct.* **2021**, *248*, 113221. [CrossRef]
53. Abambres, M.; Rajana, K.; Tsavdaridis, K.D.; Ribeiro, T.P. Neural Network-Based Formula for the Buckling Load Prediction of I-Section Cellular Steel Beams. *Computers* **2018**, *8*, 2. [CrossRef]
54. El-Aghoury, M.A.; Ebid, A.M.; Onyelowe, K.C. Optimum Design of Fully Composite, Unstiffened, Built-Up, Hybrid Steel Girder Using GRG, NLR, and ANN Techniques. *J. Eng.* **2022**, *2022*, 7439828. [CrossRef]
55. Kadry, A.A.; Ebid, A.M.; Mokhtar, A.-S.A.; El-Ganzoury, E.N.; Haggag, S.A. Parametric study of Unstiffened multi-planar tubular KK-Joints. *Results Eng.* **2022**, *14*, 100400. [CrossRef]

designs

MDPI

Article

Prospects of Triangular Modular Structures for Roadside Service Buildings

Konstantin Samoilov [1,*], Bolat Kuspangaliyev [1], Gaukhar Sadvokasova [2] and Nurlytan Kuanyshbekov [2]

1 Institute of Architecture and Construction, Satbayev University, 22 Satbayev Str., Almaty 050000, Kazakhstan
2 International Education Corporation/KazGASA, 28 Ryskulbekov Str., Almaty 050043, Kazakhstan
* Correspondence: samconiv@mail.ru; Tel.: +7-701-357-68-93

Abstract: The need for a relatively quick solution to the problem of providing highways with roadside service facilities necessitates the development of a series of appropriate standard projects. To increase the efficiency of these series, it is advisable to carry out the interconnection of space-planning solutions based on a particular module. Taking into account the variety of planning and landscape characteristics of the sites for the placement of objects of the mainline service, it seems advisable to choose as a module not a square or rectangular, but a triangular configuration, which allows in most cases to harmoniously block the modules. The proposed roof module in the form of a "regular" triangle facing the tetrahedron has a structural basis in the form of a single-tier rod spatial plate. The principal space-planning solutions of all four dozen objects from the approved nomenclature of the mainline service performed in the process of analyzing the possibilities show the real possibility of solving the development tasks on the basis of this system. The use of the proposed modular system makes it possible to successfully solve a number of tasks to reduce the harmful impact on the environment and effectively use renewable energy sources. The work is devoted specifically to the field of design.

Keywords: roadside service buildings; modular buildings; sustainable architecture; spatial grid plates; reusable structures; base module; collapsible buildings

Citation: Samoilov, K.; Kuspangaliyev, B.; Sadvokasova, G.; Kuanyshbekov, N. Prospects of Triangular Modular Structures for Roadside Service Buildings. *Designs* **2022**, *6*, 90. https://doi.org/10.3390/designs6050090

Academic Editors: Joshua M. Pearce and Tiago Pinto Ribeiro

Received: 20 June 2022
Accepted: 28 September 2022
Published: 2 October 2022

1. Introduction

Modern facilities of the mainline service demonstrate a wide variety of planning solutions and architectural and artistic forms. The problem that has arisen at this stage of roadside service development (including in Kazakhstan, where there are a number of promising opportunities in the construction sector, per B.Torgautov, A. Zhanabayev, A. Tleuken, A. Turkyilmaz, M. Mustafa, F. Karaca, F. [1], who point to challenges and opportunities in the construction sector of Kazakhstan in the aspect of a closed-cycle economy), the problem of accelerated, almost one-time construction of a large number of facilities puts forward a number of peculiar tasks that have not yet received appropriate research work (the relevance of the development of architecture of roadside service facilities in Kazakhstan is emphasized N.N. Kuanyshbekov, A.K. Tuyakaeva [2]).

The objects of roadside service naturally began to arise with the development of communication routes. With the advent of motor transport, the nomenclature of objects has increased somewhat, but the functional specifics of their work has not changed much. Accordingly, space-planning solutions, having a centuries-old practice of application and improvement, have a significant degree of study. The totality is formed in the mutual influence of several aspects important for the ongoing research. So, the general assessment of the roadside service system is covered in the works T. Cui, Y. Ouyang, Z.J.M. Shen [3], where the design of a reliable placement of an object under the risk of failures is considered. D. Ettema, T. Gärling, L.E. Olsson, M. Friman and S. Moerdijk analyze the satisfaction of Dutch drivers with roadside service [4]; M.K. Bostani, F. Hashemzehi and M.R. Anvari evaluate roadside service centers on the example of a separate highway [5]; Z. Dvorak, E. Sventekova, D. Rehak, Z. Cekerevac [6] analyze the quality of the most important elements of infrastructure in transport; H.Z. Rahman, A. Andreas, D. Perwitasari, J.S. Petroceany [7] paid attention to the development of the typology of the social infrastructure of roadside stations; M.M. Hasan, A. Alam, A.M. Mim, A. Das [8] analyzed the features of determining the level of satisfaction of users with road services; the economic assessment of the construction of roadside service facilities was carried out by such researchers as O. Makovetskaya-Abramova, A. Ivanov, Y. Lazarev, M. Shakhova, and A. Rozov [9].

Certain types of roadside service facilities and the specifics of their location are considered in various regulatory documents and research papers (A. Hurley, J.A. Jakle, K.A. Sculle [10–13] actualized the problem of roadside restaurants, taking into account the peculiarities of road transport; M. Magdic, P. Sjöstrand [14] drew attention to the key role of gas stations; K. Wolfe, R. Holland, J. Jeff Aaron [15] conducted marketing of agricultural products sold in roadside stores; K. Shanahan drew attention to the degree of correspondence between the advertising on roadside billboards and the attributes of motels in demand by tourists [16]; M. Kendrick analyzed the history of the development of models [17]; Y.W. Wang and C.R. Wang calculated the optimal scheme of the location of passenger vehicle refueling stations [18]; L. Henderson [19] drew attention to the change in the range of services provided by motels; A.F. Al-Kaisy, Z. Kirkemo, D. Veneziano and C. Dorrington examined the use of recreation areas on rural highways [20]; K.J. Sheng, A.S. Baharudin, K. Karkonasasi [21] studied the system for determining the location of a car service station in case of a breakdown; I. Xanthopoulos, G. Goulas, C. Gogos, P. Alefragis and E. Housos drew attention to the relationship between optimizing energy consumption and user satisfaction with roadside recreation areas [22]; A. Plovnick, A. Berthaume, C. Poe, T. Hodges [23] studied the problem of designing and operating sustainable recreation areas in the transport system; P. Karanja, C.W. Gathitu [24] illustrated the strategy of placing gas stations on a concrete example of an important highway; M. Rubeis, S. Groves, T. Portera and G. Bonaccorsi discuss the prospects for further development and the feasibility of operating roadside service stations [25]; D. Green, P. Roper, L. Steinmetz, L. Latter, K. Lewis, D. Gaynor [26] detailed the issue of equipment of recreation areas for heavy vehicles; A. Quito drew attention to the prospects for the development of roadside chapels [27]).

Special attention is paid to the points where alternative fuels are refueled (I. Capar, M. Kuby, V.L. Leon, Y.J. Tsai [28,29], S.H. Chung, C. Kwon [30], S.F. Bhatti, M.K. Lim, H.Y. Mak [31], M. Ghamami, A. Zockaie, Y.M. Nie [32], T.H. Tran, T.B.T. Nguyen [33]).

Accordingly, the potential of the modularity of structures, which is realized in objects of different volume and shape, looks promising (G.Angelucci, F.Mollaioli, R.Tardocchi [34]). The rapid construction of a large number of typologically different objects, technologically linked into a roadside service complex, suggests the possibility of using end-to-end modularity (A. Subbotin, S. Grigoryan [35], K.I. Samoilov [36]).

However, despite the considerable degree of study of certain aspects of the problem of roadside maintenance, the problem of interaction of objects in the complex has not been studied enough. Particularly noteworthy is the idea of unification and standardization of roadside service facilities, which provides a quick visual identification of the function of each object in the process of moving along the highway, as well as an understanding of its planning features, which makes it convenient to navigate in the service process.

The scientific novelty of the study is as follows:

- for the first time, the possibility of spatial-planning linking of typologically diverse roadside service facilities with a large-sized structural module is demonstrated;
- for the first time, the possibility of forming roadside service complexes based on multi-configuration blocking of modules, taking into account both the uniqueness of the typology of an individual object and the complex as a whole is shown;
- for the first time, a triangular module was used to solve the above problems.

Thus, the proposed material shows a set of new scientific results that determine the possibility of effective use of a modular system based on a triangular rod spatial plate with appropriate planning adaptation of the entire nomenclature of individual objects and roadside service complexes.

At the beginning, the specifics of buildings and structures that solve individual tasks that form a set of roadside service needs are considered. Then, the application of the triangular module is justified and the possibility of an integrated approach to solving these problems based on this configuration is shown. The advantages of the proposed system from the point of view of the organization of production, construction, and operation, including issues of energy efficiency and environmental protection, are indicated below. In conclusion, the prospects for further study of this topic are shown.

2. Materials and Methods

The research method used in the work is the general scientific dialectical way of cognition from observation through generalizations to practice. In its context, the following are applied:

- an integral-differential approach, which allows dividing the array of roadside service facilities on the principle of similarity of typological forms into many components, combining them into a set reflecting the possibility of planning unification;
- a formal approach that allows to trace the development of various typological forms of roadside service facilities;
- an iconographic approach that allows to explore the manifestation of the features of various historical prototypes in the modern layout of roadside service facilities;
- a structural-semiotic approach that allows modeling the development and prospects of the planning unification of roadside service facilities.

The methodology determined the techniques and sequence of work: selection and analysis of literature; field surveys; differentiation of the array of data obtained; comparison and analysis of options; formation of a unified layout design model based on a triangular module.

Having the character of mass use, road service objects are organically predisposed to typification and unification. On the one hand, this ensures their rapid visual recognition in conditions of the complexity of high-speed traffic, and on the other hand, it allows them to be built optimally quickly, taking into account the working methods. The need to improve roadside services and increase traffic flow adds to the need to expand existing single or complex structures. This naturally actualizes the need for modularity of elements, which makes it possible to increase the structure using, among other things, the initial design solutions. Among the nomenclature of roadside service facilities, the most dynamic element is filling stations of various types of fuel. Checkpoints and individual parking lots are also interesting from this point of view.

The nomenclature of roadside facilities for servicing vehicles includes: refueling with various types of fuel, washing motorcycles and cars of various sizes, service stations. Gas stations differ both in the types of fuel being refueled and in the specifics of their work. Automatic self-service filling stations are gradually becoming more widespread. Externally, they are usually a small canopy over the control panel. The main problem of common solutions is the lack of comfort during use, since the minimized surface of the canopy protects not only the car being refueled, but in most examples even the process control panel from precipitation. Moreover, if this is just an inconvenience for urban conditions, then it is a critical disadvantage for highways. Large gas stations are usually characterized by a fairly wide canopy, usually uniform for the complex, providing partial or complete shelter of the refueling car from atmospheric precipitation. In the operator's pavilion located in the immediate vicinity, there are sometimes small shops of related products and buffets of tonic drinks, as well as toilets, which imply the use of not only staff, but also road users.

Trading operations for refueling and selling products in the store, depending on the size of the complex, are carried out by one or more people. One or several people are also engaged in servicing the refueling columns themselves, for whom an appropriate set of rooms is provided in the structure of the complex: dressing rooms with showers, a recreation room. The complex also has an administration office, rooms for engineering support systems, and storerooms for technical, technological, and cleaning equipment. The architectural and artistic solution of the canopy reflects the preferences of customers interpreted by the author of the project or the corporate identity of the supplier of a particular type of fuel. The technological peculiarity of gas as a type of automobile fuel has determined the spatial solution of the corresponding gas stations, which are either included in the complex of conventional diesel-gasoline gas stations in the form of a separate structure or as independent complexes with an appropriate set of main and auxiliary rooms. The main technological feature of gas filling stations is the ground location of tanks, which ensures high explosion and fire safety of the complex. Sometimes, there are solutions with an underground location of gas tanks, but this entails the need to ensure the constant operation of a special ventilation system. Such gas stations operate both in normal and automatic mode according to the self-service system. Electric filling stations usually operate in automatic mode using a self-service system. The main problem of the formation of such structures as elements of roadside service is to provide leisure time for drivers and passengers during the refueling process, which takes from one and a half to several hours. This entails the need to create an appropriate pavilion with a set of services for short-term rest. This practice is almost not found now.

Car washes vary both by system of operation and by means of transport. Self-service car washes are often formed for passenger cars and motorcycles, which are covered posts partially fenced off by stationary or transformable partitions with control panels for pipelines of hose supply of water, detergents, compressed air, as well as waste disposal suction. Depending on the number of washing stations and placement, the structure includes an operator's pavilion with an appropriate set of main and auxiliary rooms. If a small structure of this type is included in the complex of a large car wash for trucks and buses, then the operator's pavilion is usually not made, being combined with the corresponding rooms of the main car wash. Quite often, self-service car washes are combined in one facility with automatic contactless car washes. Since the process of contactless washing involves the in-line execution of several technological processes, the pavilion itself has a longer length than for a conventional washing, providing through movement even of a passenger car, which determines the peculiarity of its position in the layout of the complex. Depending on the customer's wishes, during the contactless washing process, he can either stay in the car or stay outside. For the convenience of visual control of the process, in some examples, stained glass glazing of the washing line room is used. For trucks and buses, automatic open portal-type sinks are often used, moving along the vehicle. The limitation of the use of such sinks is their convenience only for the relatively warm season, which determines their location in a complex with closed sinks for year-round operation. The most common because of the convenience of use are pavilion sinks for cars, trucks and buses. They are mostly with a dead-end arrival or through passage. The main room of the sink is usually divided by stationary or mainly transformable curtain partitions. Auxiliary rooms are located compactly from above as an independent floor or on the mezzanine, in the middle part, on the side or symmetrically. They include a complex of rooms for customers with a room for receiving orders, a rest room, a buffet and toilets; administration—with one or more offices. A group of rooms for staff is specially planned: a control room, dressing rooms with toilets, showers and storerooms of clean and used workwear; a meal room with a built-in kitchen, storerooms of cleaning and technological equipment, cleaning materials. There is also a group of rooms for engineering support systems, panel rooms, ventilation chambers, a heat point, a water measuring unit, pumping stations. In combination with various types of sinks, there are recycling water treatment facilities. They are located in various places: in the basement space directly under the washing posts, on the side or behind in an annex or in a free-standing structure.

A separate, less often built-in, facility is a car service station. Usually, technological operations are performed by specialized personnel. However, there are self-service stations for minor repairs. Usually, the complexes are pavilions with the main room, fenced off by stationary partitions for repair posts, having independent dead-end entrances and exits to each post. The doors in the partitions provide a consistent connection of the repair rooms with each other. The repair room has a technologically determined height that ensures the maintenance of the car on the lift. Adjacent to the repair posts are storage rooms for spare parts, parts, consumables, tools and technological equipment. The complex of rooms for repair personnel, located on the same level with the repair posts or on the mezzanine, includes dressing rooms with toilets and showers, storerooms of clean and used workwear, a meal room with a built-in kitchen, a recreation room, a classroom. The administrative and clerical part includes: a document processing room, a rest room, a buffet, toilets, a control room, administration offices. Depending on the layout, the premises of engineering support systems are located compactly or dispersed.

The nomenclature of traffic control structures includes: points of traffic control services, checkpoints, points of dimensional and weight control of vehicles.

Primary traffic control is carried out at small points of traffic police services. They are usually located near populated areas. Relative to the highway, posts are of island and coastal types. The structure of such points usually includes a small canopy and partially isolated areas for the placement of duty officers. Sometimes, the item includes a separate heated pavilion, which houses a toilet, duty rooms and utility rooms. To improve visual control, the duty rooms in such pavilions are sometimes located on the second level. At the same time, the first level is used mainly as a covered parking place for a special car on duty. For extended sections between settlements, it is advisable to use structures with an expanded composition of premises. In the case of an island position on the highway, in a two-story pavilion on the ground floor, there are operator traffic control rooms, rooms for checking documents of detained cars, and toilets. Parking places for special cars on duty with the possibility of direct or rotary exit to the highway in the right direction are arranged under the end canopies. In the sequel, already in an open area, detained cars are parked in the direction of direct traffic. On the sides under the canopies, there are lanes of slow-motion free passage of vehicles. The second floor includes symmetrically located offices of operational visual control, oriented to the incoming section of the highway, staff wardrobes with toilets and showers, the office of the head of the shift on duty, a study room, a meal room with a built-in kitchen, a room for administratively detained drivers and passengers, storerooms of special and cleaning equipment, technical communication rooms.

Checkpoints with controlled passage are provided for entry and exit to toll sections of highways. At the same time, each lane has a barrier and a payment machine under the canopy. Vehicles stop for a short time only for carrying out payment transactions. The facility is located across the highway and has duty personnel pavilions located along the coastal or island scheme. In one- or two-storey pavilions, there are operator's rooms, control rooms, staff wardrobes with toilets and showers, meal rooms with built-in kitchens, the office of the shift supervisor, storerooms of special and cleaning equipment, rooms of technical means of communication and alarm systems. The pavilions are directly adjacent to the canopies for parking special cars with a direct or rotary exit in the appropriate direction. In a simplified version, the pavilions of the personnel of such checkpoints are small buildings with two or three small rooms only for operational duty officers.

Large structures with enhanced operator control of each lane are provided for specialized control of travel to the relevant road sections or entry-exit to the relevant zones. At the same time, vehicles stop at the post for the time necessary for conducting an external inspection of the rolling stock and checking the documents of the drivers. In accordance with this, each lane has a separate pavilion for accommodating an operational duty officer and a shift worker, as well as a toilet. The room on duty has a height that allows you to see the vehicle through the stained glass to the full height of the road dimension—4.5 m. In the underground level, there are rooms with anti-aircraft openings for visual inspection of the bottom of vehicles. Further, on top of the room on duty, there are bridges for inspection of the upper part of the car. All three levels of each pavilion are connected by a stairwell. The underground observation rooms through the gallery and observation bridges are directly sequentially connected to each other and to the administrative pavilion located in the middle (island version) or to the side (coastal version). This ensures an operational visually isolated passage of personnel to a specific post. In this pavilion, on the ground floor, there are duty rooms, offices for paperwork, toilets. The second floor is occupied by visual control rooms of access areas, staff wardrobes with toilets and showers, a meal room with a built-in kitchen, a classroom, technical rooms for special communications and alarm systems. Storerooms of special and cleaning equipment, warehouses of workwear, and technical rooms of engineering support systems are located in the basement. The complex includes covered parking lots for special cars with the possibility of direct or rotary exit in the appropriate direction, as well as open parking lots for detained vehicles.

A typical example of such a structure is the complex on the Narol-Naroda Road in Gujarat. It was built according to the project of the company "Archohm" (architect S. Gupta, 2010) [37]. Having an area of 3860.0 sq.m, it is planned to represent 16 (8 + 8) control and toll collection posts connected by an underground gallery with each other and with two blocks of administrative and office purpose. The buildings are elongated volumes with blind side walls and a fully glazed central cylindrical block. On the ground floor of the office block, there are: lobby, security post, elevator, waiting hall, stairs, passage, toilet, equipment storage room, garden, cash register, archive, shop, document verification room, security room. On the second floor, there are: toilet and shower for staff, locker room, elevator, passage, kiosk, cafe, guest hall, cleaning room, manager's office, women's toilet, men's toilet, staff room, classroom. On the third floor, there are: the dispatchers' pavilion, the exploited sections of the roof. The posts have sufficient height for lateral visual control. An iconic element of the architectural and artistic solution are awnings with geometry that varies depending on the weather and the location of the sun. The dominant color of the complex is red. This color has blind sections of the walls of the administrative and office pavilions, awnings and tubular bumpers stretched to the full height of the truck.

The auxiliary object of traffic control complexes are the points of weight and dimensional control of vehicles. They can be formed either in an open version or in the form of a pavilion with canopies. When closed, they include canopies with equipment for parametric control, as well as a pavilion for staff accommodation and registration of relevant documents. The pavilion's premises include toilets, storerooms for special and cleaning equipment, rooms for engineering support, and communication systems.

The nomenclature of roadside service facilities for drivers and passengers includes: public transport stops, retail and catering outlets, public toilets and showers, laundries, medical and rescue service points, wellness pavilions, picnic and recreation areas, scenic areas, playgrounds and pavilions, roadside temples of various denominations, heating points, post offices, motels, warehouses, supermarkets.

The difference between public transport stops in the city and on intercity routes is based on a much longer waiting time and a stronger impact of precipitation and wind. In addition, unlike city stops, suburban stops, due to the open space around, practically do not block the view of suitable transport. Accordingly, for city stops, it is necessary to ensure "transparency" that creates end-to-end visibility. For suburban stops, however, a larger canopy and the creation of a partially enclosed pavilion with blind rear and side walls are advisable. Roadside retail and catering outlets have different capacities depending on the intensity of cargo-passenger flow on the highway. Usually, they consist of a dining room with several tables and an adjacent kitchen with a cutting room and a dishwasher. Free access to goods is provided in the trading floor, some small goods are sold from behind the counter. The premises include a toilet for visitors and staff, an administrative office, and a number of utility rooms.

In order to avoid duplication of structures, ensuring the possibility of using the services of travelers moving in opposite directions leads to the need to create crossings across the highway. At the same time, parking is arranged on both sides, and the bridge in some places is used as a passing shopping or dining hall.

Public toilets are a very important element of roadside service. In most cases, these are small container-type structures. However, this is not very convenient, since, for example, when a regular bus stops, a queue of passengers is created, which is critically unacceptable. Accordingly, it seems appropriate to create more spacious facilities, including two compartments with vestibules-washrooms, rooms of the actual restroom with ordinary booths and booths for disabled visitors, a pantry of cleaning equipment. In some cases, it is optimal to expand the composition of the premises for the organization of a shower.

The modern practice of creating roadside medical and rescue service facilities in most cases demonstrates mobile container roadside emergency medical aid stations. However, the need to improve services determines the need to create stationary facilities, including reception rooms, dressing rooms, storerooms of medicines and equipment, toilets with large vestibules for changing clothes, tool rooms, rest rooms for duty crews, dispatching rooms, staff reception rooms, administrative rooms, workwear storerooms, wardrobes with showers, and covered parking spaces for special cars. In some cases, separate dentist and therapist offices are needed. Roadside heating points are part of the medical and rescue service complexes. Currently, in most cases they are mobile in the form of equipped cars or tents. In some cases, containers are used. However, taking into account the places of possible accumulation of cars in winter known from long-term operation practice, it is advisable to form stationary heating points in addition to mobile points. As part of their premises, wardrobes with toilets and places for drying clothes, a small buffet for hot food and drinks, a rest room, a control room, a staff room and the necessary set of rooms for engineering support systems are provided.

In the most picturesque places of the route or as part of large roadside service complexes, places for organizing picnics are provided, designed for several unrelated companies of vacationers. At the same time, the possibility of placing customers both in the open air and under the canopies of free-standing gazebos and pavilions is provided. Gazebos and pavilions are designed for a different number of vacationers. Of course, it is necessary to set up toilets, places for washing dishes, as well as containers for collecting garbage and food waste. Usually, places for cooking food on an open fire with wood warehouses are organized. Electric grills with the appropriate equipment are also provided.

To ensure the spiritual needs of those passing by, roadside structures of a cult nature are formed. They have different sizes depending on local characteristics. The space-planning solution is built depending on the canons of a particular denomination. However, the general set of rooms and spaces largely coincides. These are: a common or divided into male and female parts prayer hall, lobby, clergy room, library. There are also open courtyards or spaces under canopies for pre-prayer and post-prayer concentration. The orientation of prayer halls is created depending on the canons. Historically, there was also a type of small roadside chapels or chapels, consisting of one or two rooms for one or more travelers to pray without the participation of a clergyman.

As part of roadside complexes for long-term recreation, children's playgrounds with heated pavilions are being built, in which game rooms, utility and storage rooms and toilets are arranged. Next to the pavilion, it is advisable to arrange a large canopy for games during rain or bright sun.

The roadside service system includes small post offices associated mainly with the function of receiving correspondence and providing telephone communication in various formats. Roadside laundries are an integral part of the service. These facilities provide for both self-service and taking things to the laundry for a short-term or long-term stop. As part of their premises, there is usually a lobby, a cash register, a toilet for customers, self-service laundry rooms and laundry items, laundry pantries, self-service ironing and ironing items, inventory rooms, a laundry reception point, staff wardrobes with toilet and shower, cleaning equipment storerooms, a laundry pick-up point, engineering support systems rooms.

The complex of roadside entertainment includes various sizes of summer theaters, variety shows and stage platforms. They have open and canopied spaces for spectators and artists. The composition of the stage premises sometimes includes artistic rooms with toilets and storerooms of scenery. In places of long-term recreation, it is planned to build fitness clubs with a different set of rooms for gymnastics and water-entertainment procedures. The premises of such facilities may include: a lobby, a reception desk with a cash register and a utility room, a men's and women's department (wardrobe, linen closet, toilet, vestibule, shower room, corridor, steam room, gym, equipment pantry, coaching room, toilet), a hall with a swimming pool (swimming bath, bubble baths, children's splashing pool), a bar for light drinks, rooms of engineering systems, a playground for recreation and gymnastics under a canopy.

Along with gas stations, the most important object of roadside service are motels. They are mainly solved as one- or two-storey buildings with rooms representing an entrance hall, a bathroom and a room with a built-in kitchen. Sometimes, a bathroom with a kitchen is located in the back at the back wall, and the entrance is organized directly into the room. Three- or four-storey motels are quite rare. The main planning requirement for the models is the organization of parking the client's car opposite the entrance to his room or in the immediate vicinity of it. Some models are formed from detached houses with parking lots. Then the administration room is located in a separate pavilion. Sometimes, a vacation in motels turns into a kind of ethnographic attraction. As part of the motels, recreation areas with swimming pools are provided.

Ensuring the function of responsible storage of goods of various sizes is implemented by roadside warehouses. Usually these are large rooms, some of which are occupied by racks. Administrative and technical premises are located in the side parts. The buildings have numerous gates for the arrival of customer transport.

The largest objects of roadside service are supermarkets. The main requirement for the location of the store is the presence of a multi-seat parking lot, which determines its placement usually in the depth of the designated area. The most common is the rectangular shape of the plan, but there are other solutions. Their premises usually include: a lobby with one or two entrances with storage chambers; trolley placement zones; flow-distribution zones with areas for short-term rest of visitors; information desk; security point; currency exchange offices; branches of banks and telephone companies; pharmacy, perfume, jewelry, book and newspaper kiosks; workshops for small express repairs of shoes, bags, umbrellas, watches, phones and gadgets; buffets for light snacks; bars for soft drinks; public toilets; cash register; the main trading hall with shelving, display, counter and container-stand placement of goods. In the auxiliary compartment, there are: auxiliary and storage rooms with cold storage rooms; administrative and household premises for personnel; a reception area for goods with places for unloading trucks; a zone for pre-sale preparation of goods; a zone for storing recycled containers; a zone for storing used packaging, garbage and waste; the parking and maintenance room for electric cars, the room for cleaning machines, mechanisms and inventory, as well as the premises of engineering support systems (ventilation chambers, pumping, panel, heating points, communication nodes). Part of the auxiliary and technical rooms is located on the mezzanine.

One of the most interesting examples is the Repsol gas station, built in Madrid under the project of "Foster + Partners" in 1998). Despite the 20 years that have passed since its commissioning, its volumetric and spatial solution, implying the possibility of significant expansion while preserving the figurative and stylistic character, continues to attract the attention of researchers [38–40]. A special aspect is the figurative recognition of the object. The conceptual prototype of such an umbrella solution is, for example, Pegasus filling stations designed by E.F. Noyce in 1960 for the company "Mobile", where round umbrellas are used (Red Hill Filling Station, Leicestershire, UK) [41]. At the same time, the umbrella idea for gas stations itself has a longer history—for example, the classic gas station designed by A. Jacobsen in 1937. The continuation of the story is the Munich gas station with translucent membrane awnings, designed in 2006 by T. Frank and T. Probst, the firm

"Frank and Probst architects". In general, as for membranes and awnings, there is a wide field for modularity. An example is the structures supplied by the company "Guanzhou Faith Trass awning" for road service facilities [42]. There are not only gas stations, but also checkpoints and individual parking. The company "Travelpark Ulyanovsk" offers to implement large modules in the complexes of roadside service of the Ulyanovsk region [43]. A typical complex consists of seven volumes sequentially located parallel to the highway for various purposes, having the same spatial solution. The outer contour has a beveled elongated U-shape, forming a roof, rounded into blind end walls. The side surfaces have solid stained glass glazing. A typical complex includes a refueling station for several types of fuel with eight lanes (including buses and trucks—a kind of "fuel hypermarket"), retail outlets, catering, a motel, warehouses. The refueling unit itself is longer to provide convenient through passage along the complex. Each block has its own color, providing quick visual recognition of the function, which allows users to conveniently park in close proximity to the selected object. This, of course, improves the orientation conditions of drivers when driving on the highway.

These examples show the prospects of using a modular system for roadside service facilities, as they allow you to vary the modularity not only depending on the size of the roadside complex, but also take into account a variety of natural and climatic conditions of the region. In accordance with the regulations of some countries [44], these facilities, depending on the volume and range of services, can be divided into four categories. So, the objects of category "D" include: gas stations, toilets, points of sale. There is no gas station in the objects of category "C", but a food point and parking are added. The nomenclature of objects of category "B" includes: a gas station, a motel, a toilet, a shower, points of trade and catering, a service station, a car wash, a medical center, a parking lot and a picnic area. The largest nomenclature has objects of category "A", which include: gas stations, a motel, a toilet, a shower, points of trade and catering, a medical center, a service station, a car wash, a parking lot, a shopping and entertainment area, a picnic area. A separate object is the "Heating point". As an addition, you can also specify the need to equip checkpoints of road safety and control services. It is also advisable to include in the nomenclature, for example, objects of category "A", small structures for performing religious rites of the main religious denominations. A separate category of objects are points of ordinary and special traffic control, located, depending on the specific situation, at a distance of 50–70 km from each other, as well as in close proximity to large settlements. And, of course, it is necessary to build supermarkets mainly in the suburbs, which will ensure satisfaction of the demand not only of users of the transcontinental highway, but also of citizens, since due to good transport infrastructure (access roads, large parking lots) it will allow for prompt delivery of a wide range of goods.

Taking into account the significant number of objects and the expediency of their rapid construction, one of the possible ways to implement this is the unification of space-planning and design solutions of objects based on a modular system. As the basis of this system, it is proposed to use the construction of a rod spatial plate on angular supports. Similar constructions, the idea of which is partly based on the research of the "Geodesic Dome" by R.B. Fuller (1947), have been actively studied and widely used since the second half of the last century [45].

Usually these are one- or two-tiered structures with parallel belts formed by triangles, squares or hexagons. The support of the structural plates is nodular or with a developed core capital. Sometimes there is a perimeter support in the form of the same spatial lattice structure. There is also a solution with cable-stayed fastening of plates with supports brought up. However, the optimal solution of the tightness of the fastening unit from the point of view of isolation from precipitation is difficult here. The structure is usually formed by rods of one-dimensional length from pipes or rolled profiles of various cross-sections, depending on the location of the plate in the structure. The nodal connection is welded or bolted. Elements in the form of a group of plates, a ball or a polyhedron are used as a basis. In the vast majority of cases, the plate has a square or rectangular shape. There are variants with a triangular or hexagonal shape, depending on the structure of the actual lattice. Single or interlocked modules usually form a covering of rectangular buildings and structures.

The use of rectangular configurations for plates, for example, with a belt structure of regular triangles leads to the appearance of an additional standard size of the rod and the corresponding nodal element, which reduces the effect of uniformity of the design. This situation occurs in complexes of diverse objects, which include objects of road service. The basis of most of these facilities is a developed roof with canopies that provide comfortable boarding and disembarking of drivers and passengers of various vehicles (buses, trucks, cars) in places of road control, refueling with various types of fuel and recreation of road users. For this whole group of objects, it is proposed to use a modular single-tiered, triangular spatial plate formed on the basis of a tetrahedron. According to the regulatory recommendations [46], it is possible to use triangular-shaped plates with a discharged internal structure, when the discharged lower grid forms hexagonal cells in plan.

This is effectively used in areas of open canopies. However, in the areas where heated pavilions are located, a standardly filled lower belt allows you to fix on it (in the nodes) the suspensions of the insulated upper unused ceiling, which removes the problem of forming an independent overlap of the heated pavilion. At the same time, technologically justified episodic movements of personnel for the control of cable systems, pipelines and air ducts are provided along the appropriate bridges, also fixed to the nodes of the lower belt of the structural plate, but on top of the rod system. Given the absence of suspended equipment at the road service facilities under consideration, this solution can be successfully implemented, giving significant savings in the field of basic load-bearing structures.

The basic size is the width of two-lane traffic, 7.5 m (3.75 $\times$ 2), provided by a common tunnel size of 9.0 m. This corresponds to a "right triangle" with a side of 14.0 m. In it, the median/bisector/height is ~12.12 m (14.0 $\times$ 0.866). Accordingly, the columns located with these axial dimensions provide a full-fledged two-lane passage and sidewalks/sidewalks of ~1.56 m (taking into account the diameter of the columns of 0.4 m is 1.36 m for free passage). That is, the module is formed by rods with a conditional (calculated axial) length of 2.0 m. The lower belt is 7 triangles on the side (14.0 m), the upper belt is 8 triangles on the side (16.0 m). The height of the spatial plate will be 2.0 $(2/3)^{1/2}$ = 1.63 m, which with a margin, according to the ratio of span and section, provides spatial rigidity for this type of structures (Figure 1). Cables and pipelines of engineering support systems are conveniently located in the resulting space from the point of view of access and control. The height from the roadway is 6.0 m, which corresponds to the tunnel dimension of 5.0 m and ensures the placement of the necessary signaling devices and road signs above it. For some objects, the height can be changed according to technological parameters, which will be discussed later. The area of the room under the roof of the module is 84.0 sq.m (Figure 2).

The regional winter temperature regime allows you to leave the metal structures of the module open. The heated parts are formed as independent blocks that are not connected to the main supporting structure. It is the triangular shape of the module that makes it convenient from an architectural and planning point of view to block it in configurations corresponding to the landscape relief and functional planning features of a particular construction site.

The placement of related engineering and technological structures (underground or ground-based fuel tank farms for liquid or gaseous mixtures, transformer, pumping, fire-fighting and technological tanks, sewage treatment plants, wind generators, masts and poles of outdoor lighting, communication systems, etc.) is determined in each specific case and does not affect the space-planning solution of the object.

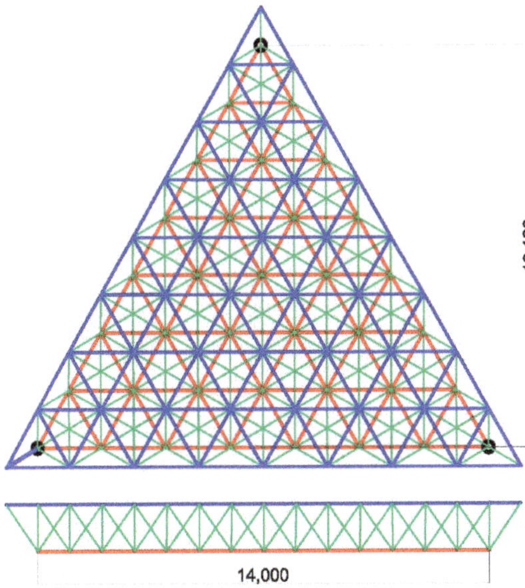

Figure 1. The scheme of the modular structural plate: a plan, a section (Dimensions are given in mm).

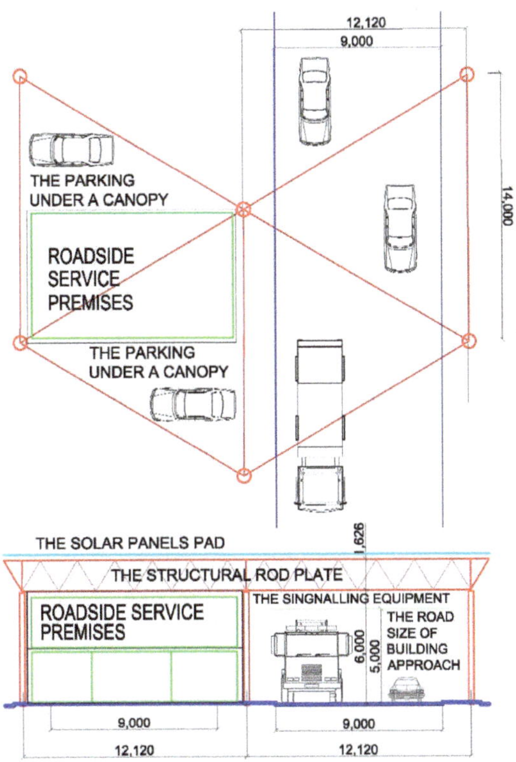

Figure 2. Module parameters: a plan, a section (Dimensions are given in mm).

Various quantitative and configurative combinations of these modules make it possible to provide an optimal architectural and planning solution for any object from the proposed nomenclature in both single and interlocked locations. The unification of the main load-bearing structures gives an advantage in ensuring the regulated quality of construction and installation work in the absence of an accessible developed construction and technical base and highly qualified personnel along most road sections. Equally promising from the point of view of quality is the possibility of centralized manufacturing of the main elements, the convenience of packaging and delivery of elements of two standard sizes (rods with nodal elements and racks), as well as the relative simplicity of their uniform assembly and subsequent installation of a large-sized structure. Depending on the accepted technological scheme of the organization of construction and production of works, section-by-section pre-assembly with delivery of individual sections in finished form is possible.

The uniformity of the space-planning appearance of structures is also of significant importance, which ensures rapid visual recognition of objects by drivers from distant perception plans in a tense situation of high-speed traffic. However, in the field of architectural and artistic solutions, there is the possibility of color-graphic differentiation of objects, for example, by regions of the region within the framework of stylistic unity and a kind of "corporate identity" of the highway. This is not only geographically oriented, but also important from a cultural and propaganda point of view.

2.1. A Set of Modular Roadside Service Facilities

In the proposed set of modular objects, the height to the bottom of the structural coating plate in accordance with the planning and technological requirements is 4.5 m (for small single-storey structures with a side entrance of buses or trucks); 6.0 m (for travel canopies, two-storey pavilions and religious buildings); 7.5 m (for special control points, warehouses and supermarkets). According to this parameter, the proposed sets of modules for 39 types of objects are grouped (Figure 3). The sequence of objects shown in the figure corresponds to the number of interlocked modules (incrementally). All objects, except for traffic control service points located across the highway and entry and exit to special sections (positions 24, 29, 32, 33, 38), are placed on the side of highways.

It is advisable to place all objects in pairs or in a staggered order. The exception is religious buildings (positions 12, 13, 14, 15, 21, 27) and supermarkets (position 40). Depending on the specific conditions, it is advisable to place them singly, providing, due to road interchanges at different levels, the possibility of entry and departure in any direction. In addition, religious buildings are oriented in accordance with the canonical requirements of each of the religions. At the same time, it is advisable to avoid the layout of complexes in which the canonical direction of one denomination has a view of the temple of another denomination. In the vicinity of religious buildings, it is advisable to place pavilions of public toilets and showers for pre-prayer procedures (position 5).

The objects have the following operationally and technologically sound spatial planning solution, linked to the adopted modular system.

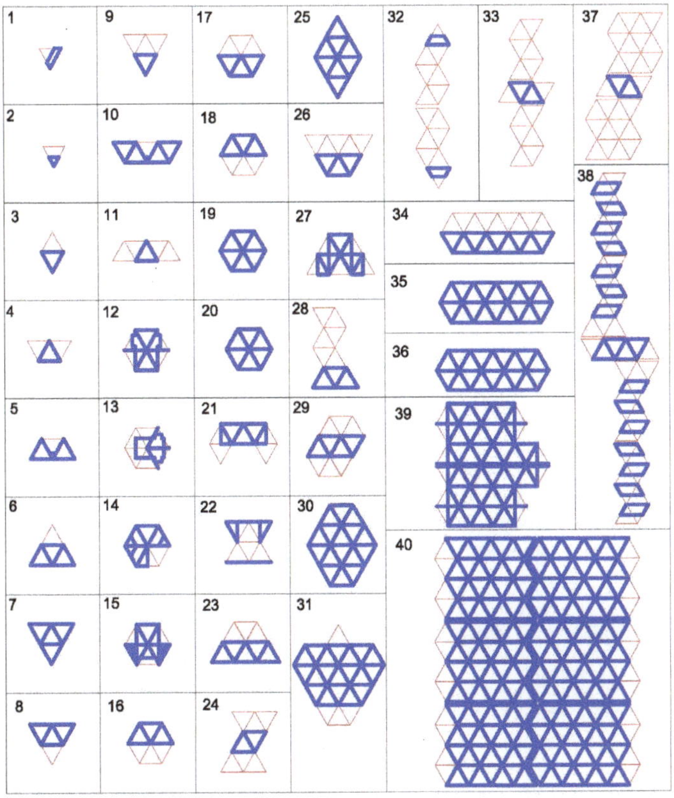

Figure 3. Modular buildings—layout schemes: 1—public transport stop; 2, 9, 26, 28—car refueling; 3, 4, 23—point of retail trade and public catering; 5, 10—public toilet and shower; 6, 8—medical center; 7, 25, 37, 39—car wash; 11—picnic area; 12, 13, 14, 15, 21, 27—temples of various denominations; 16—children's play pavilion; 17—heating point; 18—post office; 19, 30—car service station; 20—laundry; 22—scene; 24, 29, 32, 33, 38—point of traffic control; 31—Wellness Pavilion; 34, 35—motel; 36—warehouse; 40—supermarket (drawing of the authors).

Consider the planning solutions of such objects as the Public transport stop, the Automatic gas station, the Catering point, the Public toilet, the Public toilet and shower, the Picnic area and the Wellness pavilion (Figure 4). The public transport stop (Figure 4(1)): 1 module (50% canopy + 50% heated contour), dimensions 12.12 × 14.0 m (in axes), height to the bottom of the structural plate 4.5 m. The structure is a one-block one-storey structure: a platform under a canopy (42.0 sq.m) and a pavilion of a comfortable waiting area glazed from the inside in the direction of suitable transport, consisting of a hall and a toilet (42.0 sq.m).The automatic car refueling (self-service) gas or gasoline-diesel, or electric—(Figure 4(2)): 1 module (70% canopy + 30% heated contour), dimensions 12.12 × 14.0 m (in axes), height to the bottom of the structural plate 6.0 m. The structure is a one-block one-storey structure: a platform under a canopy (59.0 sq.m) for two refueling cars, a through passage, as well as a pavilion with a two-part isolated technical room (25.0 sq.m) and an external panel for self-service. The catering point (16 seats) (Figure 4(3)): 3 modules (2 canopies + 1 heated), dimensions 12.12 × 28.0 m (in axes), height to the bottom of the structural plate 4.5 m is a single-block one-story structure: central heated pavilion (84.0 sq.m) with a dining room for four four-seater tables, a kitchen, a dishwasher, a utility room, a wardrobe for staff, a toilet for visitors and staff. On the sides of the pavilion, there are canopies (84.0 sq. m each) for eating and relaxing in the warm season. The public toilet

(Figure 4(4)): 3 modules (0.8 canopy + 2.2 heated), dimensions 12.12 × 28.0 m (in axes), height to the bottom of the structural plate 4.5 m is a single-block single-storey structure with a central canopy (70.0 sq.m.) and side pavilions (91.0 sq.m + 91.0 sq.m), in which there are male and female public toilets (vestibule-washroom, a restroom with 6 ordinary booths and 1 for disabled visitors, a storage room for cleaning equipment). Having a connection with both branches, there is a cash register in the middle. The public toilet and shower (Figure 4(5)): 5 modules (0.8 canopy + 4.2 heated), dimensions 12.12 × 42.0 m (in axes), height to the bottom of the structural plate 4.5 m is a one-block single-storey structure with a central canopy (70.0 sq.m.) and side pavilions (175.0 sq.m + 175.0 sq.m.), in which there are male and female public toilets (vestibule-washroom, a restroom with 6 ordinary cabins and 1 for disabled visitors, a pantry of cleaning equipment) and showers (vestibule, the actual shower room with 6 ordinary closed booths and 2 closed booths for disabled visitors). Having a connection with both branches, there is a cash register in the middle. The picnic area (Figure 4(6)): 5 modules (80% canopy + 20% heated contour), dimensions 12.12 × 42.0 m (in axes), height to the bottom of the structural plate 4.5 m. The structure is a one-block, one-story structure with side canopies (168.0 + 168.0 sq.m) for isolated eating and recreation of two companies of clients at the same time, as well as a central pavilion (84.0 sq.m), divided into two parts. In each of them there is: a storeroom of inventory, a kitchen, a utility room and a toilet. The fitness complex (Figure 4(7)): 23 modules (17% canopy + 83% heated contour), dimensions 60.60 × 56.0 m (in axes), height to the bottom of the structural plate 6.0 m. The structure is a single-block structure with a canopy for short-term car parking in front of the main entrance. The pavilion (1596.0 sq.m.) includes: a lobby, a reception with a cash desk and an utility room, a men's department (a wardrobe, a linen closet, a toilet, a vestibule, a shower, a corridor, a steam room, a gym, a storage room, a trainer's room, a toilet), a women's department (a wardrobe, a laundry closet, a toilet, a vestibule, a shower room, a corridor, a steam room, a gym, a storage room, a trainer's room, a toilet), hall with swimming pool (swimming bath −25.0 × 7.0 m; two bubble baths, children's a splashing pool), a light drinks bar, a heating point, a water measuring unit, a recreation area under a canopy.

Consider the planning solutions of such objects as the Point of medical and rescue service, the Car Service station, the Car wash and the Point of weight and dimensional control (Figure 5). The medical and rescue service point (Figure 5(1)): 4 modules (25% canopy + 75% heated contour), dimensions 24.24 × 28.0 m (in axes), height to the bottom of the structural plate 6.0 m. The structure is a single-block structure with a canopy (84.0 sq.m) for two special cars and a two-storey pavilion (504.0 sq.m). On the ground floor, there are: two reception rooms, two dressing rooms, a corridor, a heating point with a water measuring unit, a medicine pantry, a storage room for inventory, men's and women's toilets with large vestibules for changing clothes, an instrumental, a storage room for cleaning equipment. On the second floor (level +3000), there are: a corridor, a rest room for duty crews, a control room, a server room, a panel room, a staff meal room, an administrative room, a pantry of clean workwear, a women's wardrobe with toilet and shower, a men's wardrobe with a toilet and a shower, a pantry of used workwear, a utility room. The pavilion has one internal and two external stairs leading to an open parking of special cars. The car service station (4 posts) (Figure 5(2)): 6 modules (100% heated contour), dimensions 24.24 × 28.0 m (in axes), height to the bottom of the structural plate 6.0 m. The structure is a single-block structure with a one-storey two-light side section (252.0 sq.m) and a two-storey side pavilion (504.0 sq.m) in accordance with the technical and technological parameters of the equipment. In the one-story part, in an open space, technologically separated, if necessary, by inventory blind and mesh partitions, there are 4 posts for repair and maintenance of passenger cars (3 posts), trucks and buses (1 post). Each post has an independent entrance and exit in opposite directions. The two-storey pavilion houses: on the ground floor—a room for customers, a utility room, a corridor, a toilet, a storage room for cleaning equipment, a tool room, a storage room for consumables, a storage room for equipment, a heating point with a water measuring unit, a storage rooms, a panel room; on the second floor (level +3000), male and female wardrobes with showers, storerooms of clean and used workwear, toilets, an office, a meal room with a buffet, a classroom with inventory. The two-storey pavilion has one internal and one external staircase. The car wash (3 posts—3 cars, 1 bus or truck) (Figure 5(3)): 4 modules (100% heated contour), dimensions 24.24 × 28.0 m (in axes), height to the bottom of the structural plate 6.0 m. The structure is a single-block structure. In the central, two-light part (252.0 sq.m), there are washing posts, technologically shielded by curtains. The posts have independent entrances and exits in opposite directions. In the side two-storey pavilion (168.0 sq.m), there are rooms for customers, toilets, a water metering unit, a heating station, a staff room, a storage room for inventory, a stairwell—on the first floor; hall, toilets, men's wardrobe with shower, women's wardrobe with shower, office, utility room, stairwell—on the second floor. The point of weight and dimensional control of vehicles (2 posts) (Figure 5(4)): 8 modules (75% canopy + 25% heated contour), dimensions 36.36 × 35.0 m (in axes), height to the bottom of the structural plate 7.5 m. The structure is a single-block structure. In the central pavilion, there are: vestibule, men's toilet, reception room, operator's room, document storage room, server room, vestibule, women's toilet, reception room, operator's room, water metering unit, heat point. On both sides, under the canopies, there are zones of dimensional and weight control with observation bridges.

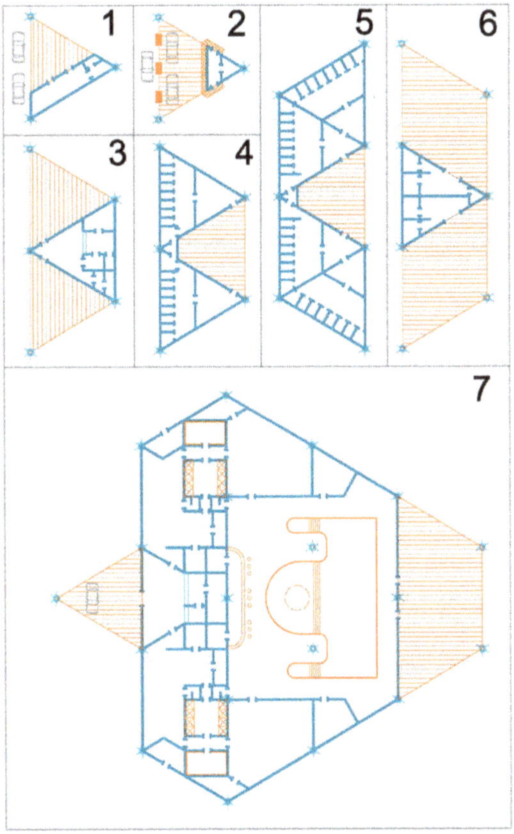

Figure 4. Planning solutions: 1—the Public transport stop; 2—the Automatic gas station (self-service); 3—the Catering point (16 seats); 4—the Public toilet; 5—the Public toilet and shower; 6—the Picnic area; 7—the Wellness pavilion.

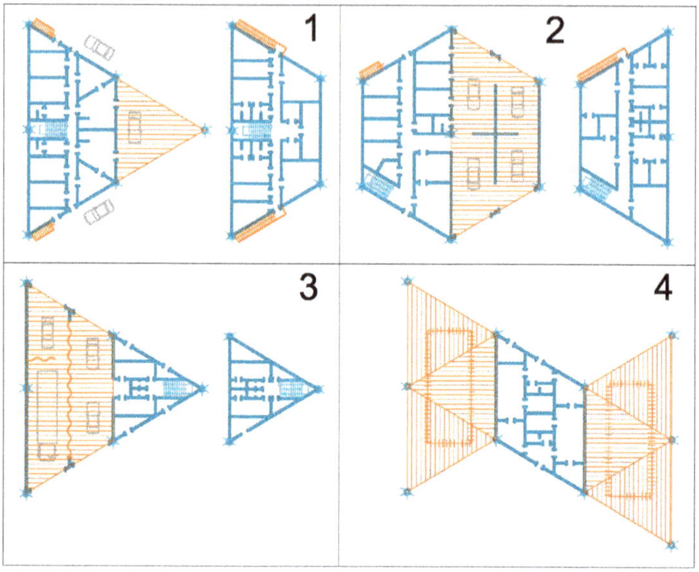

Figure 5. Planning solutions: 1—the Point of medical and rescue service—1st, 2nd floors; 2—the Car Service station—1st, 2nd floors; 3—the Car wash (4 posts)—1st, 2nd floors; 4—the Point of weight and dimensional control.

Consider the planning solutions of such objects as the Heating point, the Children's play area, the Post office, the Laundry, the First aid post, the Auditorium, the Retail and catering point, the Car refueling, the Car refueling and the Retail point (Figure 6). The heating point for drivers and passengers (Figure 6(1)): 6 modules (50% canopy + 50% heated contour), dimensions 24.24 × 28.0 m (in axes), height to the bottom of the structural plate 6.0 m. The structure is a one-block one-storey structure with canopies (252.0 sq.m) for parking cars (4 posts), trucks and buses (1 post), as well as a pavilion (252.0 sq.m) with a rest room, a panel, a heating point, a water measuring unit, a storage room for cleaning equipment, a hall, a dining room, a buffet, a dishwasher, a buffet utility room, male and female bathrooms, including dressing rooms, drying rooms, and a toilet. The children's play pavilion (Figure 6(2)): 6 modules (50% canopy + 50% heated contour), dimensions 24.24 × 28.0 m (in axes), height to the bottom of the structural plate 4.5 m. The structure is a one-block, one-story structure with canopies (252.0 sq.m) for outdoor games and recreation and an enclosed room with two game rooms, a buffet, buffet utility rooms, a corridor, men's and women's toilets. The post office (Figure 6(3)): 6 modules (50% canopy + 50% heated contour), dimensions 24.24 × 28.0 m (in axes), height to the bottom of the structural plate 6.0 m. The structure is a one-block one-story structure with canopies (252.0 sq.m) for short-term parking of cars. In the heated part (252.0 sq.m), there are: a hall, a storage room for inventory, a men's toilet, a corridor, a server room, a heating point, two special storerooms, a vestibule, a water measuring unit, a storage room for special inventory, an operator's room (4 posts), a pantry, a women's toilet, a storage room for cleaning equipment. The laundry—(Figure 6(4)): 6 modules (100% heated contour), dimensions 24.24 × 28.0 m (in axes), height to the bottom of the structural plate 4.5 m. The structure is a one-block one-storey building (504.0 sq.m), including: a lobby, a cash register, a toilet for customers, a self-service laundry room, a pantry for washing products, a self-service ironing room, an inventory room, a corridor, a laundry room, a men's staff wardrobe with a toilet and a shower, a women's staff wardrobe with a toilet and a shower, a laundry room, a detergent pantry, an ironing room, a corridor, a cleaning equipment pantry, a laundry pick-up point, a heat point, a water meter unit. The medical center (Figure 6(5)): 4 modules (25% canopy

+ 75% heated contour), dimensions 24.24 × 28.0 m (in axes), height to the bottom of the structural plate 6.0 m. The structure is a one-block, one-storey structure with a canopy (84.0 sq.m.) for short-term parking of a car and a platform in front of the entrance, and a block of premises including: a lobby, toilets for visitors, a reception, a document storage, a dentist's office, a dentist's utility room, a dentist's medicine pantry, a corridor, toilets for staff, a therapist's medicine pantry, a therapist's utility room, a water meter unit, a heat point. The auditorium/the stage area (Figure 6(6)): 6 modules (83% canopy + 17% heated contour), dimensions 24.24 × 28.0 m (in axes), height to the bottom of the structural plate 6.0 m. The structure is a one-block, one-story structure with an indoor auditorium for 400 seats, a stage box (14.0 × 14.0 m—in axes), side heated pavilions (42.0 + 42.0 sq.m) with a vestibule, toilet and an artistic room. In the corners of the hall, in closed rooms, there are sound and panel rooms. The point of trade and catering (40 seats) (Figure 6(7)): 8 modules (37% canopy + 63% heated contour), dimensions 24.24 × 42.0 m (in axes), height to the bottom of the structural plate 6.0 m. The structure is a one-block, one-story structure with a canopy (252.0 sq.m) for parking 1 bus or truck and 4 passenger cars. In the pavilion (420.0 sq.m), there are: a shopping and dining room, a shop, a toilet, a technical room, a corridor, a wardrobe, a toilet, a shower room, a kitchen with a transfer room, a distribution area, a dishwasher, a pantry, a technical room. The gas car refueling (Figure 6(8)): 4 modules (75% canopy + 25% heated contour), dimensions 24.24 × 28.0 m (in axes), height to the bottom of the structural plate 6.0 m. The structure is a one-block one-story structure with a canopy (252.0 sq.m), providing refueling on four columns, and a pavilion (84.0 sq.m) with an operating room, an operator's room, a toilet, a utility room, a technical room. The electric car refueling (Figure 6(9)): 8 modules (63% canopy + 37% heated contour), dimensions 24.24 × 42.0 m (in axes), height to the bottom of the structural plate 6.0 m. The structure is a one-block, one-story structure with a canopy (420 sq.m) for nine refueling stations for electric vehicles and hybrids. Taking into account the real time of operational (from 40 min) and full (from 75 min) electric refueling for the leisure of drivers and passengers in the immediate vicinity of cars, in the pavilion (252.0 sq.m) there is: a customer rest room, a buffet, a dishwasher, a utility room, an electrical switchboard, a toilet, a water meter and a heating point. The retail point of sale (Figure 6(10)): 2 modules (50% canopy + 50% heated contour), dimensions 24.24 × 14.0 m (in axes), height to the bottom of the structural plate 4.5 m. The structure is a one-block, one-storey structure: a platform under a canopy (84.0 sq.m) with the possibility of parking four cars in two rows and a heated pavilion (84.0 sq.m) with a sales hall, a utility room, a corridor and a toilet.

Consider the planning solutions of such objects as the Checkpoint, the Service station, and the Traffic control point (Figure 7). The coastal-type checkpoint (6 + 6 posts) (Figure 7(1)): 14 modules (86% canopy + 14% heated contour), dimensions 98.96 × 21.0 m (in axes), height to the bottom of the structural plate 6.0 m. The structure is a structure, according to antiseismic measures, divided into two symmetrical blocks. It is located across the highway and is designed to control entry and exit to special (including paid) sections. It provides passage through barriers with control panels under canopies (504.0 + 504.0 sq.m) of cars in twelve lanes in both directions. Control and support services are located in side two-storey pavilions (168.0 sq.m each). In the pavilions, there are: an operator's room, a corridor, a toilet, a technical room, a hall, a shower room, a toilet, a staff room, an office. The height of the floor is 3.0 m. The connection between the floors is carried out by an external staircase. From the outside of the pavilions under canopies (21.0 + 21.0 sq. m), it is possible to park a car of control services and special services with the possibility of direct access to the highway in the appropriate direction. The car service station (8 posts) (Figure 7(2)): 16 modules (100% heated contour), dimensions 48.48 × 42.0 m (in axes), height to the bottom of the structural plate 6.0 m. The structure is a single-block structure with a one-storey two-light central part (840.0 sq.m) and two-storey side pavilions (504.0 + 504.0 sq.m) in accordance with the technical and technological parameters of the equipment. In the central part, in an open space, there are eight posts for repair and maintenance of passenger cars (4 posts), trucks and buses (4 posts). Each post has an independent entrance and exit in opposite directions. In the two-storey side pavilions, there are rooms for customers, toilets, administrative, household and sanitary rooms for staff, storerooms of cleaning equipment, warehouses of parts, tools, materials, cleaning equipment, rooms of engineering support systems (ventilation chambers, pumping, panel, heating points, communication nodes). The height of the floor is 3.0 m. Each pavilion has one internal and one external staircase. The point of traffic control services (Figure 7(3)): 10 modules (60% canopy + 40% heated contour), dimensions 36.36 × 35.0 m (in axes), height to the bottom of the structural plate 6.0 m. The structure is a single-block structure with side canopies (252.0 + 252.0 sq.m) for controlled passage of four traffic lanes in both directions. In the central two-storey pavilion (672.0 sq.m), on the ground floor, there are: a document processing room, a corridor, a toilet, a control room, a water meter unit, a document processing room, a corridor, a toilet, a control room, a heat point. On the second floor (level +3000), there are: a corridor, a men's toilet, a men's shower room, a room for detainees, an office, an inventory room, a server room, a meal room, a men's wardrobe, an office for operational visual control, an office, an inventory room, a women's wardrobe, a women's toilet, a women's shower room, a staff training room, a shift supervisor's office, an office for operational visual control. The pavilion has one internal and two external stairs. At the ends, under the canopies, there are parking lots for special cars with the possibility of direct access to the highway in the appropriate direction.

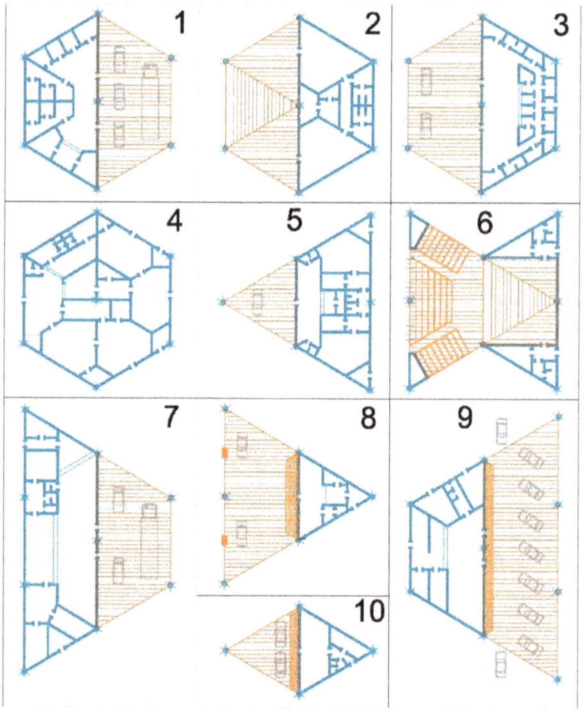

Figure 6. Planning solutions: 1—the Heating point; 2—the Children's play area; 3—the Post office; 4—the Laundry; 5—the First aid post; 6—the Auditorium; 7—the Retail and catering point; 8—the Car refueling (gas); 9—the Car refueling (electric); 10—the Retail point.

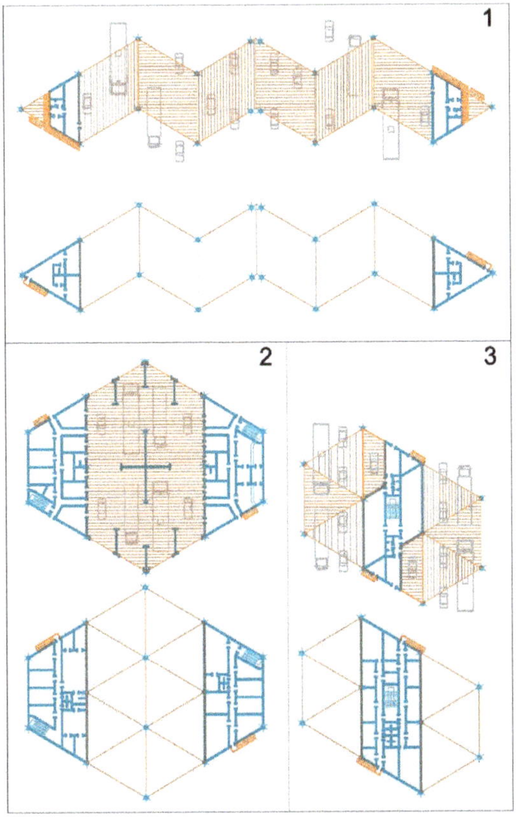

Figure 7. Planning solutions: 1—the Checkpoint (6 + 6 posts)—1st, 2nd floors; 2—the Service station (8 posts)—1st, 2nd floors; 3—the Traffic control point—1st, 2nd floors.

Consider the planning solutions of such objects as the Car wash and the Car refueling (Figure 8). The car wash (6 posts—4 cars, 2 buses or trucks) (Figure 8(1)): 8 modules (100% heated contour), dimensions 48.48 × 28.0 m (in axes), height to the bottom of the structural plate 6.0 m. The structure is a single-block structure. In the central two-light part (504.0 sq.m), there are washing posts, technologically shielded by curtains. The posts have independent exits in opposite directions; for trucks and buses, the passage is through. In the side two-storey pavilions (168.0 + 168.0 sq.m), there are: on the ground floor—a client room, a control room, a pantry, a stairwell, a heating point, an electric switchboard, an inventory room, a detergent pantry, a water meter unit, a stairwell, a pantry; on the second floor—toilets, a men's and women's wardrobe with showers and drying rooms, an office, a hall, storerooms. The petrol-diesel car refueling (Figure 8(2)): 9 modules (67% canopy + 33% heated contour), dimensions 48.48 × 28.0 m (in axes), height to the bottom of the structural plate 6.0 m. The structure is a one-block one-storey structure with a canopy (504.0 sq.m), providing refueling for ten columns, and a pavilion (252.0 sq.m) with a client and a mini-store of related products, an operator's room, a utility room, showers, a passage, a toilet, a staff room, a pantry, a water measuring unit and a heating point.

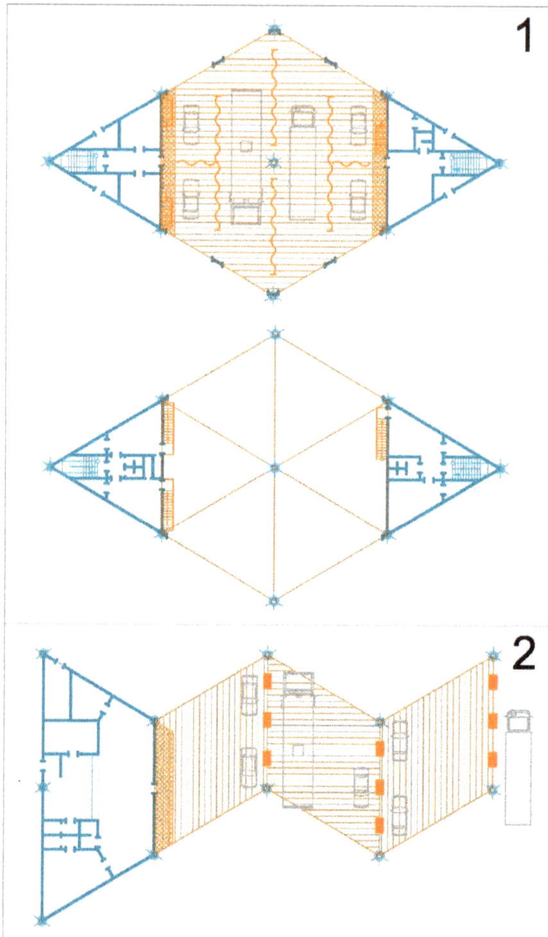

Figure 8. Planning solutions: 1—the Car wash (6 posts)—1st, 2nd floors; 2—the Car refueling (petrol-diesel).

Consider the planning solutions of such objects as the Checkpoint and the Self-service car wash (Figure 9). The island-type checkpoint (6 + 6 posts) (Figure 9(1)): 16 modules (88% canopy + 12% heated contour), dimensions 88.84 × 35.0 m (in axes), height to the bottom of the structural plate 6.0 m. The structure is a structure divided into three blocks, located across the highway, in accordance with antiseismic measures, for controlling entry and exit to special (including paid) sections. It provides passage through barriers with control panels under canopies (504.0 + 504.0 sq.m) of cars in twelve lanes in both directions. Control and support services are located in a central two-storey pavilion (168.0 sq.m on each floor) with canopies. The pavilion has: on the ground floor—a corridor, a heat point, a control room, a corridor, a water measuring unit, a control room; on the second floor (level +3000), a corridor, a men's wardrobe with toilet and shower, a server room, a shift supervisor's office, a corridor, a women's wardrobe with toilet and shower, a meal room, an inventory room. The two-storey pavilion has one internal and two external stairs. From the outside of the pavilion under canopies (84.0 + 84.0 sq.m), it is possible to park cars of control services and special services with the possibility of direct or rotary exit to the highway in the right direction. The self-service car wash (Figure 9(2)): 28 modules (93% canopy + 7%

heated contour), dimensions 88.84 × 49.0 m (in axes), height to the bottom of the structural plate 4.5 m. The structure is a complex of three blocks, the dimensions of which take into account regional antiseismic measures. In the central part there is a heated pavilion with an operator's room, a toilet, a corridor and technical rooms. Canopies are located at the ends of the pavilion. On both sides, there are symmetrically arranged canopies, under which there are 24 car and motorcycle washing stations using the self-service system. In the alignment of the main supports, there are panels with hoses for water supply for washing, liquid detergents and compressed air for drying and cleaning, as well as cabinets with cleaning materials. The posts have independent departures. Each post is shielded by curtains.

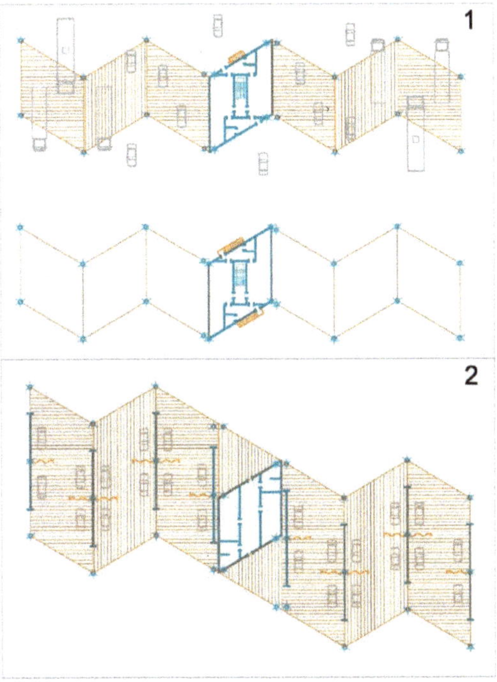

Figure 9. Planning solutions: 1—the Checkpoint (6 + 6 posts)—1st, 2nd floors; 2—the Self-service car wash (24 posts).

Consider the planning solutions of such objects as the Contactless car wash and the Warehouse (Figure 10). The contactless car washing (4 posts) (Figure 10(1)): 46 modules (13% canopy + 87% heated contour), dimensions 38.36 × 70.0 m (in axes), height to the bottom of the structural plate 7.5 m. The structure is a complex of two symmetrical blocks. In each of them, there are: a car washing line (reception post, washing post, drying post), a corridor, an operator's room with a toilet, a technical room, an inventory storeroom, an electric switchboard, a complex washing line for buses and trucks. The warehouse—(Figure 10(2)): 18 modules (100% heated contour), dimensions 24.24 × 70.0 m (in axes), height to the bottom of the structural plate 7.5 m. The structure is a one-block, one-story structure. The premises include a storage area, an administrative room, a corridor, a toilet, a storage room for inventory, a heating point, an electrical switchboard, a charging room for electric cars, a water meter unit.

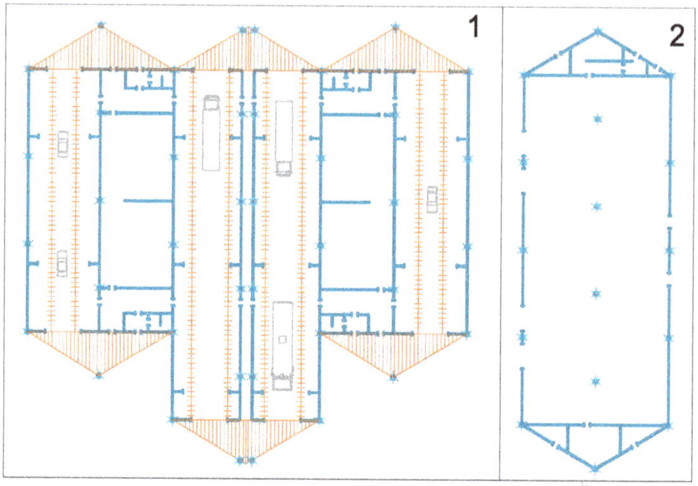

Figure 10. Planning solutions: 1—the Contactless car wash (4 posts); 2—the Warehouse.

Consider the planning solutions of such objects as the highly comfortable motel with parking under a canopy (Figure 11) and the ordinary motel (Figure 12). The luxury motel (10 apartments with individual parking site) (Figure 11): 18 modules (50% canopy + 50% heated contour), dimensions 24.24 × 70.0 m (in axes), height to the bottom of the structural plate 6.0 m. The facility is a single-block structure with a high canopy for parking ten customer cars opposite the entrance to the corresponding room (588.0 sq.m.). In the two-storey part (588.0 + 588.0 sq.m), there are 10 highly comfortable duplex rooms. The height of the floor is 3.0 m. On the ground floor, there are: a lobby (two-light), a registration service, a toilet, a utility room, a water measuring unit, a panel room, a heating point, a storage room for inventory, a storage room for clean linen, a storage room for dirty linen, rooms. The set of rooms consists of: 6 pieces of type A rooms (ordinary)—a living room, a bathroom, a staircase, a bedroom with a bathroom, a bedroom with a bathroom: 2 pieces of type B rooms (central)—a living room, a bathroom, a staircase, a bedroom, a bedroom with a bathroom, a bedroom with a bathroom and a loggia: 2 pieces of type C rooms (at the end)—a living room, a bathroom, a bedroom, a staircase, a bedroom with a bathroom, a bedroom with a bathroom, a loggia. The motel (36 rooms) (Figure 12): 18 modules (100% heated contour), dimensions 24.24 × 70.0 m (in axes), height to the bottom of the structural plate 6.0 m. The structure is a single-block structure in the form of a two-storey volume (1176.0 + 1176.0 sq.m) with a corridor layout scheme. On the ground floor, there are: a porch, a two-light lobby, a registration service, a toilet, a utility room, a corridor, a stairwell, a storage room of inventory, a utility room, a panel room, a storage room of street cleaning equipment, a corridor, a stairwell, a storage room of equipment, a water meter unit, a heating point, 10 rooms with a bedroom, a bathroom and a built-in kitchen, 10 rooms with a bedroom and a bathroom. On the second floor (level +3000), there are: a corridor with a central gallery, a stairwell, a pantry of clean linen, a pantry of dirty linen, a stairwell, a pantry of furniture, a pantry of inventory, 10 rooms with a bedroom, a bathroom and a built-in kitchen, 10 rooms with a bedroom and a bathroom.

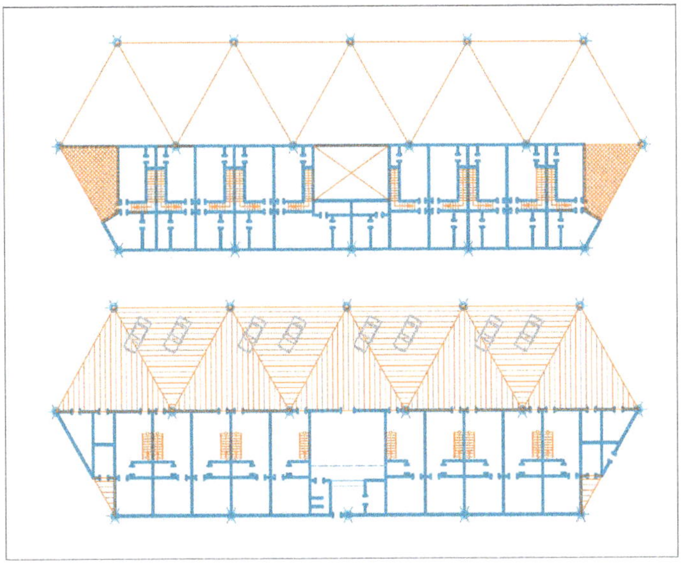

Figure 11. Planning solutions: the Highly comfortable motel (10 apartments) with parking under a canopy—1st, 2nd floors.

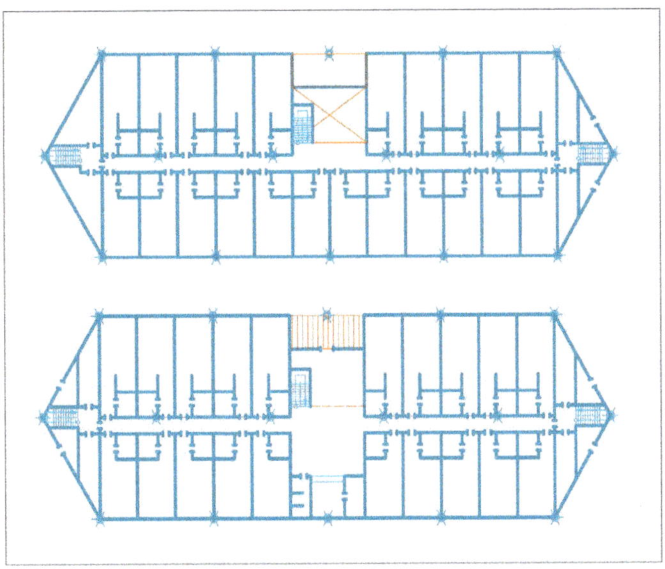

Figure 12. Planning solutions: the Ordinary Motel (24 rooms)—1st, 2nd floors.

Consider the planning solution of such an object as the Special traffic control point (Figure 13): 40 modules (55% canopy + 45% heated contour), dimensions 210.04 × 49.0 m (in axes), height to the bottom of the structural plate 7.5 m. The structure is a complex of five blocks, the dimensions of which take into account regional antiseismic measures. In the central part, there is an administrative four-level building of 6 blocks. In the basement (level −4950) there is a water measuring unit, a heat point, an electrical switchboard and passages with viewing chambers at each control post. In the side volumes, there are storerooms of cleaning equipment, storerooms of clean and used workwear, storerooms of appliances and tools. On the ground floor, duty rooms, document processing offices, men's and women's toilets, inventory storerooms are located in a symmetrical layout. Under the canopies, there are parking lots for special cars on duty with the possibility of direct or rotary exit in the right direction. On the second floor (level +3300), there are: male and female wardrobes with showers, toilets, a meal room, a classroom, a room for special equipment, an office of the head of the duty shift, offices of the visual control officers. On the third level (level +5.200), there are exits to the platform for passage to the observation bridges at each inspection post. The connection between the levels is carried out by two stairwells. In the canopies adjacent to the central block, there are two lanes of emergency free passage. On both sides, pavilions of special control posts are linearly placed under paired canopies. Each post is a three-level structure. In the basement (level −4950), there is a special control camera with a hole in the ceiling that provides inspection of the bottom of the passing vehicle control. All pavilions are sequentially connected to each other and to the central block to ensure optimal passage of personnel to the appropriate post. On the ground floor there is a high (4.6 m) duty room with stained glass windows on the side of the passage, a toilet, a shift staff room. On the third level (level +5.200), there are pavilions of stairwells that provide staff exits to observation bridges. All observation bridges are connected in series with each other and with the central unit.

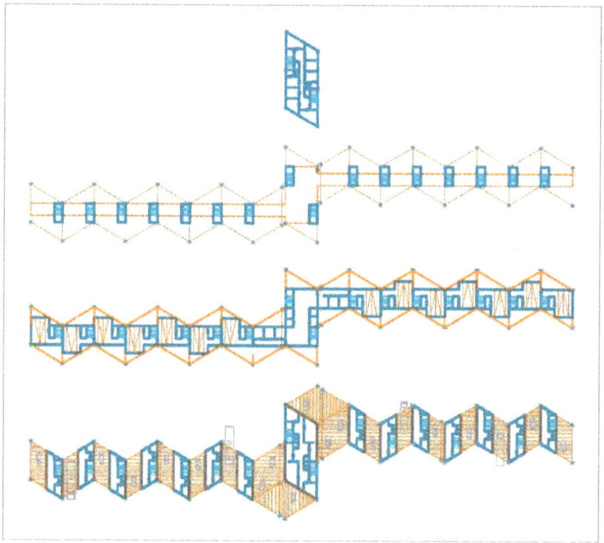

Figure 13. Planning solutions: the Special vehicle control point (levels 1, 2, 3, 4).

Consider the planning solution of such object as the Supermarket (Figure 14): 204 modules (12% canopy + 88% heated contour), dimensions 149.44 × 128.0 m (in axes), height to the bottom of the structural plate 7.5 m. The structure is a complex of six blocks, the dimensions of which take into account regional antiseismic measures. The overall dimensions are determined from the most common standardized supermarket parameters ~90 × 140 m. The heated contour (15,120.0 sq.m) accommodates: a lobby with storage chambers; trolley placement areas; flow-distribution zones with areas for short-term rest of visitors; an information desk; a security point; currency exchange offices; branches of banks and telephone companies; a pharmacy, a perfume, an jewelry, book and newspaper kiosks; workshops for small express repairs of shoes, bags, umbrellas, watches, phones and gadgets; buffets for light snacks; bars for soft drinks; public toilets; a cash register; a main trading hall with shelving, a display, counter and container-stand placement of goods.

In the two-level auxiliary compartment (mezzanine at level +4.500), there are: auxiliary and storage rooms with cold rooms; administrative and household premises for personnel; goods reception area with unloading places for trucks; goods pre-sale preparation area; a storage area for recycled packaging; a storage area for used packaging, a garbage and a waste; a parking and maintenance of electric cars, cleaning machines, mechanisms and inventory, as well as engineering support systems (ventilation chambers, pumping, panel, heating points, communication nodes). A part of the auxiliary and technical rooms is located on the mezzanine, where internal and external stairs lead. The canopies located on the sides (2016.0 sq.m) provide for the possibility of covered parking of cars (48 seats), supermarket personnel, and special services of legal, financial, sanitary-epidemiological, fire control.

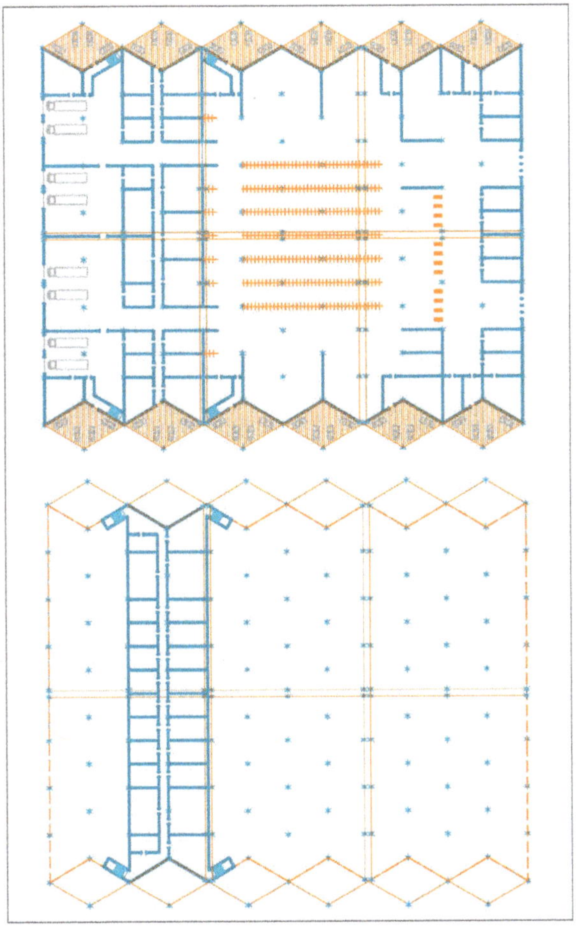

Figure 14. Planning solutions: the Supermarket—1st, 2nd floors.

A complex of appropriate buildings is designed to fulfill the religious needs of drivers and passengers. Consider planning solutions of such objects as: the Mosque, the Buddhist temple, the Protestant temple, the Orthodox temple, the Catholic temple, the Synagogue (Figure 15). The Mosque (Figure 15(1)): 8 modules (50% canopy + 50% heated contour), dimensions 24.24 × 42.0 m (in axes), height to the bottom of the structural plate 6.0 m. The structure is a one-block, one-story structure with a heated part (336.0 sq.m), a two-part (male and female parts) domed prayer hall with a mihrab niche, a covered courtyard, a reading room, a library, an imam's office, and servants' rooms located in the middle. Under the canopy (336.0 sq.m.), there are places for pre-prayer and after prayer concentration and rest. Orientation of the mihrab canonically to the Kaaba in Mecca city. The Buddhist temple (Figure 15(2)): 7 modules (43% canopy + 57% heated contour), dimensions 24.24 × 42.0 m (in axes), height to the bottom of the structural plate 6.0 m. The structure is a one-block, one-story structure with a prayer ("golden") hall placed in the middle of the heated part (336.0 sq.m), covering its gallery, a reading room and a concentration room. Under the side canopies (252.0 sq. m), there are places for pre-prayer and after prayer concentration and rest. The building is preceded by a courtyard with pagodas, the actual entrance is in the form of a canonical "Middle Gate". The entrance to the temple is canonically from the East side. The Protestant church (Figure 15(3)): 6 modules (41% canopy + 59% heated contour), dimensions 24.24 × 28.0 m (in axes), height to the bottom of the structural plate 6.0 m. The structure is a one-block, one-story structure with a heated part (294.0 sq.m), a prayer hall with an altar part placed in the middle, a block with a library, a hall and a utility room located on the side. Under the front and side canopies (210.0 sq.m), there are places for pre-prayer and after prayer concentration and rest. The altar is canonically oriented to the East. The Orthodox church (Figure 15(4)): 6 modules (25% canopy + 75% heated contour), dimensions 24.24 × 28.0 m (in axes), height to the bottom of the structural plate 6.0 m. The structure is a one-block one-story structure with a heated part (378.0 sq.m) placed in the middle of a prayer hall with a dome, a three-apse altar part and side chapels, a porch, an office, a reception room, a shop, a utility room, a pantry. Under the side canopies (126.0 sq.m), there are places for pre-prayer and after prayer concentration and rest. The altar is canonically oriented to the East. The Catholic church (Figure 15(5)): 6 modules (50% canopy + 50% heated contour), dimensions 24.24 × 28.0 m (in axes), height to the bottom of the structural plate 6.0 m. The structure is a one-block, one-story structure with a prayer hall in the heated part (252.0 sq.m) placed in the middle with a dome over the transept, a three-apse altar part, a narthex, an office and a utility room located on the sides. Under the front and side canopies (252.0 sq.m), there are places for pre-prayer and after prayer concentration and rest. The altar is canonically oriented to the East. The Synagogue (Figure 15(6)): 6 modules (33% canopy + 67% heated contour), dimensions 24.24 × 28.0 m (in axes), height to the bottom of the structural plate 6.0 m. The structure is a one-block, one-story structure with a heated part (336.0 sq.m) in the middle of a two-part (male and female parts) prayer hall, lobby, clerical offices located on the sides and auxiliary rooms. Under the front and side canopies (168.0 sq.m), there are places for pre-prayer and after prayer concentration and rest. The orientation of the altar is canonically on the city of Jerusalem.

From the same rod elements, it is possible to form supports for information boards both in front of individual structures (for example, fuel filling stations; indication of type and prices) and in front of complexes of roadside service facilities. The lattice structure also provides convenient access for technical personnel to service the back of the LED panel. The dimensions of the supports can be different depending on the specific tasks—including the possibility of installing sufficiently large screens. An example of the solution of a rod structure as a support for an information board is shown in the Figure 16.

The proposed modularity allows for both single and lockable placement of various objects as part of complexes of various categories. Blocking allows in some cases to reduce the size of buildings due to the possibility of excluding some duplicated sanitary or engineering premises. At the same time, blocking can be performed both directly, forming a single volume (for small objects), and through a seam (for large objects), ensuring compliance with antiseismic measures. The configuration of the lock is determined by the individual landscape features of the territory of the complex and the creative preferences of the authors of the project.

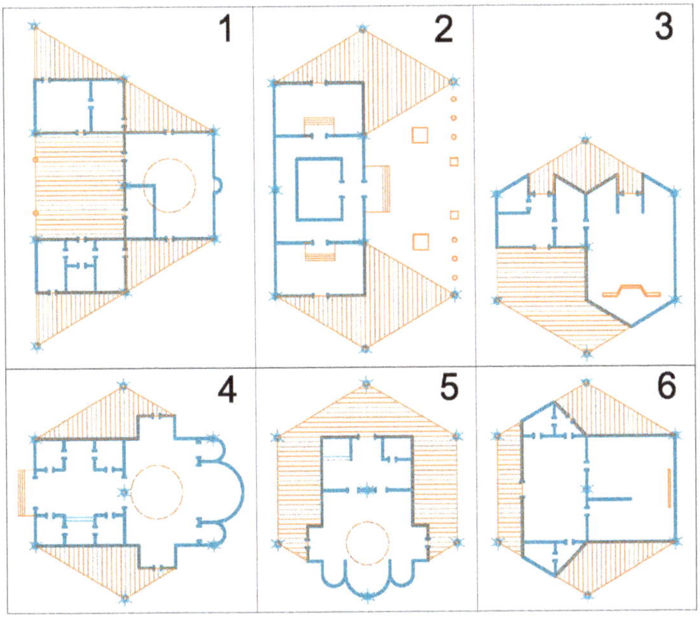

Figure 15. Planning solutions: 1—the Mosque; 2—the Buddhist temple; 3—the Protestant temple; 4—the Orthodox temple; 5—the Catholic temple; 6—the Synagogue.

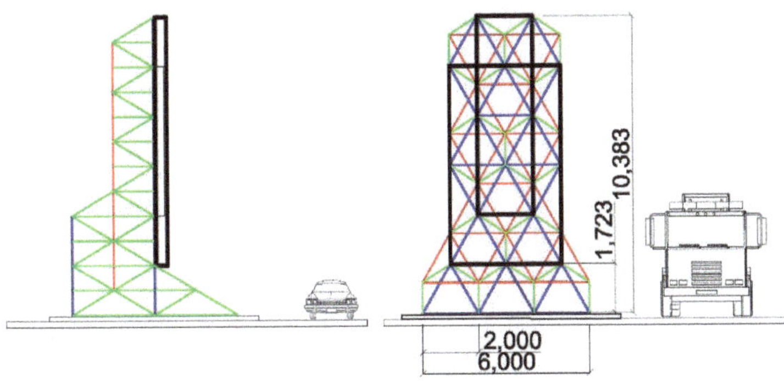

Figure 16. The Diagram (side view, front view) of the rod structure as a support for the information board (Dimensions are given in mm).

2.2. Ensuring Energy Efficiency and Reducing the Harmful Environmental Impact of Road Service Facilities

The formation of a complex of maintenance facilities along highways, as an essential part of the system of sustainable development of territories, implies the implementation of a set of measures to improve processes in all aspects, from creation to operation. The proposed modular system for the formation of objects for various purposes based on a triangular cell with an overlap of a core spatial plate makes it possible to improve a number of indicators of the complexes being created according to the basic parameters of sustainable development.

Firstly, it is the level of production. It is the uniformity of structural elements, a small number of standard sizes and the need for mass production that immediately improve a number of indicators related to energy efficiency. Centralized mass production is naturally more economical than scattered piece production in all parameters in terms of unit output.

Moreover, importantly, significantly stricter product quality control is provided at all stages of manufacturing.

Secondly, transportation of building elements and parts. The presence of only three main standard sizes (rod, nodal polyhedron, rack) allows them to be compactly packaged and tightly placed on vehicles for transportation to the installation site. Under certain conditions, it becomes possible to partially assemble, for example, pyramids on specialized sites and package several pyramids already nested in each other for transportation. This reduces the number of necessary vehicles and leads, among other things, to a reduction in the harmful impact on the environment due to exhaust emissions. At the same time, fuel is saved.

Thirdly, the construction of groups of modules on various dispersed sites. The uniformity of the assembly processes of the main elements of high factory readiness makes it possible to significantly reduce the construction time of the main structures due to the acquisition and use of skills by teams of installers. The need to use cranes is limited by a small set of works on lifting the modular element of the structural plate mounted at the bottom to the design position. The main volume of installation work does not require crane equipment. Accordingly, it becomes possible to use the in-line method of constructing objects based on the use of highly specialized construction and installation teams. This, as the successful practice of domestic construction in the second half of the last century has shown, significantly reduces the time of construction and installation work, leading, accordingly, to the saving of energy resources. A significant advantage of using a modular canopy system is that after the installation of the coating, all other construction and finishing work on the creation of pavilions with rooms for various purposes is carried out in relatively comfortable conditions under the roof, which protects builders from precipitation and excessive insolation. It also reduces energy consumption, improves the quality of work and reduces their time. The reduction of time is achieved, among other things, due to the possibility of organizing round-the-clock three-shift work, because the core structure of the canopy allows you to harmoniously place in it and move, as necessary, floodlight lighting of both the entire construction site and its individual sections.

Fourth, the materials proposed for use for the structures of the built-in heated pavilions are: steel racks and beams of the main frame, steel bent profiles and corrugated sheets for floors and staircase structures, sandwich panels for exterior walls, magnesite and gypsum cardboard plates for partitions with a frame of bent thin-sheet steel profiles and mineral wool filling, as well as ceiling linings (in rooms with high humidity or high requirements for fire-fighting indicators, special surface finishing is performed),wood-fibrous plates for the underlying layers of floors, extruded foam and foam insulation for thermal insulation of non-modular sections of structures and pipelines of engineering systems, metal-plastic windows with double-glazed filling, which entail a high degree of industry in terms of of manufacturing, packaging, warehousing, delivery, and installation, ensuring high energy efficiency in these areas. In addition, energy-consuming "wet processes" at the construction site are almost completely excluded (the exception is the device of monolithic reinforced concrete foundations. However, this can also be significantly reduced by switching to prefabricated monolithic or fully prefabricated reinforced concrete foundations of factory manufacture).

Fifth, the possibility of active use of mainly solar and to a small extent wind energy, as well as biofuels, which makes it possible to compensate for some of the total costs of electric energy complexes. The most common way of using solar photo energy generators is associated with the maximum filling of the roof space, leaving only technologically necessary passages. Sometimes it is filling the adjacent territory with generators (as, for example, at one of the Beijing gas stations). An additional source of saving artificial lighting in the interior spaces of the pavilions is the use of transparent photogenerators on the roof. In addition to the solution of replacing part of the electricity costs for lighting, photogenerators are also included in the system of operation of gas station pumps.

Gradually, a kind of "direct" use of the electricity generated by photogenerators for refueling electric vehicles is becoming more widespread. However, a purely technical problem is the time to refuel the batteries of an electric car or hybrid, which is three hours for a 70-km trip. This limits the use of such refueling stations on highways, where refueling time is essential. Accordingly, the scope of their application so far extends to parking for employees of various enterprises and educational institutions who have the opportunity to leave the car for several hours. Sometimes, these are systems for individual parking at residential buildings. At the same time, the supply of electricity from photogenerators is included in the general electrical supply system of the house, from which the electric car or hybrid is then charged. The modular electric filling stations offered for roads provide for the presence of a special pavilion for leisure activities by drivers and passengers of the corresponding vehicles.

In the aspect of the modular system proposed for roadside service facilities, it can be noted that an important advantage of this module is the possibility of placing on the roof, using the same (although, of course, smaller cross-section) core and node elements, solar panels that can be optimally oriented based on a structurally secured three-sector direction. The approximate area of the active surface of stationary batteries is 144.0 sq.m on each module. As an addition to the electrical supply system of the facility, it is recommended to place weathervane wind generators, which are fixed in the area of the module supports. Naturally, when blocking groups of modules through a seismic seam with paired columns, the weathervane wind generator is placed only on one of the columns. It is advisable to place small vertical wind turbines on side platforms, as it is used, for example, at the previously considered SEPSA gas station, Adanero (architect "Saffron Brand Consultants + Malka + Portus", 2015). Large-blade wind turbines are optimally located on relatively isolated sites. The sites for the placement of wind generators, of course, are fields for photo generators. The location of large-bladed wind generators next to road maintenance complexes and the device of their illumination at night makes it possible to improve the visual orientation of drivers, who from a sufficiently long distance will be able to see the presence of the object of interest to them.

The sixth position of increasing energy efficiency and compensating for the harmful impact on the environment of roadside service complexes is a system for collecting and cleaning rain and meltwater, as well as domestic sewage effluents for use for watering green spaces in the structure of the complex. An important component in this area is also the system of circulating water supply of vehicle washing points operating in the system of sewage treatment plants.

Some indicators can also be improved by organizing a system of differentiated collection and packaging of food waste, household and technological garbage, and used containers and packaging.

3. Results

The need for a relatively quick solution to the problem of providing all highways in the country with roadside service facilities in accordance with the directives of the nomenclature specified, taking into account historical experience, necessitates the development of a series of relevant standard projects. In order to increase the efficiency of these series, it is advisable to link up space-planning decisions based on one or another module. Considering the variety of planning and landscape characteristics of the areas of placement of objects for mainline maintenance, it seems appropriate to choose as a module not a square or rectangular, but a triangular configuration, which in most cases allows harmoniously blocking the modules. The proposed roofing module in the form of a "regular" triangle extending to a tetrahedron has a structural basis in the form of a single-tier rod spatial plate. Half a century of extensive practice in the use of such structures (mainly with a cell in the form of an equilateral pyramid with a square base), including in the Central Asian region, minimizes the construction and technological difficulties in the manufacture, delivery and installation. The height characteristics of the location of the core spatial plate are linked to the normalized road dimensions and features of the technology of functioning of each of the typologically different objects. The principle space-planning solutions of all four dozen objects from the approved nomenclature of primary maintenance services carried out in the process of the analysis of opportunities show the real possibility of solving development tasks based on this system. For large-sized complexes, in accordance with normalized anti-seismic measures, seismic cutting was performed by arranging the pair supports of the modular plate. The use of the proposed modular system allows us to successfully solve a number of tasks to reduce the harmful effects on the environment and to effectively use renewable energy sources.

4. Discussion

The lack of uniformity in approaches to the design and construction of roadside service facilities, pronounced in today's practice, slows down the design process and complicates construction. In these conditions, it seems appropriate to adapt a system of design and construction based on standard series that has a long history and good practical results. One of the approaches can be the interconnection of space-planning solutions based on a particular spatial module. The advantages of a triangular module based on a structural basis of a single-tier rod spatial plate on three supports revealed in the course of research allow us to solve a number of space-planning and spatial-layout problems for individual objects and their complexes in a harmonious way. In addition, the uniformity of design solutions and construction methods in combination with other measures allows us to successfully solve a number of tasks to reduce the harmful impact on the environment and effectively use renewable energy sources. The studied topic presents a significant prospect for expansion with further research. Thus, it is promising to detail the level of standard for projects of the proposed schemes of space-planning solutions for individual objects. It is very important to consider possible options for blocking objects of various types in complexes of various sizes that have a single roof. Moreover, of course, it is advisable to consider the possibility of using small-scale panel structures for the formation of external walls and partitions of objects.

5. Conclusions

Roadside service facilities have a long history. Their space-planning specifics reflect the specifics of cargo and passenger transportation along the road near which they are located and the level of administration in the surrounding territories. A separate issue remains passenger transport. Within the framework of this service, given its specificity and the need for almost simultaneous construction of almost identical objects on vast territories, the idea of using reusable projects first, and then standard projects for this class of objects grows. The expansion of the nomenclature of primary-line service facilities was determined by the appearance and intensive development of automobile transport, which led to a significant increase in the territory occupied by each complex due to the need to provide parking and convenient maneuvering of cars of various classes. Grouped by function: car service (refueling of various types of fuel, washing of motorcycles and cars of various sizes, service stations), road control (points of traffic control services, checkpoints, points of dimensional and weight control of vehicles), passenger service (public transport stops, retail and catering points, public toilets and showers, laundries, medical and rescue service points, health pavilions, picnic and recreation areas, stage platforms, playgrounds and pavilions, roadside temples of various faiths, heating points, post offices, motels, warehouses, supermarkets), these objects are constantly integrated into large complexes or differentiated by individual functions, depending on the actual territorial conditions.

Taking into account the variety of planning and landscape characteristics of the sites for the placement of objects of the mainline service, it seems advisable to choose as a module not a square or rectangular, but a triangular configuration, which allows in most cases to harmoniously block the modules. The proposed roof module in the form of a "regular" triangle facing the tetrahedron has a structural basis in the form of a single-tier core spatial plate. The half-century-old extensive practice of using such structures minimizes construction and technological difficulties in the manufacture, delivery, and installation. The principal space-planning solutions of all four dozen objects of the nomenclature of objects of the mainline service performed in the process of analyzing the possibilities show the real possibility of solving the development tasks based on this system. The use of the proposed modular system makes it possible to successfully solve a number of tasks to reduce the harmful impact on the environment and effectively use renewable energy sources.

Author Contributions: Conceptualization, K.S., B.K., G.S. and N.K.; methodology, K.S., B.K. and G.S.; software, K.S. and N.K.; validation, K.S., G.S. and N.K.; formal analysis, K.S., B.K. and G.S.; investigation, K.S., B.K., G.S. and N.K.; resources, K.S., B.K. and N.K.; data curation, B.K. and K.S.; writing—original draft preparation, K.S., B.K., G.S. and N.K.; writing—review and editing, K.S., B.K. and G.S.; visualization, N.K. and K.S.; supervision, K.S. and G.S.; project administration, K.S. and B.K.; funding acquisition, K.S. and B.K. All authors have read and agreed to the published version of the manuscript.

Funding: This research received no external funding.

Institutional Review Board Statement: Not applicable.

Informed Consent Statement: Not applicable.

Data Availability Statement: Not applicable.

Acknowledgments: Authors express their gratitude to the Google search engine, the USA Library of Congress, the British Library, whose resources are used for the selection of theoretical material.

Conflicts of Interest: The authors declare no conflict of interest.

References

1. Torgautov, B.; Zhanabayev, A.; Tleuken, A.; Turkyilmaz, A.; Mustafa, M.; Karaca, F. Circular Economy: Challenges and Opportunities in the Construction Sector of Kazakhstan. *Buildings* **2021**, *11*, 501. [CrossRef]
2. Kuanyshbekov, N.N.; Tuyakaeva, A.K. *Aktual'nost' Razvitiya Arhitektury ob"Ektov Pridorozhnogo Servisa v Kazahstane [The Relevance of the Development of Architecture of Roadside Service Facilities in Kazakhstan]/Sbornik Nauchnykh Trudov XIII Mezhdunarodnaya Nauchno-Prakticheskaya Konferenciya Imeni V.Tatlina*; PGUiAS: Penza, Russia, 2018; pp. 95–98.
3. Cui, T.; Ouyang, Y.; Shen, Z.J.M. Reliable facility location design under the risk of disruptions. *Oper. Res.* **2010**, *58*, 998–1011. [CrossRef]
4. Ettema, D.; Gärling, T.; Olsson, L.E.; Friman, M.; Moerdijk, S. The road to happiness: Measuring Dutch car drivers' satisfaction with travel. *Transp. Policy* **2013**, *27*, 171–178. [CrossRef]
5. Bostani, M.K.; Hashemzehi, F.; Anvari, M.R. Evaluation of roadside service centers. Case study: Zahedan to Khash Road. *Ukr. J. Ecol.* **2017**, *7*, 374–381. Available online: https://www.researchgate.net/publication/322752928_Evaluation_of_roadside_service_centers_Case_study_Zahedan_to_Khash_Road (accessed on 15 January 2022). [CrossRef]
6. Dvořák, Z.; Sventekova, E.; Řehák, D.; Čekerevac, Z. Assessment of Critical Infrastructure Elements in Transport. *Procedia Eng.* **2017**, *187*, 548–555. [CrossRef]
7. Rahman, H.Z.; Andreas, A.; Perwitasari, D.; Petroceany, J.S. Developing a typology for social infrastructure (Case study: Roadside station infrastructure). *EDP Sci.* **2019**, *276*, 02020. [CrossRef]
8. Hasan, M.M.; Alam, A.; Mim, A.M.; Das, A. Identifying User Satisfaction Level of Road services: A Focus on Rajshahi City Bypass Road, Bangladesh. *Transp. Res. Procedia* **2020**, *48*, 3132–3152. [CrossRef]
9. Makovetskaya-Abramova, O.; Ivanov, A.; Lazarev, Y.; Shakhova, M.; Rozov, A. Economic assessment of construction of the roadside service facilities. *E3S Web Conf.* **2020**, *157*, 04035. [CrossRef]
10. Guidelines for Service Stations. RTS 13. Land Transport Safety Authority: Wellington, New Zealand, 2001; 42p. Available online: https://www.nzta.govt.nz/assets/resources/road-traffic-standards/docs/rts-13.pdf (accessed on 15 January 2022).
11. Hurley, A.; Jakle, J.A.; Sculle, K.A. Fast Food: Roadside Restaurants in the Automobile Age. *J. Am. Hist.* **2001**, *87*, 1576. Available online: https://www.researchgate.net/publication/275844041_Fast_Food_Roadside_Restaurants_in_the_Automobile_Age (accessed on 15 January 2022). [CrossRef]
12. Filling Station Developments Access- & Building Line Management Technical Overview & Guidelines. Department of Police, Roads and Transport. Free State Province. 22p. Available online: http://www.energy.gov.za/files/PPA-Campaigns/free-state/Filling-Station-Developments-Access-and-Building-Line-Management.pdf (accessed on 15 January 2022).

13. Rest Areas: Mounting Costs and Increased Expectations Create the Perfect Opportunity for Exploring New Public Private Partnerships/Virginia Department of Transportation/Prepared by Ken Winter. June 2008. 21p. Available online: http://vtrc.virginiadot.org/rsb/RSB17.pdf (accessed on 15 January 2022).
14. Magdic, M.; Sjöstrand, P. The Petrol Station—A Hot Spot along the Road. Master's Thesis, IA7400/Department of Informatics School of Economics and Commercial Law University of Gothenburg, Gothenburg, Sweden, 2002; 81p. Available online: https://gupea.ub.gu.se/bitstream/2077/4587/1/magdic_sjostrand.pdf (accessed on 15 January 2022).
15. Wolfe, K.; Holland, R.; Jeff Aaron, J. Road Side Stand Marketing of Fruits and Vegetables. The University of Georgia. Center for Agribusiness and Economic Development. College of Agricultural and Environmental Sciences. September 2002; 42p. Available online: https://sustainagga.caes.uga.edu/content/dam/caes-subsite/sustainable-agriculture/documents/CR-02-09.pdf (accessed on 15 January 2022).
16. Shanahan, K. The degree of congruency between roadside billboard advertisements and sought attributes of motels by US drive tourists. *J. Vacat. Mark.* **2003**, *9*, 381–395. Available online: https://www.researchgate.net/publication/247764549_The_degree_of_congruency_between_roadside_billboard_advertisements_and_sought_attributes_of_motels_by_US_drive_tourists (accessed on 15 January 2022). [CrossRef]
17. Kendrick, M. Roadside Motels. Society of Architectural Historians Archipedia. Available online: https://sah-archipedia.org/essays/CA-01-ART-01 (accessed on 15 January 2022).
18. Wang, Y.W.; Wang, C.R. Locating passenger vehicle refueling stations. *Transp. Res. Part E Logist. Transp. Rev.* **2010**, *46*, 791–801. [CrossRef]
19. Henderson, L. America's Roadside Lodging: The Rise and Fall of the Motel. Historia 2010. 11p. Available online: https://www.eiu.edu/historia/2010Henderson.pdf (accessed on 15 January 2022).
20. Al-Kaisy, A.F.; Kirkemo, Z.; Veneziano, D.; Dorrington, C. Traffic Use of Rest Areas on Rural Highways. *Transp. Res. Rec. J. Transp. Res. Board* **2011**, *2255*, 146–155. Available online: https://www.researchgate.net/publication/272774521_Traffic_Use_of_Rest_Areas_on_Rural_Highways (accessed on 15 January 2022). [CrossRef]
21. Sheng, K.J.; Baharudin, A.S.; Karkonasasi, K. A Car Breakdown Service Station Locator System. *Int. J. Appl. Eng. Res.* **2016**, *11*, 11037–11040. Available online: https://www.researchgate.net/publication/311795116_A_Car_Breakdown_Service_Station_Locator_System (accessed on 15 January 2022).
22. Xanthopoulos, I.; Goulas, G.; Gogos, C.; Alefragis, P.; Housos, E. Highway Rest Areas simultaneous energy optimization and user satisfaction. In Proceedings of the 20th Pan-Hellenic Conference on Informatics, Patras, Greece, 10–12 November 2016; pp. 1–4. [CrossRef]
23. Plovnick, A.; Berthaume, A.; Poe, C.; Hodges, T. Sustainable Rest Area Design and Operations. U.S. Department of Transportation. Federal Highway Administration. October 2017 (DOT-VNTSC-FHWA-17-20 FHWA-HEP-18-006). 31p. Available online: http://www.bv.transports.gouv.qc.ca/mono/1204435.pdf (accessed on 15 January 2022).
24. Karanja, P.; Gathitu, C.W. Strategic Location Considerations for Fuel Filling Stations along Thika Super Highway-Kenya. *Int. J. Acad. Res. Bus. Soc. Sci.* **2018**, *8*, 220–230. Available online: https://hrmars.com/papers_submitted/4460/Strategic_Location_Considerations_for_Fuel_Filling_Stations_along_Thika_Super_Highway-Kenya.pdf (accessed on 15 January 2022). [CrossRef]
25. Rubeis, M.; Groves, S.; Portera, T.; Bonaccorsi, G. Is There a Future for Service Stations. // Boston Consulting Group. 12 July 2019. Available online: https://www.bcg.com/publications/2019/service-stations-future (accessed on 15 January 2022).
26. Green, D.; Roper, P.; Steinmetz, L.; Latter, L.; Lewis, K.; Gaynor, D. Guidelines for the Provision of Heavy Vehicle Rest Area Facilities. Edition 1.1. Austroad Research Report AP-R591-19. July 2019. 58p. Available online: https://austroads.com.au/__data/assets/pdf_file/0025/160648/AP-R591-19_Guidelines_for_the_Provision-of_HVRA_Facilities-1.1.pdf (accessed on 15 January 2022).
27. Quito, A. Roadside chapels gain traction in Western Europe as church attendance declines. QZ. 22 February 2020. Available online: https://qz.com/1805065/autobahnkirche-the-surprising-rise-of-roadside-chapels-in-europe/ (accessed on 15 January 2022).
28. Capar, I.; Kuby, M. An efficient formulation of the flow refueling location model for alternative-fuel stations. *IIE Trans.* **2012**, *44*, 622–636. [CrossRef]
29. Capar, I.; Kuby, M.; Leon, V.J.; Tsai, Y.J. An arc cover-path-cover formulation and strategic analysis of alternative-fuel station locations. *Eur. J. Oper. Res.* **2013**, *227*, 142–151. [CrossRef]
30. Chung, S.H.; Kwon, C. Multi-period planning for electric car charging station locations: A case of Korean expressways. *Eur. J. Oper. Res.* **2015**, *242*, 677–687. [CrossRef]
31. Bhatti, S.F.; Lim, M.K.; Mak, H.Y. Alternative fuel station location model with demand learning. *Ann. Oper. Res.* **2014**, *230*, 105–127. [CrossRef]
32. Ghamami, M.; Zockaie, A.; Nie, Y.M. A general corridor model for designing plug-in electric vehicle charging infrastructure to support intercity travel. *Transp. Res. Part C Emerg. Technol.* **2016**, *68*, 389–402. [CrossRef]
33. Tran, T.H.; Nguyen, T.B.T. Alternative-fuel station network design under impact of station failures. *Ann. Oper. Res.* **2018**, *279*, 151–186. [CrossRef]
34. Angelucci, G.; Mollaioli, F.; Tardocchi, R. A New Modular Structural System for Tall Buildings Based on Tetrahedral Configuration. *Buildings* **2020**, *10*, 240. [CrossRef]

35. Subbotin, A.; Grigoryan, S. Road service facilities for improving the ecological state using the PPP mechanism. *E3S Web Conf.* **2019**, *91*, 08007. Available online: https://www.e3s-conferences.org/articles/e3sconf/pdf/2019/17/e3sconf_tpacee2019_08007.pdf (accessed on 15 January 2022). [CrossRef]
36. Samoilov, K.I. The triangular module for roadside service buildings/Education materials. Almaty. 2020. 120p. Available online: https://znanio.ru/media/samoilov-ki-the-triangular-module-for-roadside-service-buildings--education-materials---almaty-2020-120-p-2711058 (accessed on 15 January 2022).
37. Toll Plaza/Archohm.—ArchDaily: The World's Most Visited Architecture Website. 26 January 2013. Available online: https://www.archdaily.com/322127/toll-plaza-archohm (accessed on 15 January 2022).
38. Repsol Service Stations.—Foster + Partners. Available online: https://www.fosterandpartners.com/projects/repsol-service-stations/ (accessed on 15 January 2022).
39. Parasol Paraíso (Repsol Filling Stations, Spain and Portugal).—The Beauty of Transport: Transport Design, Transport Architecture, and Transport's Influence on Art and Culture. 24 February 2016. Available online: https://thebeautyoftransport.com/2016/02/24/parasol-paraiso-repsol-filling-stations-spain-and-portugal/ (accessed on 15 January 2022).
40. "Foster + Partners" Repsol service station.—Divisare: The Atlas of Contemporary Architecture. 23 June 2016. Available online: https://divisare.com/projects/320787-foster-partners-nigel-young-repsol-service-station (accessed on 15 January 2022).
41. Spin Me Right Round (Red Hill Filling Station, Leicestershire, UK)—The Beauty of Transport: Transport Design, Transport Architecture, and Transport's Influence on Art and Culture. 21 August 2013. Available online: https://thebeautyoftransport.com/2013/08/21/spin-me-right-round-red-hill-filling-station-leicestershire-uk/ (accessed on 15 January 2022).
42. Shell Shape High-Speed Way Entrance Tent PTFE Membrane Structure/Guangzhou Feite Truss Tent Co., Ltd. Available online: https://www.truss-tent.com/product/shell-shape-high-speed-way-entrance-tent-ptfe-membrane-structure/ (accessed on 15 January 2022).
43. Ulyanovskie Dorogi "Ukrasyat" 3h-Zvezdnymi Gostinicami I Mnogotoplivnymi Zapravkami [Ulyanovsk Roads will be "Decorated" with Three-Star Hotels and Multi-Fuel Gas Stations]. Finance. 18 April 2017. Available online: http://1ul.ru/finance/biznes/news/ulyanovskie_dorogi_ukrasyat_3h_zvezdnymi_gostinicami_i_mnogotoplivnymi_zapravkami/ (accessed on 15 January 2022).
44. Standart Respubliki Kazakhstan 2476-2014 «Dorogi Avtomobilnye Obshego Polzovania. Trebovania k Obiektam Dorojnogo servisa i ih Uslugam» [Standard of the Republic of Kazakhstan 2476-2014 "Public Roads. Requirements for Road Service Facilities and Their Services"]. Astana, Komitet Tekhnicheskogo Regulirovaniya i Metrologii Ministerstva po Investiciyam i Razvitiyu Respubliki Kazahstan (Gosstandart), 2014. 106 s. Available online: https://ru.qaj.kz/upload/medialibrary/255/%D0%A1%D0%A2%D0%A0%D0%9A%202476-2014.pdf (accessed on 15 January 2022).
45. Mikhailov, V.V.; Sergeev, M.S. *Spatial Core Construction of Coatings (Structures)*; Izdatelstvo Vladimirskogo Gosudarstvennogo Universiteta: Vladimir, Russia, 2011; 56 s.
46. Rekomendatsii po Proektirovaniyu Strukturnykh Konstruktsii [Recommendations for the Design of Structural Constructions]/TSNIISK im.Kucherenko Gosstroia SSSR [Recommendations for the Design of Structural Constructions]. Moskwa: Stroiizdat. 1984. 416 s. Available online: http://www.complexdoc.ru/ntdpdf/537365/rekomendatsii_po_proektirovaniyu_strukturnykh_konstruktsii.pdf (accessed on 15 January 2022).

designs

MDPI

Article

A Systematic Review of Architectural Design Collaboration in Immersive Virtual Environments

Rongrong Yu [1,*], Ning Gu [1], Gun Lee [2] and Ayaz Khan [1]

[1] UniSA Creative, Australian Research Centre for Interactive and Virtual Environment, University of South Australia, Adelaide, SA 5000, Australia

[2] UniSA STEM, Australian Research Centre for Interactive and Virtual Environment, University of South Australia, Adelaide, SA 5000, Australia

* Correspondence: rongrong.yu@unisa.edu.au

Abstract: Emerging applications of immersive virtual technologies are providing architects and designers with powerful interactive environments for virtual design collaboration, which has been particularly beneficial since 2020 while the architecture, engineering and construction (AEC) industry has experienced an acceleration of remote working. However, there is currently a lack of critical understanding about both the theoretical and technical development of immersive virtual environments (ImVE) for supporting architectural design collaboration. This paper reviewed recent research (since 2010) relating to the topic in a systematic literature review (SLR). Through the four steps of identification, screening, eligibility check, and inclusion of the eligible articles, in total, 29 journal articles were reviewed and discussed from 3 aspects: ImVE in the AEC industry, ImVE for supporting virtual collaboration, and applications of ImVE to support design collaboration. The results of this review suggest that future research and technology development are needed in the following areas: (1) ImVE support for design collaboration, particularly at the early design stage; (2) cognitive research about design collaboration in ImVE, toward the adoption of more innovative and comprehensive methodologies; (3) further enhancements to ImVE technologies to incorporate more needed advanced design features.

Keywords: architectural design collaboration; immersive virtual environments; systematic literature review

Citation: Yu, R.; Gu, N.; Lee, G.; Khan, A. A Systematic Review of Architectural Design Collaboration in Immersive Virtual Environments. *Designs* **2022**, *6*, 93. https://doi.org/10.3390/designs6050093

Academic Editor: Farshid Aram

Received: 12 September 2022
Accepted: 8 October 2022
Published: 11 October 2022

1. Introduction and Background

Traditionally, most architects work and collaborate in face-to-face environments, and virtual collaboration only occurs occasionally, mainly during the latter design stages such as design review. Although the concept of computer-supported cooperative work (CSCW) in the design field has emerged and been extensively studied during the past few decades, in actuality, the area has not seen significant advances throughout that time. With emerging technologies such as immersive virtual environments (ImVE), architects and designers are able to collaborate virtually with more convenience and power. Since 2020, there has been an urgent global acceleration of remote working, leading to a rush of adoption of virtual technologies across most industries, including in the architecture and design sectors.

Design collaboration refers to team-based design activities working toward achieving the shared design goals. Effective collaboration during the initial design stage will lead to fewer problems during the latter, more complex design and construction stages [1]. Design collaboration with multiplicate problem-solving approaches can align different stakeholders' opinions toward a common baseline that can more optimally result in valuable project insights [2]. Tan [3] argues that design collaboration enhances reflection upon the actions of architects within a team. Citing the need for remote collaboration during pandemic times, Kim et al. [4] note that social networking services can increase idea clarification and the sharing of information to support active design collaboration. Combrinck and Porter [5] found that initial stages of design benefit directly from the collaboration between architects and end-users. Design collaboration occurring at the early design stage is significant for achieving design innovation and ultimately optimal design solutions.

Researchers have emphasised that digital modalities and collaboration affect the quality, efficiency, and accuracy of design [6–8]. Virtual collaboration environments enable distributed remote design collaboration, facilitating greater time and cost efficiency in design, and have become increasingly relevant and crucial since 2020. Early studies have explored the application of shared digital environments in design collaboration during the design phase [9–11], including the effects of those environments on designers. For example, Gu et al. [12] suggested that 3D virtual worlds support the production of considerable perceptual events during synchronous design collaboration. Recent developments in ImVE facilitate intuitive virtual interactions between designers, and also between designers and the design environments [13], leading to better spatial perception [14] that may be beneficial for the design process. ImVE refers to virtual environments in which the users can "immerse" themselves inside the computer-generated world and feel they are in fact an integral part [15]. This can be achieved by using head-mounted displays (HMD) or multiple projections [16]. Virtual reality (VR), augmented reality (AR), and mixed reality (MR) are typical categories of ImVE systems that allow different degrees of interactions between the physical and the virtual worlds supporting building projects with different complexity and diversity [17]: VR is a computing technique that immersively manipulates the user's senses to make him/her feel present in a simulated virtual environment [18,19]; AR on the other hand superimposes virtual information upon the real world in a graphical manner via digital computing platforms to deliver enhanced experiences of the real world [20]; and MR is a combination of AR and augmented virtuality (AV), with the potential to integrate virtual and augmented realities together [21].

In light of the increasing application of ImVE within the architecture, engineering and construction (AEC) industry, it is important to critically understand how ImVE supports design collaboration. This study has reviewed the recent (since 2010) research in this area, to reveal the current body of knowledge and different applications of ImVE for supporting design collaboration, as well as potential future research directions to further advance the field.

2. Research Method- Systematic Literature Review

This study adopts the research method of systematic literature review (SLR) to review current research on design collaboration in ImVE and synthesise further knowledge about the field. SLR is an authoritative procedural method that synthesises and delineates the boundaries of knowledge in a research domain [22,23]. In accordance with common SLR research conventions, this study adopted a four-stage SLR to review the published articles related to design collaboration in ImVE, comprising a widespread literature search, full text assessment, meta-synthesis, and critical content analysis. The following sections elaborate on the details and steps of SLR. The article retrieval process is shown in Figure 1 below. Through the 4 steps of identification, screening, eligibility check, and inclusion of the eligible articles, ultimately 29 articles were selected for analysis from the initial 1106 papers identified. The relatively small number of articles that were closely related to the topic shows that architectural design collaboration in immersive virtual environments has not been extensively explored in the field, and this in turn supports the needs for this review and for future research.

2.1. Selection of Databases and Literature Search

The SLR process should be reinforced by a detailed and impartial search in specifying relevant research. A choice of database selection that ensures a broad coverage of research must be identified and used. In this regard, this study uses Scopus and Web of Science (WoS), the two most significant platforms for article retrieval. These two databases offer a wide coverage of literature and are feasible for conducting organised queries. Past reviews have used similar databases for article selection in the AEC industry [24,25]. Firstly, prominent keywords related to architectural design collaboration and ImVE were identified, and in the next step, a number of keywords having similar semantic meanings were merged. The resulting search string used for this study was as follows:

[TITLE-ABS-KEY ("architectural design" OR architecture) AND TITLE-ABS-KEY ("collaboration" OR "design collaboration") AND TITLE-ABS-KEY ("virtual reality" OR "augmented reality" OR "mixed reality" OR "immers* techn*" OR "virtual environment*")]. A total of 689 articles from *Scopus* and 417 articles from *WoS* were identified via the search.

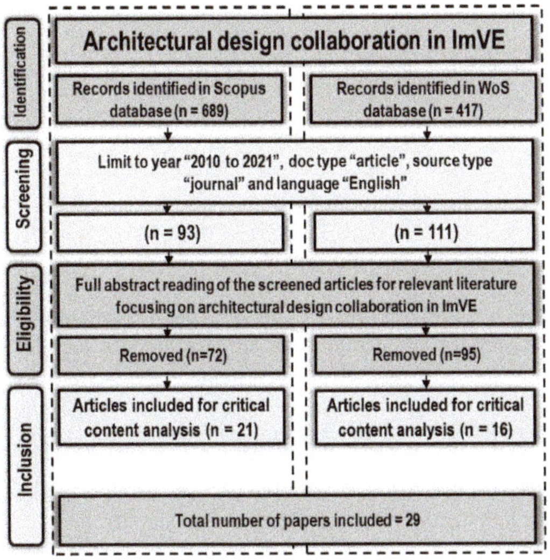

Figure 1. Article retrieval process (data was collected in March 2022).

2.2. Screening of Articles

The screening of the articles provides and constitutes the benchmark utilised in the SLR process for filtering the pool of articles. At this stage, this filtering was performed based on year, document type, source type, and the language of articles. The applied filtering procedure is as follows: [(LIMIT-TO (YEAR, "2010-present") AND (LIMIT-TO (DOCTYPE, "ar") AND (LIMIT-TO (SRCTYPE, "j") AND (LIMIT-TO (LANGUAGE, "English")]. The articles considered for this study were limited to recent ones (from 2010, in accordance with the prominent boom of ImVE). In relation to document type and source type, only peer-reviewed journals articles were considered for the study; since journal articles go through a more rigorous peer review process and provide more in-depth knowledge than other research articles such as conference papers. Finally, non-English language articles were filtered out. This stage resulted in a total of 93 articles from Scopus and 111 articles from WoS.

2.3. Eligibility and Inclusion Criteria of Articles

In this step, eligibility and inclusion criteria were developed to narrow down the articles relevant to the specific focus of this study: criteria such as abstract review, keywords, evaluation of conclusions, and results were applied to the articles from the previous step. Papers with a focus on ImVE but for health care, manufacturing, and other domains were removed at this stage. This stage resulted in a total of 21 articles from Scopus and 16 articles from WoS. Articles which were identified as duplicates in both search engines were then merged. In the end, a total of 29 articles were for this study, which is a significant number for critical content analysis. This sample size is favourable compared with other similar reviews.

2.4. Meta-Synthesis and Critical Content Analysis of Articles

Meta-synthesis refers to the analysis of metadata within the pool of articles in the SLR process [26]. It augments the SLR process by extracting the metadata from each article and is utilised to provide a basis for organising a framework for the study [27]. Information such as journal name, year of publication, focus of research, and limitations constitutes metadata and was organised in a tabular format into an Excel file. This comprehensive table can be described as an "idea thought matrix supplemented by components of analysis" [26]. The articles were then categorised further into research themes, emphasising commonalities among them to form clusters from among the retrieved articles; this process is called critical content analysis, and it facilitates establishing the current status quo and developing future trends in the domain of study [28].

3. Results

3.1. Publication Trend of Architectural Design Collaboration in ImVE

The articles in this study were limited to those from the year 2010 to present, due to the rise of ImVE only being significantly noticeable over the last decade, and this focus is more likely to deliver the latest advancements and status quo in the specific subject area. Figure 2 depicts the annual trends in research publications on the topic. From the figure, we can see that the period of time between 2010 to 2013 only has up to 3 articles per year, as this period marks the inception of ImVE in the architecture field (where architects found ImVE to be a novel design collaboration tool and could engage clients and other stakeholders in a more intuitive way). In 2014, the social media company Meta promoted its Oculus VR headset, which resulted in a substantial amount of awareness to the public. However, this did not result in increasing number of research publications in this area, due to a number of reasons; firstly, there were significant barriers encountered by the AEC industry for adopting ImVE devices including low battery life, tracking issues, interoperability concerns, and relatively high degree of skills required from users [17], and secondly the AEC industry comprises of large number of small and medium-sized enterprises (SMEs) that lacked interests in adopting ImVE due to the relatively high costs, capital required for training, and other significant obstacles [24]. A surge in the number of papers is seen from 2017 on the use of ImVE for design collaboration. This is because technological advancements seen in the latest ImVEs have resolved major prior concerns and issues. Further surge in demand resulting from the COVID-19 pandemic has brought such applications of ImVE to the broader industry. Virtual collaboration has never carried more necessity and meaning than in the recent times of the COVID-19 pandemic, which saw a general upwelling of interest among researchers since 2020 [29]. Such upward trends in this area makes this study especially timely and valuable in its analysis of the status quo, emergent themes, and future research directions for design collaboration in ImVE. Note that in Figure 2 the upward trend is not seen in the 2022 data point, as this study's data collection was only completed in March 2022.

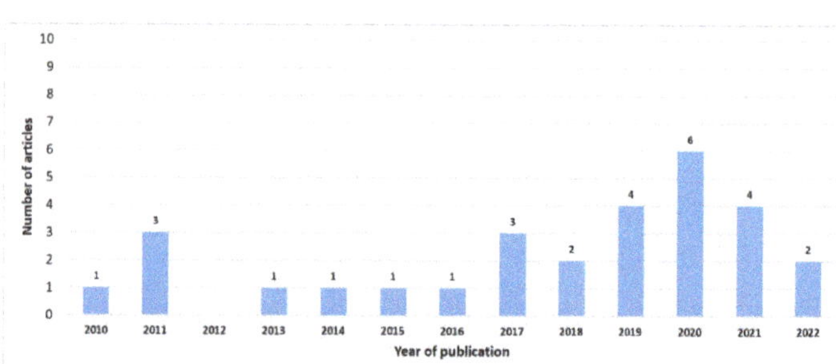

Figure 2. Annal publication trend of articles for design collaboration in ImVE.

3.2. *Distribution of Journals for Publishing the Topic*

The 29 articles included in this study were published across a total of 17 journals. Table 1 lists the journals, and the number of articles published. Six journals published at least two articles each, with the top contributing journals being *Automation in Construction* (7) and *Journal of Information Technology in Construction* (4). The analysis of journals provides a summary for researchers who may conduct similar kinds of studies. Other main journals include *Visualisation in Engineering, Frontiers in Robotics and AI,* and *Journal of Higher Education Theory and Practice.* Together, they have illustrated the applications and benefits of ImVE from a wide range of perspectives, such as automating the design collaboration process, providing better visualisations for end users, and promoting the use of information technologies within the AEC field.

Table 1 categorises the 29 articles about architectural design collaboration in ImVE that have been reviewed. The table also organises these current studies in terms of the type of technology, focus of design collaboration, main research content, and project stage in which ImVE has supported design collaboration. The results of the review are further discussed in Section 4.

Table 1. Reviewed articles on design collaboration in ImVE.

No.	Journal	Author & Year	Article Title	Type of ImVR	Focus of Design Collaboration	Main Research Content	Project Stage
1	Automation in Construction	[30]	A multi-user collaborative BIM-AR system to support design and construction	BIM and AR	Multi-stakeholder collaboration	Presented a BIM-AR system that provides the ability to view, interact with, and collaborate with 3D and 2D BIM data via AR with geographically dispersed teams.	Multiple stages
2	Automation in Construction	[31]	From BIM to extended reality in AEC industry	Extended reality	Technologies for construction collaboration between stakeholders	Explored outsourcing patterns for technologies among construction project stakeholders.	Multiple stages
3	Automation in Construction	[32]	Virtual reality applications for the built environment: Research trends and opportunities	VR	Design collaboration, multi-user virtual construction	Review paper; reviewed VR applications in AEC.	Multiple stages
4	Automation in Construction	[33]	OpenBIM-Tango integrated virtual showroom for offsite manufactured production of self-build housing	BIM, VR, and AR	Early involvement of stakeholders and end-users	Streamlined the design process and provided a pared-down agnostic openBIM system with low latency and included concurrent user accessibility.	Design stage
5	Automation in Construction	[34]	Zero latency: Real-time synchronization of BIM data in Virtual Reality for collaborative decision-making	BIM and VR	Improvement of collaboration in AEC industry	Proposed a BIM VR real-time synchronisation system based on an innovative cloud-based BIM metadata interpretation and communication method.	Design stage
6	Automation in Construction	[35]	Immersive virtual environments versus physical built environments: A benchmarking study for building design and user-built environment explorations	Immersive virtual environments	End-user involvement	Explored the use of immersive virtual environments during the design, construction, and operation phases of AEC projects.	Multiple stages

Table 1. *Cont.*

No.	Journal	Author & Year	Article Title	Type of ImVR	Focus of Design Collaboration	Main Research Content	Project Stage
7	Automation in Construction	[12]	Technological advancements in synchronous collaboration: The effect of 3D virtual worlds and tangible user interfaces on architectural design	3d virtual worlds and tangible user interface	Design collaboration	Presented and evaluated two current advancements of collaborative technologies for architectural design.	Design stage
8	Journal of Information Technology in Construction	[36]	The impact of avatars, social norms and copresence on the collaboration effectiveness of AEC virtual teams	VR	Virtual team collaboration on AEC project	Review paper; examined collaboration effectiveness of global virtual engineering project teams.	Multiple stages
9	Journal of Information Technology in Construction	[37]	Virtual Reality for the built environment: A critical review of recent advances	VR and virtual environment applications	Benefits for collaboration	Review paper; presented a classification framework to reveal the scholarly coverage of VR and virtual environment.	Multiple stages
10	Journal of Information Technology in Construction	[38]	Case studies using multiuser virtual worlds as an innovative platform for collaborative design	Multi-user virtual worlds	Collaboration between designers	Investigated the innovative use of emerging multiuser virtual world technologies for supporting human–human collaboration and human–computer co-creativity in design.	Design stage
11	Journal of Information Technology in Construction	[39]	Framework for model-based competency management for design in physical and virtual worlds	Virtual worlds	Design collaboration	Explored differences and commonalities in competencies for design in the physical and virtual worlds by examining design input, process, and outcome.	Design stage
12	Journal of Construction Engineering and Management	[24]	State-of-the-Art review on Mixed Reality applications in the AECO industry	MR	Multi-user collaboration	Review paper; reviewed MR technology applications in the AECO industry.	Multiple stages

Table 1. *Cont.*

No.	Journal	Author & Year	Article Title	Type of ImVR	Focus of Design Collaboration	Main Research Content	Project Stage
13	Journal of Construction Engineering and Management	[40]	Virtual collaborative design environment: Supporting seamless integration of multitouch table and immersive VR	VR	Multi-stakeholder collaboration	Presented the design and evaluation of a virtual collaborative design environment.	Design stage
14	Applied Sciences (Switzerland)	[41]	Developing a BIM-based MUVR treadmill system for architectural design review and collaboration	BIM based Multi-user VR	High-level immersion in architectural design review and collaboration	Presented a system framework that integrates multi-user virtual reality (MUVR) applications into omnidirectional treadmills.	Design review
15	Applied Sciences (Switzerland)	[25]	End-Users' Augmented Reality utilization for architectural design review	AR	End-user involvement in design review	Investigated how the AR system affects architectural design review from users' perspectives.	Design review
16	Applied Sciences (Switzerland)	[42]	Trends and research issues of Augmented Reality studies in architectural and civil engineering education—A review of academic journal publications	AR	Collaboration between academia and practice	Review paper; reviewed AR in AEC education, with a focus on collaboration promoting optimal connection between general pedagogy and domain-specific learning.	Multiple stages
17	Journal of Engineering, Design and Technology	[43]	Multiuser immersive Virtual Reality application for real-time remote collaboration to enhance design review process in the social distancing era	Immersive VR	Collaboration in design review	Explored design review process conducted among participants remotely located.	Design review
18	International Journal of Digital Earth	[44]	Immersive Virtual Reality for extending the potential of building information modeling in architecture, engineering, and construction sector: systematic review	Building Information Modelling (BIM) and Immersive VR	Communication and collaboration in design, construction, operation, and maintenance phases	Review paper; reviewed most commonly adopted technologies, applications, and evaluation methods of VR.	Multiple stages

Table 1. *Cont.*

No.	Journal	Author & Year	Article Title	Type of ImVR	Focus of Design Collaboration	Main Research Content	Project Stage
19	Journal of Computational Design and Engineering	[45]	Evaluation framework for BIM-based VR applications in design phase	BIM and VR	Multi-user collaboration	Developed an evaluation framework for BIM-based VR applications focused on the design phase of projects.	Design stage
20	Journal of Higher Education Theory and Practice	[46]	Innovation in architecture education: Collaborative learning method through Virtual Reality	VR	Collaborative learning	Review paper; reviewed VR encounters and long-term collaborative learning approaches.	Design education
21	Construction Innovation	[47]	Using Virtual Reality to facilitate communication in the AEC domain: A systematic review	VR	Multi-stakeholder collaboration	Review paper; explored how VR has been applied for communication purposes in AEC.	Multiple stages
22	Environment and Planning B-Urban Analytics and City Science	[48]	Architectural design creativity in multi-user virtual environment: A comparative analysis between remote collaboration media	VR	Multi-user virtual environments	Investigated the affordance of multi-user virtual environments for the production of novel and appropriate solutions in remote collaboration.	Design stage
23	Frontiers in Robotics and AI	[49]	Laypeople's collaborative immersive Virtual Reality design discourse in neighborhood design	VR	Virtual participatory urban design	Protocol study; explored design communication and participation of laypeople in a virtual participatory urban design process.	Design stage
24	Advanced Engineering Informatics	[50]	Overlay design methodology for virtual environment design within digital games	VR	Design collaboration	Protocol study; explored the use of overlay design methodology for the creation of virtual environments within digital gaming contexts.	Design stage
25	Visualization in Engineering	[51]	Virtual Reality-integrated workflow in BIM-enabled projects collaboration and design review: A case study	BIM and VR	Collaboration in design review	Developed and tested a VR integrated collaboration workflow.	Design review

Table 1. *Cont.*

No.	Journal	Author & Year	Article Title	Type of ImVR	Focus of Design Collaboration	Main Research Content	Project Stage
26	Journal of Digital Landscape Architecture	[52]	Using Virtual Reality as a design input: Impacts on collaboration in a university design studio setting	Immersive VR	Student learning and group collaboration	Presented the application of immersive VR to assist landscape architecture students in design.	Design education
27	Co-Design	[53]	Enablers and barriers of the multi-user virtual environment for exploratory creativity in architectural design collaboration	Multi-user virtual environment	Architectural design collaboration	Explored the design collaboration process using multi-user virtual environment and sketching media in face-to-face and remote collaboration modes.	Design stage
28	Computers in Industry	[54]	Mutual awareness in collaborative design: An Augmented Reality integrated telepresence system	AR	Design collaboration	Proposed a new computer-mediated remote collaborative design system, TeleAR, to enhance the distributed cognition among remote designers.	Design stage
29	IEEE Transactions on Visualization and Computer Graphics	[55]	A spatially Augmented Reality sketching interface for architectural daylighting design	AR	Collaboration between designers and end-users	Presented an application of interactive global illumination and spatially augmented reality to architectural daylight modelling.	Design stage

4. Discussion

During past decades, emerging immersive virtual technologies have significantly changed the nature of collaboration in the AEC industry at various stages of a building project including design. Technological advancements create new design environments for designers and make virtual collaboration possible, which also have an impact on designers' thinking processes as well as on the design solutions they produced [56]. This section discusses the current state of theoretical and technical developments about immersive virtual technologies, and their applications in supporting design collaboration from various perspectives, as revealed from the critical review.

4.1. Immersive Virtual Technologies and It's Attributes in the AEC Industry

One of the most significant obstacles for current design technologies has been the immersion of the relevant stakeholder in the design representation during different stages of the project. Table 1 shows that the term "immersive" is increasingly being seen in design collaboration research since 2015. Prior to 2015, researchers used alternative terms such as "virtual worlds", and "multi-user virtual worlds". Despite possessing numerous advantages for streamlining the process of design, design technologies have not provided adequate immersive presence for its users. The importance of presence when visualising a design solution is significant in the AEC industry for realising the aesthetic appearance of a space, simulating the functionality of the design, and enabling users to effectively experience the place in terms of other factors such as safety and ergonomics. Advancements in digital design technologies including ImVE have supplied architects with a myriad of opportunities, for visualising the appearance and performance of their designs [57].

Recent broader adoption of computing technologies have enabled designers to more readily utilise ImVE in practice. Conceptually, an ImVE system can be predominantly classified into four categories of elements, namely devices, platforms, applications, and tools (Figure 3).

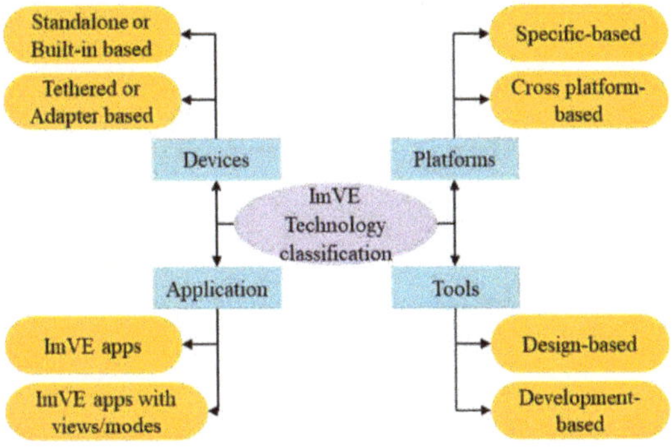

Figure 3. The conceptual model of an ImVE system.

The first category elements in the model is devices, which can be subdivided into standalone or built-in, and tethered or adapter based. The VR Quest series by Meta is a typical example of a standalone VR device, although it can also be optionally tethered to give it more rendering power. The Oculus Rift seriesis a more recent tethered models that needs to be connected to a computer through a cable. One of the earliest examples of adapter-based devices is Google Cardboard, which simply wraps around a smartphone. In relation to AR, examples such as Microsoft HoloLens act as a standalone device incorporating high-level computing capabilities. On the other hand, Holokit act as adapter based AR device using Apple ARKit enabled smartphones. There has been continuous advancement among ImVE devices with companies such as HTC, Microsoft, and Meta developing high-end devices with significant enhancements to the visual field of view, storage capacities, ergonomics, and graphics rendering to name a few. The second category of elements relates to the software platforms that serve as the basis for devices to work; such platforms can be subdivided further into specific and cross platforms. Companies such as Meta, HTC, and Magic Leap provide their own proprietary software platforms for their devices to function, while cross platforms (sometimes referred to as open platforms) allow different hardware devices to function on the same software platform. Steam VR, Windows Mixed Reality (WMR), and WebXR are examples of such open or cross platforms. The third category pertains to ImVE applications, which are subdivided into ImVE applications alone and ImVE applications with views and modes. The difference between those two relates to their real-world integration and communication. Most typically, VR applications are without real-world integration, while AR/MR applications possess real-world interaction capabilities. Finally, the fourth-category elements relate to ImVE tools that are either design based or development based. Tools such as Tilt Brush, Quill, and Aero are design oriented and are common among designers such as architects. Development tools such as Unity and Unreal Engine are significant tools for creating VR and AR visualisations respectively.

Based on data from the online market data portal Statista, the market size of ImVE is predicted to increase vastly in the coming years, from 30.7 billion USD in 2021 to 296.9 billion USD in 2024. This is indicative of the future demand for ImVE and its related technologies and applications across many industries. Figure 4 shows the main ImVE types and their usage comparison data obtained from Statista for actual 2020 and predicted 2024. From the figure, we can see that in 2020, all VR-related technologies were utilised more than AR-related technologies. Among them the VR standalone HMDs were the most popular immersive technology (accounted for 43.76%). However, for 2024 the prediction is that AR-related technologies combined will be utilised more than VR combined, and among the AR technologies the AR standalone HMDs will have the highest usage (accounted for 31.28%). AR represents the superimposition of the virtual information over the real world to construct the synthetic environment aiming to enrich reality [58]), and the growing trend of research shifting away from VR and toward AR/MR can also be noticed in the ImVR literature that was reviewed (Figure 5). In the figure, we can see that although most of the reviewed articles focused on VR-related ImVE technologies, in recent years, there is a visible increase of AR/MR utilisation in design collaboration. The growing application of AR technologies for design collaboration may due to the fact that AR technologies were based on early VR technologies that had been developed to extend VR technologies, focusing on the capability of augmenting the real world with virtual information, allowing real and virtual information to coexist and at a same time making user interactions more intuitive through references to reality [59].

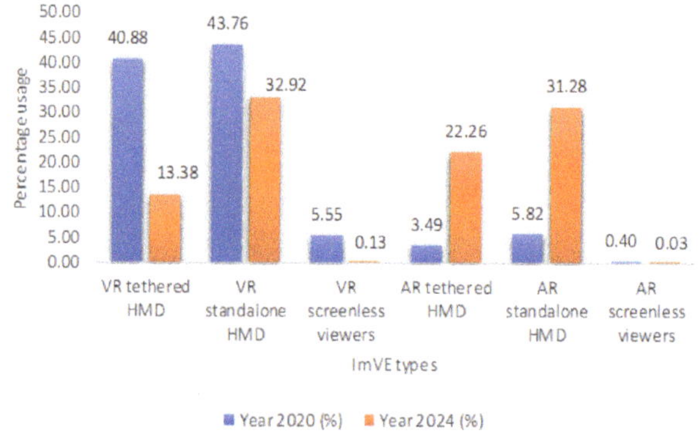

Figure 4. Main ImVE types and their usage comparison (Source: Statista).

Some of the main attributes of ImVE include presence, immersion, and interactivity [60,61]. Presence describes the complete feeling of being in an immersive simulated space, wherein the user is psychologically immersed in the virtual environment in a manner that they temporarily escape their real world [62–64]. The immersion level is affected by the sophistication of the simulation in terms of the quality of its visual representation, consistency, freedom of movement of the user, and physical interaction/feedback functionality within the ImVE. The interactivity level indicates to what extent a user is able to alter the ImVE in real-time [65]. Additionally, immersive AR technologies are able to augment real world spaces with overlayed virtual information in a manner such that real and virtual information can coexist simultaneously to enable enhanced intuitive user interactions [59].

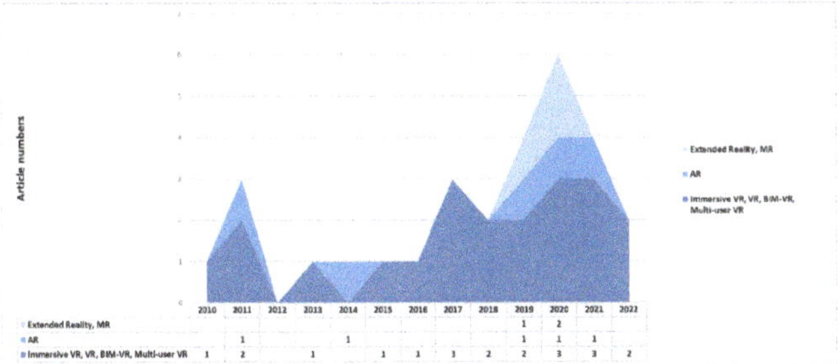

Figure 5. Types of ImVE technologies the review articles focused on.

4.2. ImVE in Supporting Virtual Collaboration

Virtual collaboration tools currently in use within the AEC industry largely focus on design review or construction scheduling, such as Unity Reflect Review, Resolve, Trezi, Fuzor, and BIM 360. A few recently emerging tools are intended for use in the design ideation stage, including Wild, Mindesk, and Arkio. For design ideation purposes, most tools support importing 3D models from commonly used architectural design software such as Revit and SketchUp. Mindesk and Arkio support the use of Grasshopper, which is a popular parametric design tool for design ideation in architecture. Arkio provides functions including real-time Boolean operations, sun studies, smart guides for geometry alignment and creation, integrated street maps, etc., which can support a range of design ideation purposes. For design collaboration, Wild provides native digital sketching tools for both ideation and as (speech to text) annotation for design review. Mindesk allows teams to collaborate on the same parametric model in multi-user VR sessions. Arkio on the other hand focuses on supporting multi-user model modification during design collaboration. All of the aforementioned virtual collaboration tools enable access via VR including desktop VR options that are convenient for users without headsets. Wild and Arkio also allow mobile access for users. Another virtual collaboration tool Hyve 3D, is primarily focused on providing 3D sketching in immersive design environments, to support both ideation and remote collaboration. Hyve 3D requires a Macbook, iPad Pros, and a 4K projector but a VR headset is not needed to have the immersive experience. Table 2 summarises the main ImVE used for virtual design collaboration in the AEC industry. In addition, generic communication and collaboration tools are also used in the AEC practice, including Teams, Zoom, Slack, etc. Some of those generic tools also provides certain visual collaboration functions, for example, Miro has a 2D digital whiteboard for supporting brainstorming, Asana can assist with project workflow planning with a visual timeline and calendar, and as Jira can be linked to Navisworks and allow the display of 3D models. From the table, we can see that there are recently developed design collaboration tools utilising ImVE technologies, some of which are focused on application at the early design stage. However, they are not yet widely adopted throughout the AEC industry. Furthermore, the review results suggest that academic research has fallen behind the technological advancement, and there is a lack of understanding of how the recently developed ImVE technologies support design collaboration, particularly at the early design stage.

Table 2. Virtual design collaboration tools in the AEC industry (Source: respective official website of each individual platform).

Virtual Design Collaboration Tools	Project Stages	Collaboration Means	Interoperability	Accessibility	Design Support
The Wild (https://thewild.com/)	Multiple stages	Annotation (speech to text) for design review	Revit, SketchUp, BIM 360	VR, desktop or mobile	Sketching, Inspection of object BIM data
IrisVR (joined Wild) (https://irisvr.com/)	Multiple stages	Optimised *Revit* to VR workflow, multi-user meetings, collaborative tracking, annotation, other user controls, BIM coordination.	Revit, Sketchup, Navisworks, Rhino	VR or desktop	Inspection of elements, tape measure, scale model mode, sun studies
Generic communication tools such as Zoom (https://zoom.us/), Teams (https://www.microsoft.com/en-au/microsoft-teams/log-in (accessed on 12 September 2022)), Slack (https://slack.com/intl/en-au/in (accessed on 12 September 2022)), Miro (https://miro.com/), Asana (https://asana.com/).	Multiple stages	Multi-user meeting, 3D model viewing, 2D shared white board, visual timeline, calendar, chat box	-	-	3D model viewing, 2D drawing sharing.
Fuzor (https://www.kalloctech.com/)	Multiple stages with a focus on construction	Multi-user view	Revit, Rhino	VR, desktop or mobile	Flythrough and walkthrough videos
BIM 360 (https://www.autodesk.com/bim-360/in (accessed on 12 September 2022))	Multiple stages with a focus on construction	Multi-user model modification	BIM	Desktop	Model modification
Hyve3D (https://www.hyve3d.com/)	Ideation	Multi-user model modification, sketching	Can import model from Revit, SketchUp, but primarily focus on sketching	Need iPad Pro, Macbook Pro, and 4K projector, but does not require a headset	3D sketching simultaneously in complementary (orthogonal) representations.

Table 2. *Cont.*

Virtual Design Collaboration Tools	Project Stages	Collaboration Means	Interoperability	Accessibility	Design Support
Arkio (https://www.arkio.is/)	Ideation, modelling	Multi-user model modification, sketching	Revit, SketchUp, BIM 360, Rhnio	AR/VR or mobile	Parametric design, sketching, real-time Boolean operations and parametric volumes, sun studies, PC spectator mode, see inside with sections, smart guides, integrated street map, and instantly enable 3D buildings from OpenStreetMap, etc.
Mindesk (https://mindeskvr.com/)	Ideation, design review	3D/surface modelling, multi-user model modification, navigation, selection, review and co-presence and use of body language for communication	Revit, Rhino, Grasshopper, Solid works	VR or desktop	Parametric design, Unreal Studio and OBS Studio for VR
Theia BigRoom (upcoming) (https://www.theia.io/bigroom/in (accessed on 12 September 2022))	Design review	Multi-user meeting, 3D sketching tools, whiteboard drawing tools, ideation boards, task lists and post-it notes	BIM	VR, desktop or mobile	Unreal engine, interactive sun + sky.
Unity Reflect review (https://unity.com/products/unity-reflect-reviewin (accessed on 12 September 2022))	Design review	Walkthroughs in VR and AR, annotation, and filter BIM data to effectively communicate design intent to stakeholders	Revit, SketchUp, BIM 360, Navisworks, Rhino	AR/VR, desktop, or mobile	Sun studies, overlay models in 1:1 AR at scale (marker-based or tabletop)
Resolve (https://www.resolvebim.com/)	Design review, facility management	Annotation (speech to text), multi-user meeting	BIM	VR or desktop	Annotate by measuring and sketching, issue tracking integration with BIM 360, inspect BIM properties
Trezi (https://trezi.com/)	Manufacturing	Multi-user meeting	BIM	VR or desktop	Model modification, design review

4.3. Applications of ImVE in Supporting Design Collaboration

ImVE provides opportunities for designers and other stakeholders to work collaboratively in a shared virtual environment. In the AEC industry, ImVE combining with standard design and collaboration platforms specific to the sector such as BIM have significantly enhanced communication and collaboration across design, construction, operation, and maintenance stages [44]. This study reviewed recent research on design collaboration in ImVE. As shown in Figure 6, 15 out of 29 articles focus on the design stage, 10 articles discuss collaboration across multiple stages including design of a building project, and 4 articles focus on design review. Collaboration research focussing on the design stage in ImVE covers a wide range of topics including the development of various VR environments or frameworks to support design collaboration [40,45], cognitive exploration on how designers collaborate in VR environments [49,50], and in an education context, virtual design studio through VR collaboration [46,52]. Articles focussing on design review discuss various directions including developing a BIM-based multi-user VR system for architectural design review and collaboration [41], investigating how AR affects architectural design reviews based on the user's perspectives [25], and integrating VR into collaboration workflow [51]. For articles focussing on collaboration in ImVE across multiple stages or in the general AEC context, some present relevant ImVE applications [44,47] and other advance technical developments of ImVE, especially by combining with BIM, to facilitate collaboration [30,35]. The review of the 29 articles has identified 3 main application areas of ImVE in supporting design collaboration: (1) design review with end-users and other stakeholders in ImVE; (2) visual data in BIM-based ImVE for supporting design collaboration; and (3) cognition and education in collaborative ImVE. These three application areas will be further discussed as follows.

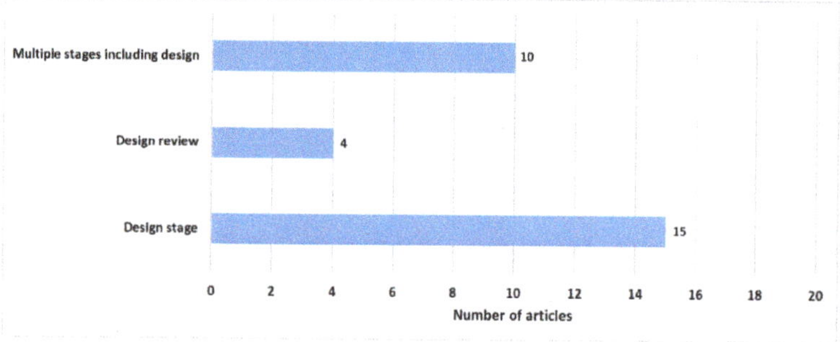

Figure 6. Number of articles focusing on different project stages.

4.3.1. Design Review with End-Users and Other Stakeholders in ImVE

In current practice, architects have embraced ImVE in the participatory process to engage with end-users and to obtain their feedback. For instance, Pour Rahimian et al. [33] consider VR an effective tool for end-user engagement due to the advanced visual communication of building design, and a dynamic feedback initiator. Whether by simulating the realistic scale of the building, or functional aspects of the design, or other end-users experiences such as safety, ImVE has enabled telepresence for different stakeholders including clients, end-users and authorities [56]. ImVE can be intimidating initially for new users especially outside the professional project team, as it involves various hardware and software setup. However, the benefits of ImVR are enormous, as it can leverage the showcase of virtual models to clients in a superior way [66], enabling them with a more thorough understanding of the design and better align the business- client requirements. As a result, design professionals including architects are increasingly using VR for design showcase [67].

ImVE delivers spatial information and at the same time allows collaborative communication. In the AEC disciplines, the goal of including clients during the design process beyond the final showcase—clients who often have limited spatial comprehension and limited specialist knowledge—is to improve both design quality and client satisfaction [33,68], and has gained increasing popularity in the sector. Similarly, Heydarian et al. [35] also find that ImVE is an effective tool in the design phase of a building project in terms of acquiring performance feedback from end users. These benefits are evident in all types of ImVE. For example, VR has been considered to engage clients to the project in an inclusive way [35], and could be more effective than the traditional approach for design review [43]. Lee et al. [25] suggest that AR is effective in reviewing the visual elements of a building and leads to a higher degree of satisfaction in terms of user experience. AR could be an effective platform to investigate evaluate the appearances of a virtual model and to examine the user experience in the design review process [25]. Sheng et al. [55] present an application of an interactive AR application to architectural daylight modelling in which designers and end users can review and have their input on the daylighting design. The involvement of end users leads to higher performing design and end user satisfaction. MR has shown potential in increasing the spatial understanding of end users [69], to allow the interpretation and translation of virtual content through wearable devices [70]. Enabling an end user to experience a realistic and immersive visualisation of a building project, has a different effect on an end user's cognition, and can reveal new possibilities. Unlike AEC professionals, end users are unable to be effectively related to two-dimensional drafting documents. Standard BIM models, despite having multiple advantages over drawings, still lack in leveraging the experiences of the end users to critically understand the design. It is therefore necessary to utilise BIM combing with immersive technologies in participatory design to adequately support end users' decision-making processes [71]. Combining BIM and immersive technologies can also bring many other benefits. For example, the ability to allow collaborators to interact with BIM models without being physically together at different stages of a building project, can be achieved via VR. Zaker and Coloma [51] further suggest that VR collaboration in the design review is beneficial for a wide range of disciplines involved.

Early design review and visualisation, optimisation of building performance analysis, and building maintenance and operations, are possible areas where MR integration can improve the delivery of the building project [24]. The rapid development in MR technologies has significant potential for the AEC industry by combining digital- and real-world information, which is beneficial for design collaboration [69]. MR collaboration can be face-to-face or remote. Face-to-face MR collaboration is achieved by using a shared coordinate system, where collaborators interact with the same set of virtual data information in person [72]. In remote MR collaboration, remote data sharing and remote collaboration are both accomplished utilising the MR platform [73]. Currently, MR collaboration has not been widely adopted within the AEC industry, and there is especially a lack of applications for remote MR collaboration, despite that such applications are likely to benefit AEC professionals to communicate and interact across distributed locations, due to its strong capability of integrating both digital- and physical-world information [24].

4.3.2. Data Visualisation in BIM-Based ImVE for Supporting Design Collaboration

BIM combining with ImVE enable collaboration among multiple parties via both two dimensional and three-dimensional models [74]. Data visualisation in BIM-based ImVE can effectively support and improve design collaboration; in particular, the data visualisation in immersive BIM-based VR environments during the design process can potentially facilitate a deeper understanding among collaborators [75], which can make real-time visualisation and communication more accurate during design [76,77]. Additionally, studies suggest that the barriers in BIM-VR data exchange need further exploration, which may otherwise limit applications of VR within the AEC industry [44]. For example, Du et al. [34] introduced a real-time synchronisation system for BIM-based VR, which is cloud-based, and updates changes to the BIM model and VR model simultaneously to facilitate effective data exchange.

BIM-based AR enriches the real world with digital data by providing more dynamic outcomes with real-time visualisations through a seamless management process [78], hence BIM-based AR environments have potential to benefit design collaboration, by supporting richer interactions and media representations [79]. Visualising and sharing of information by multiple users through AR can augment the real environment by embedding relevant digital data for supporting design decision-making. Overall, by combining the real and the digital, AR can improve designers' information processing and communication [80] and offer subsequent benefits in terms of project visualisation, monitoring, and control [81]. One example of a BIM-based AR system was developed by Garbett et al. [30], which allows distributed teams to view, interact and collaborate on both 3D and 2D BIM data via AR. The other immersive technology—MR—can also be used to enhance BIM model information and its visualisation, feeding mixed-reality representations back to the original BIM model during the design process [82]. BIM-MR integration can potentially further enhance the visualisation of the BIM model, along with supporting more context-aware interactions between the designers and between the designers and the design environment [33]. However, one of the main limitations in BIM-MR integration is the amount of data and details of a BIM model, which is at times not required and not appropriate for an effective corresponding MR application. Therefore, timely data-keeping, and dropping of non-required BIM data, should be addressed in future studies to improve the performance of BIM-based MR applications for design collaboration. Future data storage technologies supported by cloud computing will also help addressing this issue, and can expand the potential utilisation of BIM-based MR applications, to include large-scale collaborative projects, which tend to produce significant amounts of data. Data storage and transfer, together with other issues such as the accuracy of spatial registration, user interface, and multi-user collaboration have been suggested to be key areas for future MR research [24].

4.3.3. Design Cognition and Education in Collaborative ImVE

The exploration of design perception, design physiology, and design cognition and neurocognition in a collaborative environment can generate the knowledge needed in order to support improved design patterns, creativity, and reasoning among multiple users to support their designing and collaboration [83]. Research shows that ImVE aids designers' cognitive processes such as those related to working memory, design data search and access, spatial cognition, and attention allocation. Particularly, ImVE has a positive effect on users' perception and memory [75,77]. Hermund et al. [84] determined that immersive VR representations during the design process were less demanding on designers' cognitive load than traditional 3D visualisation on desktop computers. Design is not a linear process, since designers typically formulate a design problem and develop a design solution in parallel [85]. Studies indicate that ImVE can potentially lead to higher performance of designers particularly in problem finding [86], which can have positive effect in both the problem and solution spaces. In another study focusing on collaboration, Hong et al. [53] developed a multi-user virtual environment and track users' problem-solving measures in a shared design setting. Their results suggest positive effective of ImVE especially in increasing inspiration for new approaches to problem-solving among design collaborators.

Advances in design computing and cognition research have provided a number of methodological approaches to studying both human–human and human–agent communications and interactions in ImVE [38]. For example, Roupe et al. [40] developed a virtual collaborative design environment that enhances the communication, collaboration, understanding, and knowledge sharing of participants. They conducted two collaborative design workshops to explore the collaboration behaviours of designers in a virtual collaborative environment, utilising direct documented observations and semi-structured interviews with participants, to explore their experiences and views about their collaborative design processes. Leon et al. [1] developed a pre-BIM conceptual design stage protocol for cognitive design studies in collaborative virtual environments, and its coding scheme included team formation, introduction of the brief, discussion of project requirements, solution synthesis and brainstorming, solution evaluation, consensus, and final solution. For design collaboration in ImVE, communication tools have mainly consisted of a few forms of text-based tools, voice chat tools, visual sharing tools and avatars, which can potentially improve the communication efficiency of design collaboration in the AEC industry [47]. In one example, based on participatory observation in a virtual collaboration environment it has been found [36] that the use of avatar movement is effective for communicating non-verbal information that enhances the effectiveness of collaboration. Other studies focusing on various communication approaches during collaboration in ImVE, include Kim et al. [50] which applied an overlay design methodology in studying virtual environments based on protocol studies of participants' collaborative design process, and found that method can effectively assist with communication, doubling the collaboration segments among team members and reducing the overall time needed to complete their design compared to use of traditional design method. Another example is Wang et al. [54] development of a computer-mediated remote collaborative design system TeleA. Via measurement of participants' physiological movements such as gestures, facial expressions, fine motor movements and bodily postures and distances, as well as observation of their emotional states, that study suggested that developed system had a positive effect on designers' communication and collaboration, especially for distributed cognition and mutual awareness.

Furthermore, previous studies have also explored design collaboration in ImVE in the context of architectural education focusing on teaching and learning. For instance, Rauf et al. [46] discussed the application of VR in architectural education to support collaborative learning, and identified the effect of VR applications from various collaborative levels including student-instructor, student-client and student-industry. Similarly, Diao and Shih [42] reviewed AR applications in the broader AEC education and suggested that collaborative learning enabled by AR can promote the connection between general pedagogy and domain-specific learning. George et al. [52] studied students' design processes in ImVE and found that VR is effective in enhancing students' understanding of design decisions by assisting them with rapid design prototyping.

Among various design studies, especially cognitive design studies on collaboration techniques and frameworks, the dominant majority have the overarching aim of streamlining the collaboration process to produce more optimal design output. ImVE has the potential to assist in producing creative design solutions during the collaboration. Particularly, design solutions appeared to be more creative in terms of both novelty and appropriateness in multi-user virtual environments than in traditional sketching environments due to the explicit communication cues for sharing the collaborative procedure and spatial information provided by the former [48]. Similarly, Chowdhury and Schnabel [49] also found that the more spontaneous exchange of visual information in a shared virtual environment is beneficial for producing optimal design solutions.

5. Conclusions and Future Research

This study has reviewed recent research on design collaboration in ImVE from the past ten years. The results demonstrate an increasing research focus on this area through the past decade. From a technical development perspective, immersive technologies are being rapidly developed and applied within the AEC industry to facilitate collaboration at various design and construction stages. Recent research on design collaboration in ImVE has focused on design review with end users and other stakeholders in ImVE, visual data/information in BIM-based ImVE in supporting design collaboration, and cognitive studies of design collaboration in ImVE. To date, current research provides us with some early understandings of design collaboration in ImVE as applied in the above-mentioned ways. However, academic research about ImVE support for design collaboration, is not keeping pace with the accelerating rate of its technological advancement. Generally there is a lack of critical understanding about design collaboration in ImVE, thus the following future research is needed.

First, additional studies regarding ImVE support for design collaboration, especially at the early design stage, are needed. Commonly used design collaboration platforms in the AEC industry such as BIM tend to be tailored to the later design and management stages rather than providing support for the conceptualising, interacting, and sharing of design concepts at the early stage [7,87]. Most virtual collaboration tools are focused on visualisation [31,88,89]. A few recent tools that support design ideation (such as Wild, Mindesk, and Arkio) provide limited design functions of sketching or parametric modelling, etc., and they have not yet been widely applied and tested throughout the design industry. Effective collaboration during the initial concept design stage will lead to fewer problems during the later stages such as developed design, detailed design, and design documentation [1]. In addition, effective collaboration is often supported by digital and network tools, for example, through social networking services for enhancing remote design collaboration [4]. There is currently a lack of critical understanding about the effectiveness of these virtual technologies and various emerging add-on tools for meeting the needs of design collaboration, especially at the early design stage [90].

Secondly, further cognitive research on design collaboration in ImVE adopting more innovative and comprehensive methodologies is needed. Current research has explored the design collaboration process in ImVE from various cognitive perspectives, including the creativity of design solutions (i.e., novelty and appropriateness) in multi-user virtual environments [48], design collaborators' problem-solving measures in multi-user virtual environments [53], and various explorations from design teaching and learning perspectives [42,46]. More comprehensive understandings about how architects collaborate in ImVE from a cognitive perspective, compared with face-to-face or more traditional computer-based collaboration, will allow us to identify the barriers of adopting ImVE for design collaboration, and designers' needs for future tools to better support design collaboration. Very recently, more innovative and comprehensive cognitive studies such as by adopting or combining design and neuroscience perspectives could possibly lead to a deeper understanding of designers' collaboration, through measurements of designers' biometric responses such as eye-tracking and electroencephalogram (EEG), to complement the existing knowledge about the impact of ImVE on designers' collaboration process.

Finally, the continuing development of new ImVE technologies with more advanced design features (such as intuitive parametric and generative design functions) is needed to better assist with designing during collaboration. Current limitations of collaborative ImVE in terms of design support include insufficient conceptual sketching and overly simplified modelling functions. The parametric design functions provided by some tools rely mostly upon external add-ons rather than within the collaboration environment itself. There is a clear need for an algorithmic design feature in collaborative ImVE [89]. Furthermore, there is also a lack of advanced design features such as those for analysing building performance, cost, and land use, which are essential for making more informed design decisions during architectural design collaboration.

Author Contributions: R.Y., conceptualization, methodology, formal analysis, original draft preparation, supervision; N.G., conceptualization, writing-review and editing, supervision; G.L., conceptualization, writing-review and editing, supervision; A.K., conceptualization, methodology, formal analysis. All authors have read and agreed to the published version of the manuscript.

Funding: This research received no external funding.

Data Availability Statement: Not applicable.

Conflicts of Interest: The authors declare no conflict of interest.

References

1. Leon, M.; Laing, R.; Malins, J.; Salman, H. Making collaboration work: Application of A conceptual design stages protocol for pre-BIM stages. *WIT Trans. Built Environ.* **2015**, *149*, 205–216.
2. Feast, L. Professional perspectives on collaborative design work. *CoDesign* **2012**, *8*, 215–230. [CrossRef]
3. Tan, L. Collaborative Cultures of Architecture Teams: Team Learning and Reflective Practice. *Des. J.* **2021**, *24*, 489–498. [CrossRef]

4. Kim, M.J.; Hwang, Y.S.; Hwang, H.S. Utilising social networking services as a collective medium to support design communication in team collaboration TT—SNS as a collective medium. *ArchNet-IJAR* **2020**, *14*, 409–421. [CrossRef]
5. Combrinck, C.; Porter, C.J. Co-design in the architectural process. *Archnet-IJAR* **2021**, *15*, 738–751. [CrossRef]
6. Froese, T.M. The impact of emerging information technology on project management for construction. *Autom. Constr.* **2010**, *19*, 531–538. [CrossRef]
7. Garber, R. *BIM Design: Realising the Creative Potential of Building Information Modelling*; John Wiley & Sons, Inc.: Hoboken, NJ, USA, 2014.
8. Succar, B. Building information modelling framework: A research and delivery foundation for industry stakeholders. *Autom. Constr.* **2009**, *18*, 357–375. [CrossRef]
9. McCall, R.; Johnson, E. Using argumentative agents to catalyze and support collaboration in design. *Autom. Constr.* **1997**, *6*, 299–309. [CrossRef]
10. Kalay, Y.E.; Khemlani, L.; Choi, J.W. An integrated model to support distributed collaborative design of buildings. *Autom. Constr.* **1998**, *7*, 177–188. [CrossRef]
11. Gross, M.D.; Yi-Luen Do, E.; McCall, R.J.; Citrin, W.V.; Hamill, P.; Warmack, A.; Kuczun, K.S. Collaboration and coordination in architectural design: Approaches to computer mediated team work. *Autom. Constr.* **1998**, *7*, 465–473. [CrossRef]
12. Gu, N.; Kim, M.J.; Maher, M.L. Technological advancements in synchronous collaboration: The effect of 3D virtual worlds and tangible user interfaces on architectural design. *Autom. Constr.* **2011**, *20*, 270–278. [CrossRef]
13. Pour Rahimian, F.; Seyedzadeh, S.; Oliver, S.; Rodriguez, S.; Dawood, N. On-demand monitoring of construction projects through a game-like hybrid application of BIM and machine learning. *Autom. Constr.* **2020**, *110*, 103012. [CrossRef]
14. Paes, D.; Arantes, E.; Irizarry, J. Immersive environment for improving the understanding of architectural 3D models: Comparing user spatial perception between immersive and traditional virtual reality systems. *Autom. Constr.* **2017**, *84*, 292–303. [CrossRef]
15. Mostafavi, A. Architecture, biometrics, and virtual environments triangulation: A research review. *Archit. Sci. Rev.* **2021**, 1–18. [CrossRef]
16. Furht, B. Immersive Virtual Reality. In *Encyclopedia of Multimedia*; Springer: Boston, MA, USA, 2008; pp. 345–346.
17. Khan, A.; Sepasgozar, S.; Liu, T.; Yu, R. Integration of BIM and Immersive Technologies for AEC: A Scientometric-SWOT Analysis and Critical Content Review. *Buildings* **2021**, *11*, 126. [CrossRef]
18. Diemer, J.; Alpers, G.W.; Peperkorn, H.M.; Shiban, Y.; Mühlberger, A. The impact of perception and presence on emotional reactions: A review of research in virtual reality. *Front. Psychol.* **2015**, *6*, 1–9. [CrossRef] [PubMed]
19. Serrano, B.; Baños, R.M.; Botella, C. Virtual reality and stimulation of touch and smell for inducing relaxation: A randomized controlled trial. *Comput. Hum. Behav.* **2016**, *55*, 1–8. [CrossRef]
20. Behzadan, A.H.; Kamat, V.R. Visualization of construction graphics in outdoor augmented reality. In Proceedings of the Winter Simulation Conference, Orlando, FL, USA, 4 December 2005.
21. Milgram, P.; Kishino, F. A Taxonomy of Mixed Reality Visual Displays. *EICE Trans. Inf. Syst.* **1994**, *E77-D*, 1321–1329.
22. Linnenluecke, K.M.; Marrone, M.; Singh, A.K. Conducting systematic literature reviews and bibliometric analyses. *Aust. J. Manag.* **2020**, *45*, 175–194. [CrossRef]
23. Pickering, C.; Grignon, J.; Steven, R.; Guitart, D.; Byrne, J. Publishing not perishing: How research students transition from novice to knowledgeable using systematic quantitative literature reviews. *Stud. High. Educ.* **2015**, *40*, 1756–1769. [CrossRef]
24. Cheng, J.C.P.C.P.; Chen, K.; Chen, W. State-of-the-Art Review on Mixed Reality Applications in the AECO Industry. *J. Constr. Eng. Manag.* **2020**, *146*, 03119009. [CrossRef]
25. Lee, J.; Seo, J.; Abbas, A.; Choi, M. End-Users' Augmented Reality Utilization for Architectural Design Review. *Appl. Sci.* **2020**, *10*, 5363. [CrossRef]
26. Lachal, J.; Revah-Levy, A.; Orri, M.; Moro, M.R. Metasynthesis: An Original Method to Synthesize Qualitative Literature in Psychiatry. *Front. Psychiatry* **2017**, *8*, 269. [CrossRef] [PubMed]
27. Hedges, L.V. Meta-Analysis. *J. Educ. Stat.* **1992**, *17*, 279–296. [CrossRef]
28. Stemler, S. An overview of content analysis. *Pract. Assess. Res. Eval.* **2000**, *7*, 17. [CrossRef]
29. Matthews, B.; See, Z.S.; Day, J. Crisis and extended realities: Remote presence in the time of COVID-19. *Media Int. Aust.* **2020**, *178*, 198–209. [CrossRef]
30. Garbett, J.; Hartley, T.; Heesom, D. A multi-user collaborative BIM-AR system to support design and construction. *Autom. Constr.* **2021**, *122*, 103487. [CrossRef]
31. Alizadehsalehi, S.; Hadavi, A.; Huang, J.C. From BIM to extended reality in AEC industry. *Autom. Constr.* **2020**, *116*, 103254. [CrossRef]
32. Zhang, Y.; Liu, H.; Kang, S.; Al-hussein, M. Automation in Construction Virtual reality applications for the built environment: Research trends and opportunities. *Autom. Constr.* **2020**, *118*, 103311. [CrossRef]
33. Rahimian, F.P.; Chavdarova, V.; Oliver, S.; Chamo, F.; Amobi, L.P. OpenBIM-Tango integrated virtual showroom for offsite manufactured production of self-build housing. *Autom. Constr.* **2019**, *102*, 1–16. [CrossRef]
34. Du, J.; Zou, Z.; Shi, Y.; Zhao, D. Zero latency: Real-time synchronization of BIM data in virtual reality for collaborative decision-making. *Autom. Constr* **2018**, *85*, 51–64. [CrossRef]

35. Heydarian, A.; Carneiro, J.P.; Gerber, D.; Becerik-Gerber, B.; Hayes, T.; Wood, W. Immersive virtual environments versus physical built environments: A benchmarking study for building design and user-built environment explorations. *Autom. Constr.* **2015**, *54*, 116–126. [CrossRef]
36. Anderson, A.; Dossick, C.S.; Iorio, J.; Taylor, J.E. The impact of avatars, social norms and copresence on the collaboration effectiveness of AEC virtual teams. *J. Inf. Technol. Constr.* **2017**, *22*, 287–304.
37. Kim, M.J.; Wang, X.; Love, P.E.D.; Li, H.; Kang, S.C. Virtual reality for the built environment: A critical review of recent advances. *J. Inf. Technol. Constr.* **2013**, *18*, 279–305.
38. Merrick, K.E.; Gu, N.; Wang, X. Case studies using multiuser virtual worlds as an innovative platform for collaborative design. *Electron. J. Inf. Technol. Constr.* **2011**, *16*, 165–188.
39. Cerovšek, T.; Zupančič, T.; Kilar, V. Framework for model-based competency management for design in physical and virtual worlds. *Electron. J. Inf. Technol. Constr.* **2010**, *15*, 1–22.
40. Roupe, M.; Johansson, M.; Maftei, L.; Lundstedt, R.; Viklund-Tallgren, M. Virtual Collaborative Design Environment: Supporting Seamless Integration of Multitouch Table and Immersive VR. *J. Constr. Eng. Mangement* **2020**, *146*, 1–31. [CrossRef]
41. Keung, C.C.W.; Kim, J.I.; Ong, Q.M. Developing a bim-based muvr treadmill system for architectural design review and collaboration. *Appl. Sci.* **2021**, *11*, 6881. [CrossRef]
42. Diao, P.H.; Shih, N.J. Trends and research issues of augmented reality studies in architectural and civil engineering education-A review of academic journal publications. *Appl. Sci.* **2019**, *9*, 1840. [CrossRef]
43. Tea, S.; Panuwatwanich, K.; Ruthankoon, R.; Kaewmoracharoen, M. Multiuser immersive virtual reality application for real-time remote collaboration to enhance design review process in the social distancing era. *J. Eng. Des. Technol.* **2022**, *20*, 281–298. [CrossRef]
44. Safikhani, S.; Keller, S.; Schweiger, G.; Pirker, J. Immersive virtual reality for extending the potential of building information modeling in architecture, engineering, and construction sector: Systematic review. *Int. J. Digit. Earth* **2022**, *15*, 503–526. [CrossRef]
45. Kim, J.I.; Li, S.; Chen, X.; Keung, C.; Suh, M.; Kim, T.W. Evaluation framework for BIM-based VR applications in design phase. *J. Comput. Des. Eng.* **2021**, *8*, 910–922. [CrossRef]
46. Rauf, H.L.; Shareef, S.S.; Othman, N.N. Innovation in Architecture Education: Collaborative Learning Method Through Virtual Reality. *J. High. Educ. Theory Pract.* **2021**, *21*, 33–40. [CrossRef]
47. Wen, J.; Gheisari, M. Using virtual reality to facilitate communication in the AEC domain: A systematic review. *Constr. Innov.* **2020**, *20*, 509–542. [CrossRef]
48. Hong, S.W.; El Antably, A.; Kalay, Y.E. Architectural design creativity in Multi-User Virtual Environment: A comparative analysis between remote collaboration media. *Environ. Plan. B Urban Anal. City Sci.* **2019**, *46*, 826–844. [CrossRef]
49. Chowdhury, S.; Schnabel, M.A. Laypeople's Collaborative Immersive Virtual Reality Design Discourse in Neighborhood Design. *Front. Robot. AI* **2019**, *6*, 1–10. [CrossRef]
50. Kim, I.; Hong, S.; Lee, J.H.; Bazin, J.C. Overlay Design Methodology for virtual environment design within digital games. *Adv. Eng. Inform.* **2018**, *38*, 458–473. [CrossRef]
51. Zaker, R.; Coloma, E. Virtual reality-integrated workflow in BIM-enabled projects collaboration and design review: A case study. *Vis. Eng.* **2018**, *6*, 1–15. [CrossRef]
52. George, B.H.; Sleipness, O.R.; Quebbeman, A. Using virtual reality as a design input: Impacts on collaboration in a university design studio setting. *J. Digit. Landsc. Archit.* **2017**, *2017*, 252–259. [CrossRef]
53. Hong, S.W.; Jeong, Y.; Kalay, Y.E.; Jung, S.; Lee, J. Enablers and barriers of the multi-user virtual environment for exploratory creativity in architectural design collaboration. *CoDesign* **2016**, *12*, 151–170. [CrossRef]
54. Wang, X.; Love, P.E.D.; Kim, M.J.; Wang, W. Mutual awareness in collaborative design: An Augmented Reality integrated telepresence system. *Comput. Ind.* **2014**, *65*, 314–324. [CrossRef]
55. Sheng, Y.; Yapo, T.C.; Young, C.; Cutler, B. A Spatially Augmented Reality Sketching Interface for Architectural Daylighting Design. *IEEE Trans. Vis. Comput. Graph.* **2011**, *17*, 38–50. [CrossRef] [PubMed]
56. Yu, R.; Gu, N.; Ostwald, M.J. *Computational Design: Technology, Cognition and Environments*, 1st ed.; CRC Press, Taylor & Francis Group: Boca Raton, FL, USA, 2021.
57. Yu, R.; Ostwald, M.J. Comparing Architects' Perceptions of the Usefulness of Digital Design Environments with their Aspirations for Sustainable Design in Australia. *J. Sustain. Archit. Civ. Eng.* **2021**, *2021*, 5–20. [CrossRef]
58. Azuma, R.T. A Survey of Augmented Reality. *Presence Teleoperators Virtual Environ.* **1997**, *6*, 355–385. [CrossRef]
59. Billinghurst, M.; Clark, A.; Lee, G. A survey of augmented reality. *Found. Trends Hum. -Comput. Interact.* **2014**, *8*, 73–272. [CrossRef]
60. Ryan, M.-L. *Narrative as Virtual Reality 2: Revisiting Immersion and Interactivity in Literature and Electronic Media*; Johns Hopkins University Press: Baltimore, MD, USA, 2015.
61. Walsh, K.R.; Pawlowski, S.D. Virtual reality: A technology in need of IS research. *Commun. Assoc. Inf. Syst.* **2002**, *8*, 20. [CrossRef]
62. Schubert, T.; Friedmann, F.; Regenbrecht, H. The experience of presence: Factor analytic insights. *Presence Teleoperators Virtual Environ.* **2001**, *10*, 266–281. [CrossRef]
63. Slater, M.; Steed, A. A virtual presence counter. *Presence Teleoperators Virtual Environ.* **2000**, *9*, 413–434. [CrossRef]
64. Sundar, S.S.; Xu, Q.; Bellur, S. Designing interactivity in media interfaces: A communications perspective. *Conf. Hum. Factors Comput. Syst. Proc.* **2010**, *4*, 2247–2256. [CrossRef]

65. Steuer, J. Defining virtual reality: Dimensions determining telepresence. *J. Commun.* **1992**, *42*, 73–93. [CrossRef]
66. Sidani, A.; Dinis, F.M.; Sanhudo, L.; Duarte, J.; Santos Baptista, J.; Poças Martins, J.; Soeiro, A. Recent Tools and Techniques of BIM-Based Virtual Reality: A Systematic Review. *Arch. Comput. Methods Eng.* **2021**, *28*, 449–462. [CrossRef]
67. Hilfert, T.; König, M. Low-cost virtual reality environment for engineering and construction. *Vis. Eng.* **2016**, *4*, 1–18. [CrossRef]
68. Asgari, Z.; Rahimian, F.P. Advanced Virtual Reality Applications and Intelligent Agents for Construction Process Optimisation and Defect Prevention. *Procedia Eng.* **2017**, *196*, 1130–1137. [CrossRef]
69. Wang, X.; Dunston, P.S. User perspectives on mixed reality tabletop visualization for face-to-face collaborative design review. *Autom. Constr* **2008**, *17*, 399–412. [CrossRef]
70. Piroozfar, P.; Essa, A.; NewSchool, E.R.P.F. The application of Augmented Reality and Virtual Reality in the construction industry using wearable devices. In Proceedings of the 9th International Conference on Construction in the 21st Century, Dubai, United Arab Emirates, 5–7 March 2017.
71. Damen, T.; MacDonald, M.; Hartmann, T.; Giulio, R.D.; Bonsma, P.; Luig, K.; Sebastian, R.; Soetanto, D. Bim based collaborative design technology for collective self-organised housing. In Proceedings of the 40th IAHS World Congress on Housing: Sustainable Housing Construction, Funchal, Portugal, 16–19 December 2014.
72. Billinghurst, M.; Kato, H.; Kiyokawa, K.; Belcher, D.; Poupyrev, I. Experiments with Face-To-Face Collaborative AR Interfaces. *Virtual Real.* **2002**, *6*, 107–121. [CrossRef]
73. Cidota, M.; Lukosch, S.; Datcu, D.; Lukosch, H. Workspace Awareness in Collaborative AR using HMDs: A User Study Comparing Audio and Visual Notifications. In Proceedings of the 7th Augmented Human International Conference 2016, Geneva, Switzerland, 25–27 February 2016; p. 3.
74. Dunston, P.S.; Wang, X.; Lecturer, S.; Program, P. A hierarchical taxonomy of aec operations for mixed reality applications. *J. Inf. Technol. Constr.* **2011**, *16*, 433–444.
75. Roupé, M.; Johansson, M.; Tallgren, M.V.; Jörnebrant, F.; Tomsa, P.A. Immersive visualization of Building Information Models. *In Living Systems and Micro-Utopias: Towards Continuous Designing, Proceedings of the 21st International Conference of the Association for Computer-Aided Architectural Design Research in Asia (CAADRIA 2016), New Delhi, India, 28–30 April 2016*; pp. 673–682.
76. Johansson, M.; Roupé, M.; Bosch-Sijtsema, P. Real-time visualization of building information models (BIM). *Autom. Constr.* **2015**, *54*, 69–82. [CrossRef]
77. Calderon-Hernandez, C.; Paes, D.; Irizarry, J.; Brioso, X. Comparing Virtual Reality and 2-Dimensional Drawings for the Visualization of a Construction Project. In Proceedings of the ASCE International Conference on Computing in Civil Engineering 2019, Atlanta, GA, USA, 17–19 June 2019; pp. 17–24.
78. Hou, L.; Wang, X. Experimental framework for evaluating cognitive workload of using AR system in general assembly task. In Proceedings of the 28th International Symposium on Automation and Robotics in Construction (ISARC 2011), Seoul, Korea, 29 June–2 July 2011; pp. 625–630. [CrossRef]
79. Wang, X.; Love, P.E.D.; Kim, M.J.; Park, C.S.; Sing, C.P.; Hou, L. A conceptual framework for integrating building information modeling with augmented reality. *Autom. Constr.* **2013**, *34*, 37–44. [CrossRef]
80. Shin, D.H.; Dunston, P.S. Technology development needs for advancing Augmented Reality-based inspection. *Autom. Constr.* **2010**, *19*, 169–182. [CrossRef]
81. Chi, H.L.; Kang, S.C.; Wang, X. Research trends and opportunities of augmented reality applications in architecture, engineering, and construction. *Autom. Constr.* **2013**, *33*, 116–122. [CrossRef]
82. Lee, J.; Kim, J.; Ahn, J.; Woo, W. Context-aware risk management for architectural heritage using historic building information modeling and virtual reality. *J. Cult. Herit.* **2019**, *38*, 242–252. [CrossRef]
83. Shealy, T.; Gero, J.; Hu, M.; Milovanovic, J. Concept generation techniques change patterns of brain activation during engineering design. *Des. Sci.* **2020**, *6*, e31. [CrossRef]
84. Hermund, A.; Klint, L.; Bundgaard, T. BIM with VR for Architectural Simulations Building Information Models in Virtual Reality as an Architectural and Urban Designtool. In Proceedings of the ACE 2018, Singapore, 14–15 May 2018.
85. Dorst, K.; Cross, N. Creativity in the design process: Co-evolution of problem-solution. *Des. Stud.* **2001**, *22*, 425–437. [CrossRef]
86. Wu, T.-H.; Wu, F.; Kang, S.-C.; Chi, H.-L. Comparison of Virtual Communication Environment for Remote BIM Model Review Collaboration. In Proceedings of the 36th International Symposium on Automation and Robotics in Construction (ISARC), Banff, AB, Canada, 21–24 May 2019; pp. 1149–1154.
87. Okeil, A. Hybrid design environments: Immersive and non-immersive architectural design. *ITcon* **2010**, *15*, 202–216.
88. Davila Delgado, J.M.; Oyedele, L.; Demian, P.; Beach, T. A research agenda for augmented and virtual reality in architecture, engineering and construction. *Adv. Eng. Inform.* **2020**, *45*, 101122. [CrossRef]
89. Castelo-Branco, R.; Leitão, A. Algorithmic Design in Virtual Reality. *Architecture* **2022**, *2*, 31–52. [CrossRef]
90. Idi, D.B.; Khaidzir, K.A.M. Critical perspective of design collaboration: A review. *Front. Archit. Res.* **2018**, *7*, 544–560. [CrossRef]

designs

MDPI

Article

Influence of Microclimate on Older Peoples' Outdoor Thermal Comfort and Health during Autumn in Two European Cities

María Teresa Baquero Larriva * and Ester Higueras García

Escuela Técnica Superior de Arquitectura, Universidad Politécnica de Madrid, Av. Juan de Herrera, 2, 28040 Madrid, Spain
* Correspondence: m.t.baquero.larriva@upm.es

Abstract: Public spaces and green areas have been proven to influence people's mental and physical health, thermal comfort being one of the main indicators. The growing trend of an ageing population globally led this research to analyse the outdoor thermal comfort of older adults in public spaces from two cities in Europe: Madrid in Spain and Newcastle upon Tyne in the United Kingdom during autumn. A mixed methodology through environmental measurements and surveys was performed in situ. In addition, the UTCI (Universal Thermal Climate Index) and PET (Physiological Equivalent Temperature) outdoor thermal comfort indices were applied. The results highlighted the risk of thermal stress and the vulnerability of this group of the population to the effects of climate on their health. Although most older people had 'neutral' thermal sensation, 86.3% of them would be at risk of cold stress in Newcastle, whilst in Madrid 31.5% would be at risk of cold stress and 35.7% of heat stress. Those results could be a starting point for the design of more comfortable and healthy public spaces that improve the quality of life of all citizens within the guidelines of active ageing and healthy cities.

Keywords: outdoor thermal comfort; older people; thermal stress; health risk; urban public spaces

Citation: Baquero Larriva, M.T.; Higueras García, E. Influence of Microclimate on Older Peoples' Outdoor Thermal Comfort and Health during Autumn in Two European Cities. *Designs* **2023**, *7*, 27. https://doi.org/10.3390/designs7010027

Academic Editor: Tomohiko Sakao

Received: 23 December 2022
Revised: 13 January 2023
Accepted: 18 January 2023
Published: 1 February 2023

1. Introduction

The Intergovernmental Panel on Climate Change (IPCC) forecasts an increase in the intensity, number, and duration of extreme weather events, so it is expected that the extreme temperature will increase [1]. This affects normal temperatures in autumn and spring; in some cases, the change in season is imperceptible [2]. Parallel to this, according to the World Health Organization (WHO), the ageing population in 2050 will triple. In addition, more than half of the world's population lives in urban habitats. By 2030, six out of ten people in the world will live in a city, of which 900 million would be older adults [3].

One of the characteristics of ageing is its diversity among people over 65 years old. Thus, health depends on various factors such as genetic predisposition, lifestyle and quality of life, and at the same time is influenced by the physical and social environment. However, statistics indicate that a large percentage of older people tend to present multimorbidity and are also more sensitive to environmental effects on health (noise pollution, heat waves, cold waves, etc.) [4]. Some physiological changes that occur with ageing affect the thermal sensitivity, perception, adaptation and preferences of older adults [5]. For instance, muscle strength, work capacity, activity level, metabolic rate, vascular reactivation, thermoregulation capacity, sweating and hydration levels decrease and affect their ability to detect and respond to temperature changes, making them vulnerable to thermal extremes [6].

Public spaces are determinants for people's health [7]. Thermal comfort depends on several factors including: geographical (latitude, altitude); microclimatic (temperature, humidity, wind, radiation); personal (activity, clothing, age, gender, state of health, among others); psychological factors (aptitude, experience, expectation, memory, etc.) and built environment factors (vegetation, shade, surface material, SVF, microclimate, etc.) [8]. Therefore, a good urban design based on bioclimatic criteria influences the quality of these spaces and the well-being of their users and the environment [9].

Most studies on thermal comfort are based on the physical and physiological characteristics of the 'type person' (a man, 35 years, 1.70 m in height, 75 kg in weight, level of shelter and standard metabolic rate [10–12], without taking into account the physiological conditions of older people or other vulnerable groups [13]. Novieto and Zhang [6] applied the IESD-Fiala model to represent the ageing human body considering the metabolic rate, heart rate and weight. They found that these factors differ between 10 and 19.2% of the characteristics of an average young person. Additionally, through sensitivity tests and simulations to establish the impact of these factors on the thermal comfort of older people, they discovered that the most influential factor was the basal metabolic rate. In a literature review of the existing literature on the thermal comfort of older adults [5], it was found that there are differences between 0.2 °C to 4 °C between the comfort ranges of the elderly and the rest of the age groups, the latter groups being more tolerant of the outside environment [5]. However, most of these studies refer to indoors and are heterogeneous in terms of methodologies, sample sizes and climatic zones, evidencing the need for more research in this area [13].

Until the early years of the twenty-first century, not much attention had been paid to the study of exterior thermal comfort and most evaluation systems had been developed for interiors, in stable conditions, that is, without considering the multiple factors that affect the urban microclimate [14,15]. Some authors have tried to adapt the interior thermal comfort indices such as PMV [16], SET [17] and ET [18]. Subsequently, other methods have been developed to establish the sensation of thermal comfort for open spaces, such as the case of PET (Physiological Equivalent Temperature) [19] and UTCI (Universal Thermal Climate Index) [20]. These are the most used in outdoor assessment [21].

Different climatic zones have specific characteristics that influence thermal comfort conditions and thermal adaptation among diverse seasons [22,23]. Thermal adaptation refers to the ability to adapt to the microclimate conditions, and this could be physical, physiological or psychological [24]. For instance, some aspects are related to cultural and social factors specific to each locality and context that affect thermal adaptation, including physiological conditions (health status, gender, age and metabolic rate), psychological issues (origin, expectation, personal experience, attitude, etc.) and clothing insulation levels [25,26].

In this context, the objectives of the present study are as follows:

- To evaluate how microclimate variables (air temperature, wind speed, mean radiant temperature, relative humidity and sky view factor) affect the thermal comfort of older people in public spaces during autumn in different climatic zones (Csa—Mediterranean Climate and Cfb—oceanic climate).
- To identify thermal comfort ranges for older adults in outdoor public spaces during autumn.
- To identify health risk via thermal stress (due to extrem cold or heat) for older adults in different climates during autumn according to PET and UTCI indexes.

2. Materials and Methods

To evaluate the influence of outdoor microclimate on the thermal comfort of older people, five public spaces were selected within two different cities corresponding to different climatic zones in Europe, Madrid (continental Mediterranean climate) and Newcastle upon Tyne (humid temperate oceanic) (Figure 1). Fieldwork was performed in the months of autumn 2018 in Madrid and autumn 2019 in Newcastle upon Tyne (September, October, and November 2019).

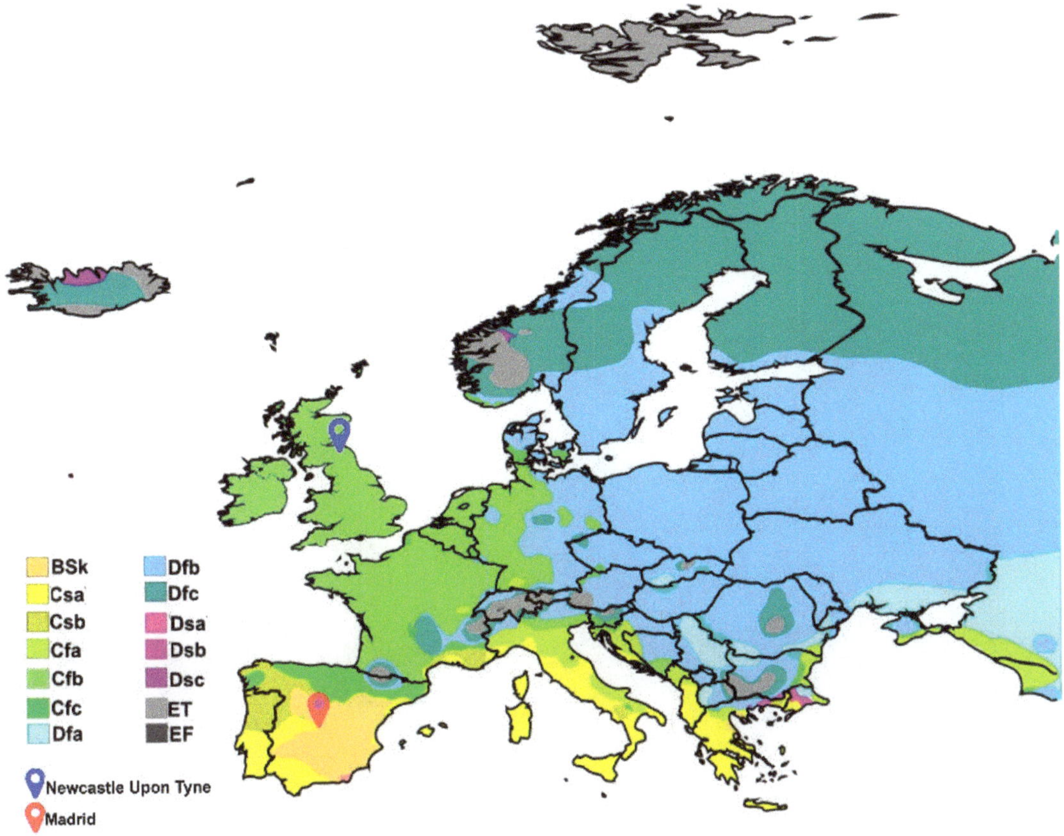

Figure 1. Case studies location in the Köppen–Geiger's climate classification map of Europe. (BSk, cold semi-arid; Csa, hot summer Mediterranean; Csb, warm-summer Mediterranean; Cfa, humid subtropical; Cfb, oceanic climate west coast; Cfc, oceanic climate subpolar; Dfa, hot summer humid continental; Dfb, warm summer humid continental; Dfc, regular subarctic; Dsa, dry and hot summer humid continental; Dsb, dry and warm summer humid continental; Dsc, dry summer, regular subarctic; ET, tundra; EF, ice cap climate).

2.1. Description of Case Studies

2.1.1. Madrid, Spain

The city of Madrid is in central Spain, at latitude 40°26′ N and longitude 3°41′ W at an altitude of 667 metres above sea level (Figure 2). According to the Köppen–Geiger climate classification, its climate corresponds to 'Csa' Mediterranean (temperate climate with dry and hot summers). In summer, it is characterised by its low average relative humidity of 37% and an average temperature of 25–32 °C. In winter, it presents a moderate–high humidity of around 71% and average temperatures of 2–11°C. The average annual temperature is around 14.1 °C [27,28].

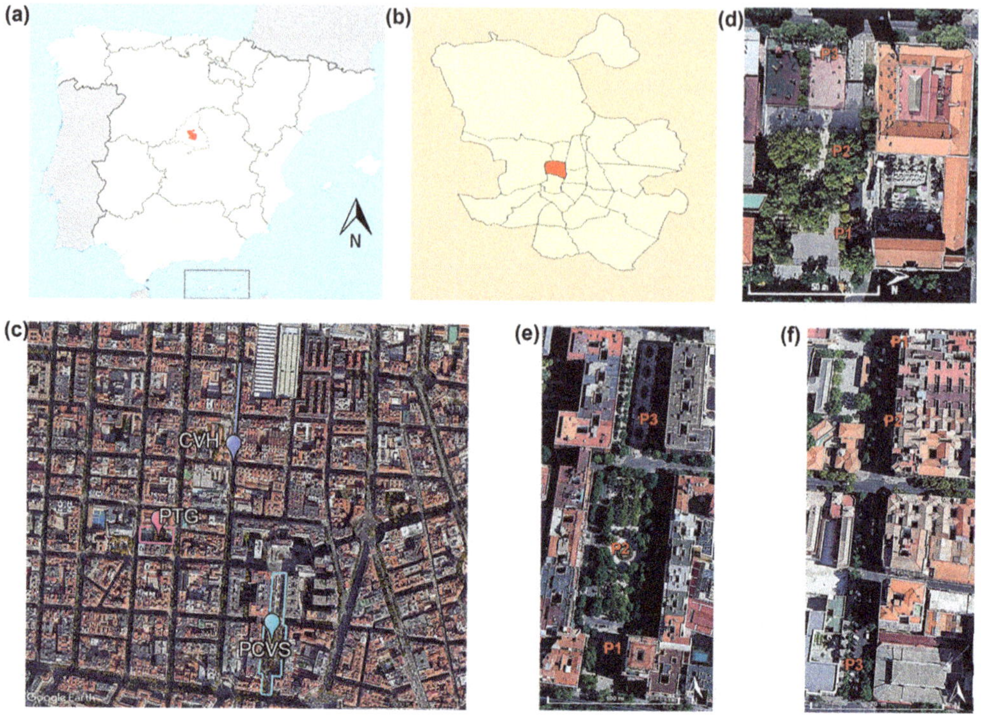

Figure 2. Case study of Madrid. (**a**) Location of Madrid in Spain; (**b**) location of Chamberi district in Madrid; (**c**) location of Park Galileo Theatre (PTG), *Conde del Valle de Schill* Square (PCVS) and *Vallehermoso* street (CVH); (**d**) aerial view of PTG and location of measurement points; (**e**) aerial view of PCVS and location of measurement points, (**f**) aerial view of CVH and location of measurement points.

A park (PTG), a square (PCVS) and a street (CVH) in a high-density, mainly residential neighbourhood (Arapiles, Chamberi) were selected as case studies due to their diverse characteristics. Figure 2 shows the location of these case studies.

2.1.2. Newcastle upon Tyne, United Kingdom

Newcastle upon Tyne is in the northeast of England, in the metropolitan borough of Tyne and Wear, situated at an altitude of 30 m above sea level, at latitude 54°58′40″ N and longitude 1°36′48″ W. Its climate corresponds to 'Cfb' humid temperate oceanic, characterised by cool summers, with abundant and well-distributed rainfall throughout the year (Figure 1). The mean temperature in summer is 15 °C and the mean in winter is 4 °C. The average annual temperature is 8.5 °C, with an average annual precipitation of 655 mm [29].

In the city of Newcastle, two public spaces are located in the centre of the city in an area of mainly commercial use (Westgate), a square (OES) and a pedestrian street (NS), and were taken as case studies because of observations of the concentration of older people. Figure 3 shows the location of those case studies.

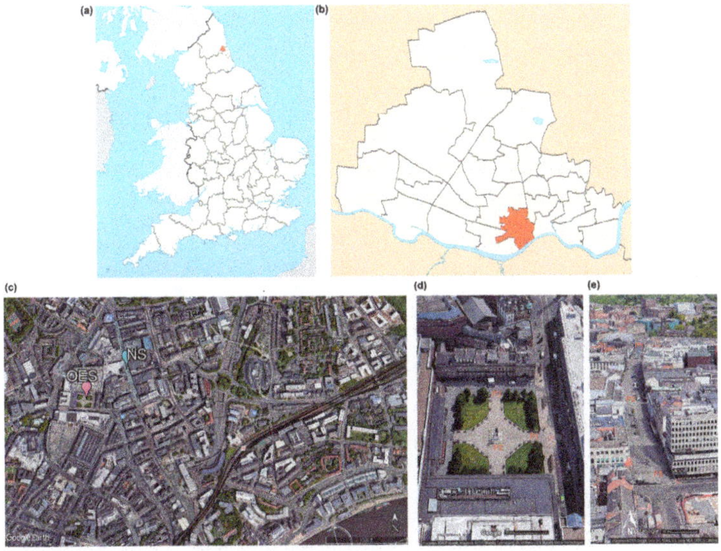

Figure 3. Case study of Newcastle upon Tyne. (**a**) Location of Newcastle in the UK; (**b**) location of Westgate in Newcastle upon Tyne; (**c**) location of Old Eldon Square (OES) and Northumberland Street (NS); (**d**) aerial view of OES and location of measurement points; (**e**) aerial view of NS and location of measurement points.

2.2. Environmental Measurements

The fieldwork was conducted one day per month, between 10:00 and 18:00, the hours for the concentration of older people in those public spaces during September, October and November (2018 and 2019) on calm weather days. Three sample points were selected for environmental measurements in each public space, where equipment was located and fixed for 15 min (i.e., it remained stationary), whilst at the same time surveys were performed for people who were around those sample points. Figure 4 presents an example of the measurement process.

Figure 4. Measurement and survey at point 2 in CVH Street, Madrid.

The environmental variables including the relative humidity (RH), air temperature (Ta), wind speed (Ws) and sky view factor (SVF) were measured at 15-min intervals at each point at 1.1 m above floor level corresponding to the centre of gravity of the human body as recommended in ISO 7726 [30]. Characterisation of the measurement equipment can be found in Table 1.

Table 1. Environmental measuring equipment specifications.

Variable	Equipment	Measurement Range	Accuracy
Ta (°C)	Thermohydrometer data logger HOBO UX100 (HOBO, MA USA)	−20–70 °C	±0.2 °C
RH (%)		1–95%	±2.5%
Ws (m/s)	Anemometer Proster Digital MS6252a (Proster, Hong Kong)	0.4–30 m/s	±2%
SVF	Rayman 1.2 software		
Tmrt (°C)			

Ta = air temperature; RH = relative humidity; WS = wind speed; Tmrt = mean radiant temperature; SVF = sky view factor.

The mean radiant temperature (Trm) that represents the combination of air temperature and short- and long-wave radiation fluxes has been calculated using Rayman 1.2 software, in which environmental variables measured in situ, such as the air temperature, relative humidity and wind speed, in addition to the geographic data of the site, date and time, are entered. This method has been used by several authors [31–33]. Similarly, for the sky-view factor (SVF) fisheye lens (180°) pictures were taken during each measurement with a Sigma 8 mm circular lens in the north direction and then calculated in Rayman 1.2 software [31,33,34].

2.3. Thermal Perception Assessment and Sample Selection

The sample was selected using simple random methods. Simultaneously with the urban environmental measurements, older people (who visually appeared to be older than 65 years of age) present (people who passed by and people who stayed) at each measured point in the assessed public spaces were asked about their willingness to answer the survey, which consisted of a two-part questionnaire.

The first part was related to personal demographics and behavioural characteristics (age, sex, level of clothing insulation, time and frequency of the visit). The second part was focused on their thermal perception through a seven-point scale of (−3. cold; −2. cool; −1. slightly cold; 0. neutral; 1. slightly warm; 2. warm; 3. hot) for thermal sensation (TSV) assessment [35] and the McIntyre three-point scale (1. warmer; 2. would not change; 3. colder) [36] for thermal preference (TP) assessment. For humidity and wind perception, a four-point scale was used (1. Very pleasant—4. Very unpleasant) [37].

2.4. Thermal Comfort Indices

Whilst the UTCI index was computed using the official program version 0. 002 [20], the PET index was derived using Rayman. The features of a 'typical adult', as used in most research, were used in the PET standard calculation (a man, 35 years old, weight 75 kg, 1.75 m tall, 0.9 clo and 80 W). In this study, the data on the average older person in the UK and in Spain has included separating men from women, as well as the appropriate degree of activity (W) and clothing (clo) corresponding to each interviewed person.

2.5. Statistical Analysis

All the collected data were processed in IBM SPSS statistics software and analysed with a significance of 5%. As the decision of which statistical test to use depends on the distribution of the data and the type of variable, a prior normality test (Kolmogoroy–Smirnov) was applied, if data were normally distributed, a parametric test such as a Pearson correlation was performed, otherwise, non-parametric tests were employed (Mann–Whitney, Kruskal–Wallis, Spearman correlation, chi-squared). The thermal sensation and thermal preference (qualitative/ordinal) were considered dependent variables for this study, whilst environmental, personal and others were independent. Additionally, linear regression was applied between thermal-sensation vote and air temperature to identify the neutral temperature at which more older people felt comfortable.

3. Results and Discussion

3.1. Outdoor Environmental Conditions

Outdoor microclimatic conditions varied across both climates. Table 2 presents the environmental measurement data from autumn 2018 in Madrid and autumn 2019 in Newcastle. As regards the average air temperature, there is a difference of 4.8 °C, with Newcastle being colder. However, Madrid presented lower minimum and higher maximum temperatures, with a greater daily thermal amplitude, and the average radiant temperature was higher in Madrid, whilst relative humidity and wind speed were higher in Newcastle.

Table 2. Summary of environmental data measured in situ in public spaces in Madrid and Newcastle upon Tyne.

City	Public Space	Ta °C	RH %	Ws m/s	Mrt °C	SVF
		Mean	Mean	Mean	Mean	Mean
Madrid	PCVS	20.9	38.3	1.5	35.0	0.04
	CVH	17.5	50.0	1.1	28.3	0.02
	PTG	19.4	50.5	1.2	30.5	0.03
	Mean	19.3	46.2	1.3	31.2	0.03
Newcastle upon Tyne	OES	13.3	60.8	1.6	22.7	0.01
	NS	15.9	54.7	1.9	20.3	0.02
	Mean	14.6	57.8	1.7	21.5	0.01

Ta = air temperature; RH = relative humidity; Ws = wind speed; Mrt = mean radiant temperature; SVF = sky view factor.

3.2. Sample Description

The sample size in both cases was similar, seventy people in Madrid and seventy-three in Newcastle upon Tyne.

In Madrid, 56% were women. Regarding the level of clothing insulation (clo), 60% corresponds to 1 clo and 34.3% to 1.5 clo, as this is considered normal for autumn (about 1 clo) [10]. The rest (5.7%) had worn around 0.5 clo, whilst in Newcastle, 47% were women, and 74% of the interviewees wore around 1 clo of clothing insulation.

No statistical relationship was found between gender or age and the level of clothing ($p > 0.05$) using the Kruskal–Wallis test. However, the older they were, the higher the level of clothing they wore, especially in the case of Madrid, with men wearing the most clothing. Figure 5 presents the sample distribution regarding gender, age and city.

3.3. Thermal Sensation of Older Adults in Outdoor Public Spaces

The thermal sensation of the people is a result of the extrinsic conditions of the place in addition to the subjective ones of the people. In this case, in Madrid, 63% of older people stated to have a 'neutral' thermal sensation, whilst 14.3% of the interviewees perceived it between 'slightly hot and hot', 22.8% between 'slightly cold and cold'. About 69% selected 'no change' as the thermal preference.

Regarding the humidity, 87% of the interviewees perceived this, whilst the perception of the wind speed was between 'pleasant' and 'very pleasant' for 83% of them.

In Newcastle, just 32% of older people had a 'neutral' thermal sensation whilst 31% of the interviewees perceived it as between 'slightly cool and cold' and just 10% between 'slightly hot and hot'.

Regarding preference, 42% preferred 'no change' and 31% would like it to be 'warmer'.

The humidity was perceived as between 'pleasant' and 'very pleasant' for 57% of older people, whilst the wind speed was pleasant for 53% of them.

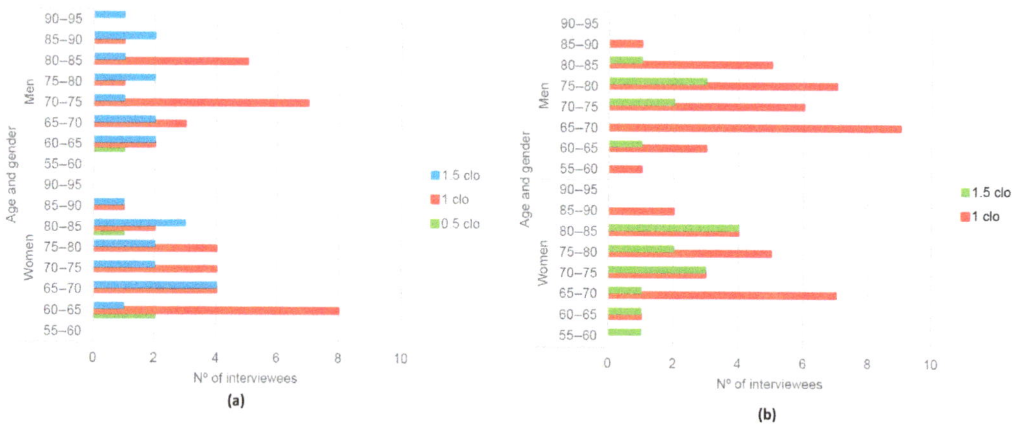

Figure 5. Sample characteristics. (**a**) Madrid; (**b**) Newcastle upon Tyne.

A chi-squared test was performed to identify the association between gender, thermal comfort and thermal preference for both cities, and results showed that there was no statistical relation. Figure 6 shows that thermal comfort votes were similar between both genders, even though in the case of Madrid a greater percentage of men felt colder, whilst in Newcastle men felt warmer than women. In the case of thermal preference, in both cities women were more dissatisfied with the thermal environment. In the case of Madrid, more women would have liked to be warmer and the same amount colder, whilst in Newcastle more women would have liked it to be warmer. These findings are in line with some studies that have found statistical differences between thermal comfort for men and women, where men were more satisfied with the thermal environment and women were more sensitive, especially in cooler conditions [38–40].

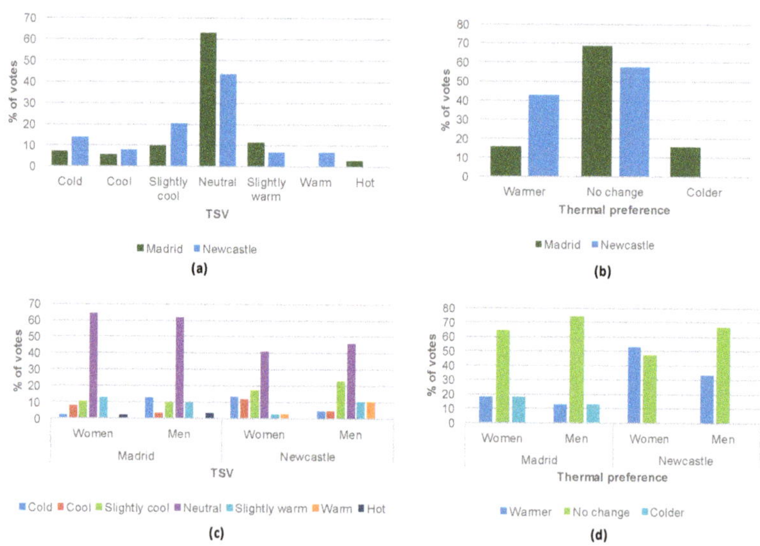

Figure 6. Thermal perception. (**a**) Thermal sensation (TSV) by city; (**b**) thermal preference by city; (**c**) thermal sensation by gender and city; (**d**) thermal preference by gender and city.

3.4. Influence of the Microclimate on Thermal Perception

To identify statistical differences between the two case studies, the t-Student test for environmental variables, the Kruskal–Wallis test for the level of clothing (clo), age and visit time, and the chi-squared test for gender and thermal comfort (TSV = 0), were applied. A summary of these tests is presented in Table 3. Significant differences between all the environmental variables were measured except for the sky view factor. Additionally, there is a distinction between activity, time and frequency of visits. Regarding thermal perception and thermal sensation, those varied among both cities, but the thermal preference and thermal acceptability did not differ.

Table 3. Statistical differences between Madrid climate (Csa) and Newcastle upon Tyne climate (Cfb).

Variable		Statistical Test Chi-Squared Sig (Bilateral)	Kruskal-Wallis Sig (Bilateral)	Student's-t Sig (Bilateral)
Microclimatic	Ta (°C)			0.000 **
	HR (%)			0.000 **
	Ws (m/s)			0.000 **
	SVF			0.080
	Trm (°C)			0.000 **
Personal	clo		0.611	
	Age		0.415	
	Gender	0.274		
	Visit time		0.001 **	
	Activity	0.000 **		
	Frequency of visit	0.000 **		
Perception	Humidity perception	0.239		
	Wind perception	0.000 **		
	TSV	0.023 *		
	PT	0.172		
Thermal index	UTCI 'Wellbeing'	0.0246 *		
	PET 'no thermal stress'	0.007 **		

Ta = air temperature; RH = relative humidity; Ws = wind speed; Mrt = mean radiant temperature; SVF = sky view factor. * The correlation is significant at level 0.05 (2 tails); ** The correlation is significant at level 0.01 (2 queues).

Although clothing insulation was not statistically different between both cities, it was higher in Madrid, and this could be attributed to the behavioural and adaptative cultural differences between both cities [25].

Additionally, the Pearson's correlation test was performed to identify the main environmental variables that influence thermal comfort for both cities (Table 4). In the case of Madrid, all the environmental variables were correlated to thermal comfort ($p \leq 0.00$) less than the SVF. Whist in the case of Newcastle, just the wind speed ($\leq$0.01) was related to thermal comfort, similar to other study findings [40]. Madrid presented a higher percentage of people in comfort (Figure 6) and the perception of ventilation was also more pleasant there.

Table 4. Correlation between thermal comfort and environmental variables for Madrid and Newcastle.

Variable	Pearson Correlation	
	Madrid	Newcastle
	Coef. Correlation	
Ta (°C)	0.14 **	0.128
HR (%)	−0.13 **	0.177
Ws (m/s)	0.10 *	−0.252 *
SVF	−0.03	−0.06
Trm (°C)	0.24 **	0.176

* The correlation is significant at level 0.05 (2 tails); ** the correlation is significant at level 0.01 (2 tails). Ta = air temperature; RH = relative humidity; SVF = sky view factor; Trmt = mean radiant temperature.

Older people there might be better adapted to the microclimatic conditions of their place of residence. Furthermore, in dense cities with heavy traffic and uneven distribution of green areas, there are differences in the microclimate between the public spaces that are worth considering (Table 2), with temperature differences of 3.4 °C, 11.7% in relative humidity and 0.4 m/s in wind conditions among the public spaces in Madrid. On the other hand, we found differences of 2.6 °C in temperature, 6.1% in relative humidity and 0.3 m/s in wind among the public spaces in Newcastle. This confirms the existence of differentiated microclimates in the city [9] which influence the thermal comfort for older people.

3.5. Neutral Temperature and Comfort Zone

Linear regressions to represent older people's mean thermal sensation (MTSV) as a function of the mean air temperature (Ta) were obtained to identify the 'neutral temperature' (Tn) (MTSV = 0) and thermal sensitivity. The two regression equations passed the goodness-of-fit (R2 > 0.5). Both variables were different among both cities; in Madrid, the neutral temperature was 20.4 °C, whilst in Newcastle it was 17 °C; a difference of 3.4 °C between both climates. Figure 7 presents these equations for both climates. The slopes represent thermal sensitivity to temperature changes; it was found to be higher in Newcastle. This could be understood, as older people in Madrid were found to be more tolerant than in Newcastle and this could be attributed to their adaptative behaviour, thus people in Madrid wore higher levels of clothing insulation [25].

There are no outdoor thermal comfort studies for older people in Madrid and Newcastle to compare our results. As a reference from the literature review, this range would be between 23.9–28.1 °C for the average adult's outdoor thermal comfort in Madrid (Csa climate). Whilst no thermal comfort zone was found in previous studies for Newcastle or similar climates [41], as a reference, we considered Nikolopoulou and Lykoudis [42], who assessed outdoor thermal comfort for different climatic zones in Europe for the average adult. They found a great variation of 10°C for neutral temperature across Europe. In Cambridge and Sheffield (Cfb climate), it was found that neutral temperatures in autumn were 23.2 °C and 16.7 °C, respectively. The one found in Newcastle was similar to Sheffield's, whilst in Athens (Csa climate) it was 19.4 °C, one degree lower compared to our findings for Madrid.

The thermal comfort zone is defined by ASHRAE as the range of air temperature where at least 80% of the space occupants are satisfied with the thermal environment. Some authors suggest that comfort zones should be considered within the TSV interval of −1 and +1 [34,43–48]. By applying these values to the linear regression equations (Figure 7), the thermal comfort zone for Madrid in autumn would be between 9.8–31.1 °C, whilst in Newcastle it would be between 12.9–21 °C.

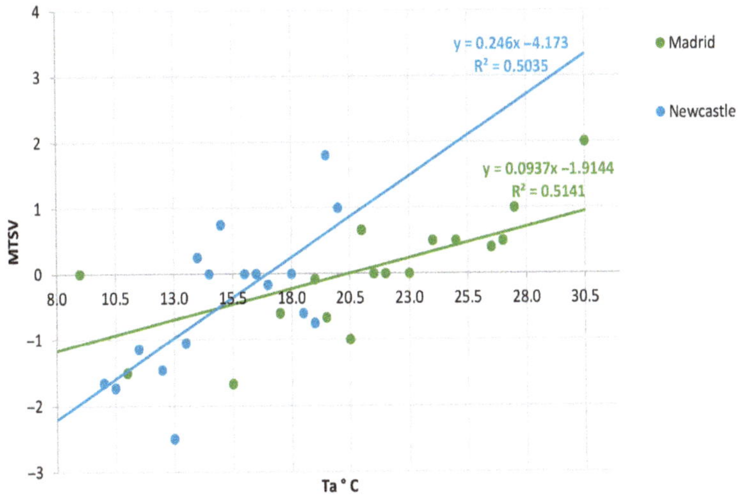

Figure 7. Linear regression air temperature vs. mean thermal sensation votes (MTSV) for Madrid and Newcastle.

The thermal comfort zone range was wider in Madrid. This could be explained because the thermal amplitude was greater during the experimental campaign in this city in autumn, with temperatures ranging from 8–31 °C. It would be necessary for people to adapt to these sudden changes in a few hours, which is even worse for the older population where adaptation is slower due to their metabolism and other physiological changes [5]. It is necessary to consider the differences between older adults in the perception of both cold and heat events during the autumn season that are becoming more usual with climate change [2], where it is necessary to know the comfort zone of these groups that are more vulnerable to both thermal extremes.

3.6. Thermal Comfort Indices (Physiological Equivalent Temperature) PET and Universal Thermal Index (UTC) vs. Thermal Sensation (TSV)

According to the PET index, it was found that in Madrid, 32.9% of older people would be in thermal comfort and 21.4% would be in 'lightly warm, light heat stress', whilst in Newcastle, only 13.70% would be in thermal comfort and 86.3% would be between 'light cold stress' (34.25%), 'moderate cold stress' (31.5%) and 'strong cold stress' (20.55%).

Regarding UTCI, in Madrid, 75.7% of the older adults interviewed were within the wellness zone (9–26 °C) whilst 17.1% were within the 'moderate heat' zone (26–32 °C). In the case of Newcastle, 68.49% were within the wellness zone (9–26 °C) and 32.51% were within the 'light cold' zone where the physiological response would be the reduction of one degree of the temperature of the skin on the hands after 120 min of exposure [49].

Furthermore, the thermal sensation and thermal preference of older people interviewed were significantly correlated with the UTCI and PET indices ($p < 0.01$). In Madrid, 60.4% of people who were in 'well-being' according to the UTCI index had 'neutral' thermal sensations. Similarly, when analyzing the PET with the thermal sensation, it was found that the highest percentages of 'neutral' responses occurred within the 'comfort: no thermal stress' zone (65.2%).

In the case of Newcastle, it was found that the 'well-being' of the UTCI index corresponded to 84.4% of the 'neutral' responses, whilst regarding the PET index, the comfort zone corresponded to just 6.3% of the 'neutral' answers and 21.4% of 'no change' answers. It is important to note that the highest percentage of neutral TSV in older people would be within 'light stress due to cold' according to the PET index (Figure 8). Although most of them claimed to perceive the thermal environment as comfortable, they could be at risk of thermal stress due to cold that could affect their health. This may be due to the loss of thermal sensitivity in both cities, especially in the case of Newcastle [37].

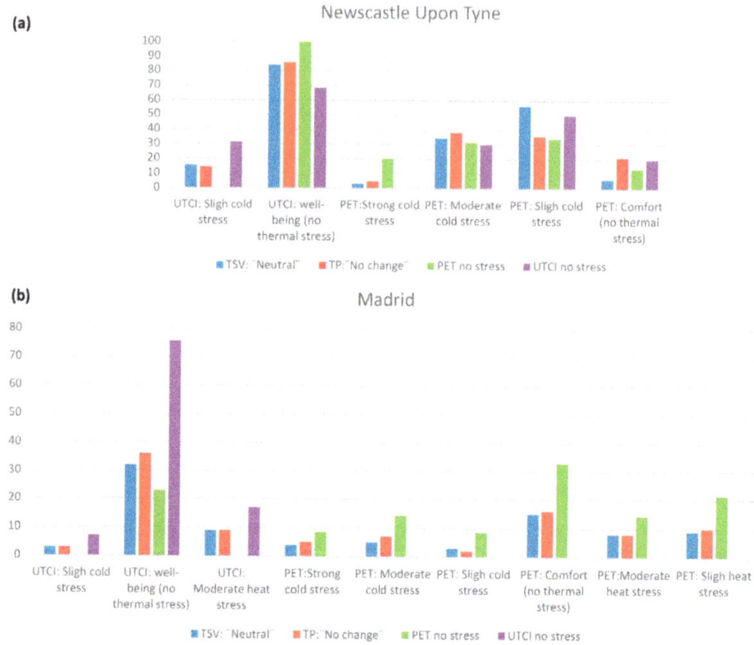

Figure 8. PET and UTCI and comparison with MTSV for (**a**) Newcastle upon Tyne and (**b**) Madrid.

The PET range for well-being (without thermal stress) was similar in both cities, being between 17.86–22.08 °C in Madrid and between 17.94–19.93 °C in Newcastle with a difference between $\pm 0.8 \pm 2.15$ °C. Just one previous study assessed the thermal comfort zone with the PET index in Madrid for the average adult in autumn [50], and it was found to be between 15.6–25.5 °C, whilst there is no evidence of similar studies for Newcastle. Figure 8 presents the PET and UTCI indices for each city and a comparison with TSV and TP.

4. Conclusions

The global trend of ageing populations and other challenges such as climate change that affect microclimate conditions for all the seasons led this research to analyse the outdoor thermal comfort of older adults in public spaces from two cities in Europe: Madrid in Spain and Newcastle upon Tyne in the United Kingdom, during autumn. As we have seen throughout this research, these are two very different cities in terms of size, density, climate, culture, etc.

The statistical influence of microclimatic variables on thermal comfort is evident in older people in both cities. Some differences in cultural behaviour and adaptation were identified. Older adults in Madrid wore higher levels of clothing insulation, were more tolerant and presented higher thermal comfort, and their thermal comfort range was wider due to the greater thermal amplitude (numerical difference between the minimum and maximum values observed during the day).

A difference of 3.4 °C was found in the neutral temperature between the cities; older people in Newcastle were shown to be more sensitive to climate changes.

In both cases, the risk of cold and heat stress was identified according to the PET index, although most of the older people perceived the environment as comfortable. The vulnerability of this population group to the effects of climate on their health is evident, making further research on the subject necessary to establish mitigation and adaptation strategies for future extreme weather and climate change scenarios across Europe.

Additionally, some final considerations can be highlighted which are important for further research:

1. A dense city has significant microclimatic differences in the variables of temperature, relative humidity, and wind, which must be assessed at the pedestrian level to be able to establish the appropriate bioclimatic recommendations for healthier urban spaces.
2. Older people, the majority in most European cities, have intrinsic differences in their perception of both cold and heat. For this reason, both the definition of the comfort zone and the strategies must be qualified by considering age groups over 65 and over 80 years of age.
3. The final thermal comfort of a group must always combine external data with the real perception of the people in situ due to the wide number of variables that affect perception.

Therefore, considering that the European population is very old, it is necessary to establish spatial recommendations to improve the microclimatic conditions of public spaces, which should include sunny areas in winter and shaded areas in summer; green areas with deciduous trees and shrubs; areas protected from the winter wind thanks to walls, furniture or small windbreaks; and areas with fountains and clear paving. But at the same time, areas that are active but not noisy should be established. Older adults constitute a significant group in many urban neighbourhoods, which need to be considered so that they have spaces appropriate to their uniqueness and are not exposed to the risks of extreme heat or cold that they are not aware of, as has been reflected in this research.

Some limitations of this study are:

- Due to the small sample size, these results can only be taken as a reference.
- There was a wide climatic variability in both cities.
- The exposure time to microclimate conditions was not considered for the acclimatisation of older people. Future research should consider this to obtain more accurate results and to avoid biases, as some authors has suggested.

However, these results could be a starting point for the design of more comfortable and healthy public spaces that improve the quality of life of all citizens within the guidelines of active ageing and healthy cities.

Author Contributions: Conceptualisation, M.T.B.L. and E.H.G.; methodology, M.T.B.L.; software, M.T.B.L.; validation, M.T.B.L.; formal analysis, M.T.B.L.; investigation, M.T.B.L. and E.H.G.; resources, M.T.B.L.; data curation, M.T.B.L.; writing—original draft preparation, M.T.B.L.; writing—review and editing, M.T.B.L. and E.H.G.; visualisation, M.T.B.L.; supervision, E.H.G.; project administration, M.T.B.L.; funding acquisition, M.T.B.L. All authors have read and agreed to the published version of the manuscript.

Funding: This work was supported by *Secretaría de Educación Superior, Ciencia, Tecnología e Innovación del Ecuador*. SENESCYT under Grant [109-2017 (SENESCYT-SDFC-DSEFC-2017-4040-O)].

Data Availability Statement: The data presented in this study are available on request from the corresponding author. The data are not publicly available due to [privacy].

Conflicts of Interest: The authors declare no conflict of interest.

References

1. IPCC. *Climate Change 2014*; Synthesis Report; Cambridge University Press: Cambridge, UK, 2014. [CrossRef]
2. Wang, J.; Guan, Y.; Wu, L.; Guan, X.; Cai, W.; Huang, J.; Dong, W.; Zhang, B. Changing Lengths of the Four Seasons by Global Warming. *Geophys. Res. Lett.* **2021**, *48*, e2020GL091753. [CrossRef]
3. Organización Iberoamericana de Seguridad Social. Boletín del Programa Iberoamericano de Cooperación Sobre Adultos Mayores. 2016. Available online: https://oiss.org/boletin-no-10-del-programa/ (accessed on 1 September 2022).
4. Comission on Social Determinants of Health. Closing the Gap in a Generation: Health Equity through Action on the Social Determinants of Health. Final Report of the Commission on Social Determinants of Health, Geneva. 2008. Available online: http://www.who.int/social_determinants/final_report/csdh_finalreport_2008.pdf (accessed on 1 September 2022).

5. Baquero Larriva, M.T.; Higueras García, E. Thermal comfort for the elderly: A systematic review of the scientific literature. *Rev. Esp. Geriatr. Gerontol.* **2019**, *54*, 280–295. [CrossRef]
6. Novieto, D.; Zhang, Y. Thermal comfort implications of the aging effect on metabolism, cardiac output and body weight. *Adapt. Chang. New Think. Comf.* **2010**, 1–9. Available online: https://acortar.link/UQQr7C (accessed on 1 September 2022).
7. Fariña, J.; Higueras, E.; Román, E. *Ciudad, Urbanismo y Salud. Criterios Generales de Diseño Urbano para Alcanzar los Objetivos de una Ciudad Saludable*; Envejecimiento Activo, Instituto Juan de Herrera: Madrid, España, 2019.
8. Erell, E.; Pearlmutter, D.; Williamson, T. *Urban Microclimate: Designing the Spaces between Buildings*, 1st ed.; Routledge: London, UK, 2011. [CrossRef]
9. Higueras, E. Urbanismo Bioclimático, GG, Barcelona. 2006. Available online: https://editorialgg.com/urbanismo-bioclimatico-libro.html (accessed on 1 September 2022).
10. Neila, J. *Arquitectura Bioclimática en un Entorno Construido*; Munilla-Lería: Madrid, Spain, 2004; ISBN 84-89150-64-8.
11. Kalmár, F. An indoor environment evaluation by gender and age using an advanced personalized ventilation system. *Build. Serv. Eng. Res. Technol.* **2017**, *38*, 505–521. [CrossRef]
12. Havenith, G.; Holmér, I.; Parsons, K. Personal factors in thermal comfort assessment: Clothing properties and metabolic heat production. *Energy Build.* **2002**, *34*, 581–591. [CrossRef]
13. Aghamolaei, R.; Lak, A. Outdoor Thermal Comfort for Active Ageing in Urban Open Spaces: Reviewing the Concepts and Parameters. *Ageing Int.* **2022**, 1–14. [CrossRef]
14. Oke, T.R. Street Design and Urban Canopy Layer Climate. *Energy Build.* **1988**, *11*, 103–113. [CrossRef]
15. Fernández, F.J.; Allende, F.; Rasilia, D.; Martilli, A. *Alcaide, Estudio de Detalle del Clima Urbano de Madrid*; Departamento de Geografía, Universidad Autónoma de Madrid: Madrid, Spain, 2016. Available online: https://acortar.link/tgKL6C (accessed on 1 September 2022).
16. Gagge, A.; Fobelets, A.; Berglund, L. A standar predictive index of human response of thermal environment. *ASHRAE Trans.* **1986**, *92*, 709–731.
17. Spagnolo, J.; de Dear, R. A field study of thermal comfort in outdoor and semi-outdoor environments in subtropical Sydney Australia. *Build. Environ.* **2003**, *38*, 721–738. [CrossRef]
18. Blazejczyk, K.; Epstein, Y.; Jendritzky, G.; Staiger, H.; Tinz, B. Comparison of UTCI to selected thermal indices. *Int. J. Biometeorol.* **2012**, *56*, 515–535. [CrossRef]
19. Mayer, H.; Höppe, P. Thermal comfort of man in different urban environments. *Theor. Appl. Climatol.* **1987**, *38*, 43–49. [CrossRef]
20. Innternational Sciety of Biometeorology. UTCI Universal Thermal Climate Index. 2004. Available online: http://www.utci.org/ (accessed on 12 July 2019).
21. Potchter, O.; Cohen, P.; Lin, T.; Matzarakis, A. Outdoor human thermal perception in various climates: A comprehensive review of approaches, methods and quantification. *Sci. Total Environ.* **2018**, *631–632*, 390–406. [CrossRef]
22. Manu, S.; Shukla, Y.; Rawal, R.; Thomas, L.; de Dear, R. Field studies of thermal comfort across multiple climate zones for the subcontinent: India Model for Adaptive Comfort (IMAC). *Build. Environ.* **2016**, *98*, 55–70. [CrossRef]
23. Frontczak, M.; Wargocki, P. Literature survey on how different factors influence human comfort in indoor environments. *Build. Environ.* **2011**, *46*, 922–937. [CrossRef]
24. Yang, W.; Wong, N.; Jusuf, S. Thermal comfort in outdoor urban spaces in Singapore. *Build. Environ.* **2013**, *2*, 426–435. [CrossRef]
25. Knez, I.; Thorsson, S. Thermal, emotional and perceptual evaluations of a park: Cross-cultural and environmental attitude comparisons. *Build. Environ.* **2008**, *43*, 1483–1490. [CrossRef]
26. Fong, C.; Aghamohammadi, N.; Ramakreshnan, L.; Sulaiman, N.; Mohammadi, P. Holistic recommendations for future outdoor thermal comfort assessment in tropical Southeast Asia: A critical appraisal. *Sustain. Cities Soc.* **2019**, *46*, 101428. [CrossRef]
27. Departamento de Producción de la Agencia Estatal de Meteorología de España y Departamento de Meteorología e Clima de Portugal, Atlas Climático Ibérico. Temperatura del Aire y Precipitación (1971–2000), Closas-Orc. 2011. Available online: http://www.aemet.es/documentos/es/conocermas/publicaciones/Atlas-climatologico/Atlas.pdf (accessed on 2 June 2020).
28. AEMET. Temperaturas Medias y su Comparación con las de los Últimos 30 Años. (Observatorio de Retiro), Banco Datos. Ayunt. Madrid Territ. y Medio Ambient. 2019. Available online: http://www-2.munimadrid.es/CSE6/control/seleccionDatos?numSerie=14020000020 (accessed on 22 March 2019).
29. Time and Date, Climate & Weather Averages in Newcastle upon Tyne, England, United Kingdom. Annual Weather Avergages Near Newcastle upon Tyne. 2015. Available online: https://www.timeanddate.com/weather/uk/newcastle-upon-tyne/climate (accessed on 29 June 2020).
30. International Standard Organization. *ISO 7726: Ergonomics of the Thermal Environment. Instruments and Methods for Measuring Physical Quantities*; International Standard Organization: Geneva, Switzerland, 1998.
31. Andrade, H.; Alcoforado, M. Microclimatic variation of thermal comfort in a district of Lisbon (Telheiras) at night. *Theor. Appl. Climatol.* **2008**, *92*, 225–237. [CrossRef]
32. Hwang, R.; Lin, T.; Matzarakis, A. Seasonal effects of urban street shading on long-term outdoor thermal comfort. *Build. Environ.* **2011**, *46*, 863–870. [CrossRef]
33. Martinelli, L.; Lin, T.; Matzarakis, A. Assessment of the influence of daily shadings pattern on human thermal comfort and attendance in Rome during summer period. *Build. Environ.* **2015**, *92*, 30–38. [CrossRef]

34. Hwang, R.; Chen, C. Field study on behaviors and adaptation of elderly people and their thermal comfort requirements in residential environments. *Indoor Air* **2010**, *20*, 235–245. [CrossRef]
35. ASHRAE. STANDARD 55—Thermal Enviornmental Conditions for Human Occupancy. 2020. Available online: https://www.ashrae.org/technical-resources/bookstore/standard-55-thermal-environmental-conditions-for-human-occupancy (accessed on 1 December 2017).
36. Mclntyre, D. A Guide to Thermal Comfort. *Appl. Ergon.* **1973**, *4*, 66–72. [CrossRef] [PubMed]
37. Larriva, M.B.; Higueras, E. Health risk for older adults in Madrid, by outdoor thermal and acoustic comfort. *Urban Clim.* **2020**, *34*, 100724. [CrossRef]
38. Karjalainen, S. Thermal comfort and gender: A literature review. *Indoor Air* **2012**, *22*, 96–109. [CrossRef] [PubMed]
39. Indraganti, M.; Rao, K. Effect of age, gender, economic group and tenure on thermal comfort: A field study in residential buildings in hot and dry climate with seasonal variations. *Energy Build.* **2010**, *42*, 273–281. [CrossRef]
40. Yao, F.; Fang, H.; Han, J.; Zhang, Y. Study on the outdoor thermal comfort evaluation of the elderly in the Tibetan plateau. *Sustain. Cities Soc.* **2022**, *77*, 103582. [CrossRef]
41. Larriva, M.B. Confort Térmico y Acústico para la Tercera edad en Espacios Públicos de la Ciudad Consolidada del Clima Mediterráneo Continental: Caso de Estudio Barrio Arapiles. Ph.D. Thesis, Universidad Politécnica de Madrid, Madrid, Spain, 2021. [CrossRef]
42. Nikolopoulou, M.; Lykoudis, S. Thermal comfort in outdoor urban spaces: Analysis across different European countries. *Build. Environ.* **2006**, *41*, 1455–1470. [CrossRef]
43. Wong, L.; Fong, K.; Mui, K.; Wong, W.; Lee, L. A field survey of the expected desirable thermal environment for older people. *Indoor Built Environ.* **2009**, *18*, 336–345. [CrossRef]
44. Bills, R.; Soebarto, V.; Williamson, T. Thermal experiences of older people during hot conditions in Adelaide. In Proceedings of the Revisiting the Role of Architectural Science in Design and Practice: 50th International Conference of the Architectural Science Association, Adelaide, Australia, 7–9 December 2016; pp. 657–664.
45. Fan, G.; Xie, J.; Yoshino, H.; Yanagi, U.; Hasegawa, K. Investigation of indoor thermal environment in the homes with elderly people during heating season in Beijing, China. *Build. Environ.* **2017**, *126*, 288–303. [CrossRef]
46. Ji, X.; Lou, W.; Dai, Z.; Wang, B.; Liu, S. Predicting thermal comfort in Shanghai's non-air-conditioned buildings. *Build. Res. Inf.* **2006**, *34*, 507–514. [CrossRef]
47. *ASHRAE Standar 55-1966*; Thermal Comfort Conditions. ASHRAE: Atlanta, GA, USA, 1966.
48. Feriadi, H.; Wong, N.; Chandra, S.; Cheong, K. Adaptive behaviour and thermal comfort in Singapore's naturally ventilated housing. *Build. Res. Inf.* **2010**, *31*, 37–41. [CrossRef]
49. Błazejczyk, K.; Broede, P.; Fiala, D.; Havenith, G.; Holmér, I.; Jendritzky, G.; Kampmann, B.; Kunert, A. Principles of the new Universal Thermal Climate Index (UTCI) and its application to bioclimatic research in European scale. *Misc. Geogr.* **2010**, *14*, 91–102. [CrossRef]
50. García, F.F.; Galán, E.; Cañada, R. Caracterización del régimen bioclimático medio del área metropolitana de Madrid, mediante la aplicación de la temperatura fisiológica (PET). *Territoris* **2012**, *8*, 83–101. Available online: https://raco.cat/index.php/Territoris/article/view/259755. (accessed on 5 February 2021).

designs

MDPI

Article

Simulation and Analysis of Thermal Insulators Applied to Post-Disaster Temporary Shelters in Tropical Countries

Bruno B. F. da Costa [1,*], Caio F. P. Silva [2], Ana Carolina F. Maciel [2], Herson D. P. Cusi [3], Gladys Maquera [3] and Assed N. Haddad [1,*]

[1] Universidade Federal do Rio de Janeiro, Rio de Janeiro 21941-901, Brazil
[2] Universidade Federal de Uberlândia, Uberlândia 38400-902, Brazil; caio.civil.ufu@gmail.com (C.F.P.S.); anamaciel@ufu.br (A.C.F.M.)
[3] Universidad Peruana Unión, Juliaca 21100, Peru; hparic@upeu.edu.pe (H.D.P.C.); gladys.maquera@upeu.edu.pe (G.M.)
* Correspondence: bruno.barzellay@macae.ufrj.br (B.B.F.d.C); assed@poli.ufrj.br (A.N.H)

Abstract: Containers are fundamental elements for the development of international trade; however, it is estimated that there are more than 17 million retired containers stacked in ports around the world. Considering the high costs involved in the process of storing, transporting, or destroying these materials, in addition to their non-degradable nature, it is urgent to develop strategies for the sustainable use of these decommissioned containers. In this context, repurposing these containers into permanent structures is becoming a predominant trend. One solution is converting steel shipping structures into habitable spaces. However, due to the urgency with which Container Houses (CHs) are demanded in case of disasters, they are usually planned to be built as quickly as possible, serving as many people as possible, and do not consider the basic principles of energy efficiency. The performance of the CHs is, then, impaired, including risks of overheating, corrosion, and rust, among others, during service, making them an even more stressful experience for their users who are already in a vulnerable situation. Therefore, the objective of this study is to compare the performance of two thermal insulators applied to a temporary shelter container designed to promptly serve vulnerable populations. The model was developed in Building Information Modeling (BIM) software and simulated in Building Energy Simulation (BES) software, aiming to obtain subsidies for its technical and economic viability analysis. The results indicated that thermal insulators are able to generate significant savings in energy consumption, with mineral wool presenting better long-term performance.

Keywords: thermal analysis; BIM; BES; temporary shelter; container housing; computational simulation; thermal insulators

Citation: da Costa, B.B.F.; Silva, C.F.P.; Maciel, A.C.F.; Cusi, H.D.P.; Maquera, G.; Haddad, A.N. Simulation and Analysis of Thermal Insulators Applied to Post-Disaster Temporary Shelters in Tropical Countries. *Designs* **2023**, *7*, 64. https://doi.org/10.3390/designs7030064

Academic Editor: Farshid Aram

Received: 28 March 2023
Revised: 17 April 2023
Accepted: 24 April 2023
Published: 9 May 2023

1. Introduction

Every year, a surprising amount of people are forced to leave their homes in search of shelter and protection due to natural disasters. According to the Norwegian Refugee Council [1], in the year 2021 alone, 23.7 million people were affected by all kinds of geophysical and climatic catastrophes, such as earthquakes, volcanic eruptions, landslides, storms, floods, wildfires, droughts, and extreme temperature events. However, contrary to the unpredictable nature of climatic events, there is a constant increase in the occurrence of armed conflicts, political persecution, and other types of violence around the world, causing a significant rise in the number of people who move to preserve their lives [2]. Last year alone, 14.4 million people were displaced due to violence, the highest number over the past ten years. Thus, in 2021, adding natural and anthropogenic causes, 38 million refugees were accounted for, in 141 countries, with an estimated financial cost of around 21 billion dollars [1].

Indeed, the economic impact of this exodus is relevant, but the social and humanitarian damage is inestimable. Refugees are often forced to move with minimal resources, depending on all sorts of emergency aid, especially food, medicine, and shelter [2]. They are entire families who have no choice but to leave their cities, their jobs, and their lives. In this context, considering that the majority of those affected reside in underdeveloped countries where resources are already scarce, the situation becomes even more critical. According to [1], 74.7% of all displacements throughout 2021 were concentrated in Sub-Saharan Africa and the East Asia and Pacific regions, followed by South Asia. In these areas alone, there are over 34 million homeless people. Therefore, providing temporary shelter is a priority [3,4].

However, due to the urgency with which they are demanded, these shelters are usually provided in a rudimentary way with people being allocated to any available covered area, such as sports facilities, churches, or hangars. These places, in addition to not offering adequate basic infrastructure, steal the privacy of families who are compelled to sleep, store their belongings, clean themselves, and live with unknown people. In turn, in cases where international humanitarian aid is provided and where victims are taken to relief camps, the most used shelter structures are tents. This is an interesting and worthy option since families often own a private space. On the other hand, this type of structure, despite offering the minimum necessary protection, lacks comfort, and appears to be a campsite not a building. As a result, the feeling of protection against adverse thermal effects, strong winds, and suffocating dust, among others, is impaired. In this context, the use of shipping containers as temporary post-disaster shelters has drawn the attention of specialists in recent years [2,4–6].

Shipping containers compose the core of the world's cargo transportation system [7]. However, the rapid dissemination of this type of structure has always presented challenges regarding the sustainability of the logistic model used [8]. This is mainly because the lifecycle of containers is generally not constrained by their effective lifespan but rather by logistical constraints.

The beginning of the export process occurs when the container is put into operation when it is sent empty to the exporter (Figure 1). The cargo is then accommodated and the container is transported to the port of origin. After the maritime transit stage, the container is unloaded at the port of destination and forwarded to the importer's warehouse. After the completion of unloading, the empty container is transported back to the port terminal. At this point, the operator faces a dilemma. Upon arriving empty at the port, the container is stored awaiting return freight or round trip. The purpose of the last one is to use the containers disembarking at the port of destination for the export of other cargo after delivery to the final customer so that the containers do not return empty to the port of origin. Nevertheless, in recent years, there has been a significant slowdown in the flow of maritime transport [9] so making round trips is no longer so trivial. In this case, if it is impossible to reuse the container in a round trip, it must be returned empty to the port of origin. Yet, since the cost of retrieving empty containers back to their origin is almost as costly as moving a fully loaded container [7], it is too expensive and manufacturing new containers is considered more economical [10].

This strategic decision has contributed to the increase in the number of unused containers stored in seaports [11], causing problems associated with the allocation of space and requiring a great effort for their reallocation [9]. Currently, it is estimated that there are tens of millions of retired containers in ports around the world [8,12]. On the other hand, manufacturing new units does not eliminate the need to end the life cycle of out-of-service containers and, in this context, there are basically two options available. The first and most obvious is recycling, given the non-degradable nature of steel [11,12]. However, melting and remanufacturing the standard 3.63 ton container requires 8000 kWh of electrical energy [13–15], so this is not the most sustainable option. Thus, the second and most recent option has been the attempt to find new market niches in which these elements can be reused. Container housing is one of the most promising [9].

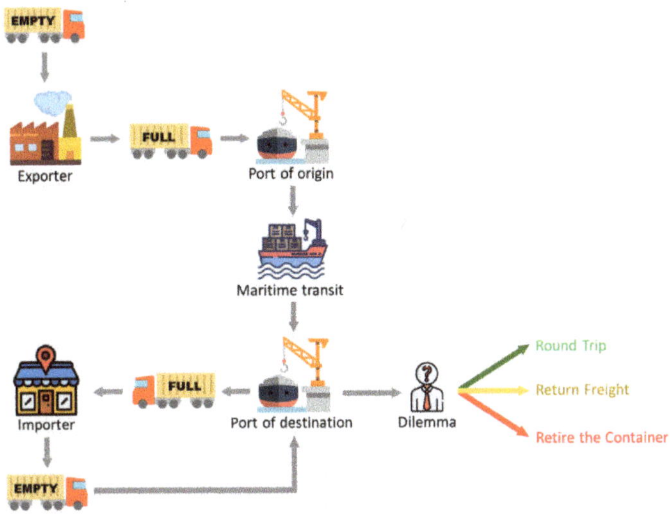

Figure 1. Export and import flow of goods using containers.

Shipping container buildings benefit from the intrinsic characteristics of modular construction [12] being economical and fast to build [16]. Furthermore, as they are made of corten steel alloy or weathered steel, the containers are highly resistant and durable, once they are designed to withstand many years in the salt air and spray on the ocean [17]. Nevertheless, regarding Container Houses (CH) for temporary shelters, portability is the most important feature of this system. In other words, the ease and speed with which they can be transported to the places where they are needed [6], providing immediate assistance to the victims. However, although the use of containers as buildings is not a recent concept, there is a gap in the technical literature available on the subject, especially concerning to their use as temporary shelters [14,16].

In fact, the technique has already been used successfully in countries, such as Korea, Japan, China, and Turkey to build relief camps for refugees and displaced populations [13]. Nonetheless, due to the urgency with which CHs are demanded in case of disasters, they are usually planned to be built as quickly as possible, serving as many people as possible, and not considering basic principles of comfort and sustainability [2]. The performance of the CHs is then impaired, including risks of overheating, corrosion, and rust, among others, during service, making them an even more stressful experience for their users who are already in a vulnerable situation.

Therefore, the objective of this study is the development of a novel standard Sustainable Container Housing (SCH) project, designed to promptly serve vulnerable populations with a focus on the analysis through computer simulation of the technical and economic viability of using two thermal insulators. The model will be developed in BIM software (Building Information Modeling) in order to allow the realization of thermo energetic simulation of the same, and the efficient survey of its cost, aiming to obtain subsidies for its technical and financial viability. It is proposed to obtain a versatile, cheap, and durable model, so that vulnerable populations can be promptly assisted in housing that, although temporary, allows them to live with dignity and comfort.

2. Theoretical Background

2.1. Modular Construction

Modular construction has become a trend in the last two decades [5,11], and the technologies associated with its application have evolved significantly in the construction industry [13]. Also known as volumetric construction, modular building system, or modular architecture [3,18], there is still no formal definition for the terminology. However, the analysis of recent studies indicates the evolution and convergence of meanings for the term. Chatzimichailidou and Ma [19] define modular construction as a process where the units are built off site and, then, transported and assembled on-site. This concept is congruent with Lacey et al. [20] and Musa et al. [18] which complement it, indicating that this is a technique applicable to volumetric units which generally constitute the building's structural element. Ye et al. [21], Rakotonjanahary et al. [22], and Hong [3] go forward describing the components that compose the modular unit and that must be integrated in order to allow the transport of a complete module to the building installation site, that is, wall solutions, ceiling, frames, electrical and hydraulic installations, HVAC, and fixed furniture. Finally, Musa et al. [18] highlight the need to consider the logistical aspects as part of the modular construction process, as the use of the method only brings the expected results if executed in accordance with rigorous planning.

In this context, the increasing interest in the subject has generated an intense registration of the advantages of modular construction in the literature. Koke et al. [12] indicate that, when compared to the traditional construction method, modularization has a smaller environmental footprint. The information is confirmed by Cao et al. [23] who conclude that a traditional residential building consumes approximately 20% more energy and 36% more natural resources during construction. Bertolini and Guardigli [7] present positive environmental results regarding the use of modular elements in the construction industry through Life Cycle Analysis (LCA).

However, currently, the main benefits related to the use of modularization concern the greater productivity made possible by the system [11,24]. According to Jeong et al. [25], about 70% of the construction work for a modular unit is conducted off site, so little work is conducted during on-site assembly. As a result, most of the construction process can be carried out using the production line model, similar to that of the automobile industry, with well-defined workstations and less need for workers to move around the factory floor. This peculiarity, in addition to significantly reducing occupational risks [26], allows for optimizing internal finishing work and even facades, decreasing the construction schedule [20,21] as they are carried out in an industrial environment and not on the construction site which is subject to various adversities [22]. In addition, the use of specialized labor in a controlled environment favors a higher quality product delivery [18,21] and less waste generation [13,20].

Despite arousing great interest and the advantages of its use, the diffusion of modular construction has faced barriers [18]. Perhaps the most relevant is the difficulty in meeting the necessary technical criteria to guarantee user comfort and energy efficiency throughout its lifespan [12]. This is because the scope of modular construction is very broad, allowing the application of a wide variety of materials and construction techniques and making it difficult to develop comprehensive technical guides [20]. Among the various existing possibilities, the container house has stood out [11].

2.2. Shipping Containers

Shipping containers are essentially large steel boxes that are used to transport cargo [27]. This element was idealized by Malcom McLean in the 1930s with the aim of rationalizing the transport of goods when he was still working as a driver of small trucks at the Port of Hoboken, in New Jersey, USA [6]. In 1958, Malcom patented containers as an "Apparatus for shipping freight" [28]. However, the product was intended to be a universal cargo transport solution [2]. After the patent registration, the possibility of transporting goods through standardized structures without the need for constant loading and unloading [6] caught the attention of the U.S. Military. The widespread use of this solution by the military influenced its acceptance by most shipping companies [28]. In this context, following the evolution of international trade, container production has expanded rapidly in recent decades [27], reaching unsustainable standards as shown in the Introduction section of this paper.

Containers use as buildings emerged as a way to reuse these structures and it is not being a new concept because only in the last twenty years has this constructive method achieved greater development [15,29]. This delay is due, at least in part, to the constant erroneous association of this kind of building with trailer houses, normally considered by the general population as unattractive and uncomfortable [18]. Currently, this solution is becoming a trend for several applications, such as hotels, healthcare facilities, low-income housing, and post-disaster settlements [11–13].

The shipping containers currently used in building construction can be divided into two main groups, the 20-ft container, with a length of 6.096 m, and the 40-ft container, with a length of 12.192 m [2,14], and both with a width of 2.438 m [28]. For architectural purposes, these containers offer a limited height, that is, 2.591 m of external height [11] which results in an internal ceiling height of just 2.385 m [28]. Although the International Residential Code (IRC) allows a ceiling height of 2.134 m [30], most national building codes require a minimum ceiling height of 2.40 m [10,11,15]. Therefore, a special subcategory of containers is more favorable, despite not being available on the market in the same quantity as the previous ones. High Cube (HC) containers have the same width and length as 20-ft and 40-ft containers but have an external height of 2.896 m, resulting in a minimum internal height of 2.655 m, attending to the national regulations requirements [9]. Table 1 presents the dimensions of the above described containers.

Table 1. Geometric characteristics of the most popular containers.

Model	External Dimensions			Minimal Internal Dimensions			
	Length (m)	Width (m)	High (m)	Length (m)	Width (m)	High (m)	Floor Area (m^2)
20-ft container	6.096	2.438	2.591	5.710	2.352	2.385	13.430
40-ft container	12.192	2.438	2.591	11.998	2.352	2.385	28.219
20-ft HC container	6.096	2.438	2.896	5.710	2.352	2.655	13.430
40-ft HC container	12.192	2.438	2.896	11.998	2.352	2.655	28.219

The shipping container standardization is a great advantage in terms of modularity, as it offers versatility in assembly options [10,11], besides facilitating the lifting, transport, and connection operations [14]. Considering a building as an articulation of properly combined spaces to meet the user needs [9], containers make it possible to arrange two or more elements in countless ways. Or even, the application of a single container holding all the infrastructure necessary for the user, which is the adopted approach in this study.

Although there are several container typologies, they must all be manufactured in accordance with the standards set by the International Organization for Standardization (ISO) and the International Convention for Safe Containers (CSC) [11,14]. Among the standards of greatest interest are ISO 668 [31], ISO 830 [32], ISO 6346 [33], ISO 1496-1 [34], ISO 1161 [35], ISO 2308 [36], and ISO 3874 [37] and also the guide provided by CSC [38]. These guidelines present the necessary specifications to ensure the uniformity of the geometric and mechanical properties of the containers for transportation purposes [9], but currently, there are no standards, guidelines, or codes for the use of containers as building materials [11]. Therefore, it is essential to know the main characteristics of these elements, so that the design and construction of the container house reach the required quality standard for use as a building.

The manufacturing process of shipping containers, following the modularization trend, is simple. This element consists of intrinsically structural components, such as corner posts, floor, and closure elements, such as walls, doors, and roofs. Once completed, they all become an integral part of the container's structural system. Initially, the walls are cut, corrugated, and welded together and then welded to the container floor. This is made up of a mesh of metal beams that will later be covered with a wooden floor [27]. The next step is to install the doors and corner posts which are welded to the walls and floor. Finally, the roof is welded, completing the container's shell [27] and giving the element a high load capacity [2,9].

In addition to the great structural support, another characteristic that differentiates shipping containers as building materials is their durability [13]. This kind of container is designed to withstand extreme weather conditions during service, that is, they are elements that spend most of their life exposed to rain, wind, and sea air [2,9,28]. For this reason, they are built in Corten steel or weathering steel, a steel resistant to atmospheric corrosion which includes alloying elements that affect the corrosion process of materials and protect the steel integrity [27,28]. Corten steel has a weather resistance of four to eight times greater than ordinary steel, in addition to good weldability and workability [27]. In this context, this material promotes a great reduction in maintenance costs when applied in buildings, since these are generally located in less aggressive environments.

2.3. *Shipping Containers for Post-Disaster Reconstruction*

The flexibility of modular construction and the benefits of using shipping containers as buildings, the so-called container architecture, have contributed to the development of the technique for various applications [11], including coffee shops, fast food kiosks, sales stands, public restrooms, hotels, residential buildings, field hospitals, information centers, leisure spaces, military barracks, scientific research laboratories, and educational buildings, among others. However, one of the most relevant applications of this technology is its use as post-disaster housing, due to its quick and easy installation [5]. More specifically, as a temporary shelter where victims are allocated before being moved to new homes [2]. In this sense, some studies have recently addressed the topic, especially in countries with a high occurrence of climate catastrophes and armed conflicts. Nevertheless, the literature is still limited.

Zafra et al. [39] applied Building Energy Modeling (BEM) to conduct a thermal performance assessment of container shelters in the Philippines. Two design models were created and simulated by changing the insulation material using the EnergyPlus engine. The authors concluded that, regardless of the design, the use of insulating materials is essential to obtain thermal comfort in containers in tropical climates. Obia [40] conducted several architectural and structural changes in temporary container shelters and concluded that all modifications were well accepted by respondents, confirming the technology's flexibility. Shen et al. [11] analyzed the effectiveness of climate-adaptive design for container buildings in three different climate zones, Stockholm, Berlin, and Rome. The results indicated that the integration of passive strategies and renewable technologies was the method that obtained the best results. Ling et al. [2] carried out a review of the feasibility of using containers as transitional shelters. The literature analysis indicated that the use of containers for this purpose has great potential, mainly due to its economic and operational advantages. However, the authors highlighted some system weaknesses that must be quickly mitigated, such as the lack of design guidelines and community acceptance. In fact, the analysis of this last topic was precisely the objective of the Wong et al.'s [41] research which sought to assess the level of acceptance of Malaysian citizens in relation to containerized houses. The used methodology was a questionnaire applied to 454 respondents. The results indicated that only 45% of the participants would consider living in a container house but that its use as a commercial establishment already has great approval.

The research of Tan and Ling [14] aimed to understand the current status of the technical aspect of the container for shelter provision. The authors found that this kind of building meets several technical criteria necessary for its use as a building, such as minimum internal area, ventilation, and fire safety. However, they concluded that further research is essential for this constructive technique to reach its full potential. Sun et al. [42] analyzed the advantages of using container construction in regions with very cold weather. The authors concluded that the system's versatility, combined with the ease of installation and customization possibilities make it adaptable to different locations. In cold weather regions where work schedules must be thought through to avoid periods when workers are exposed to extreme temperatures, being able to do most of the work inside the factory can save time and money.

Elrayies [10] analyzed the thermal comfort of a container house in a hot and humid climate region. The study was performed through computer simulation, varying the external thermal insulation materials. Materials used were mineral wool, closed-cell spray polyurethane foam (ccSPF), and straw. The author concluded that the use of insulation is essential for habitable containers and that the type of material depends on the local climate. In this case, the most effective insulator was ccSPF. Bowley and Mukhopadhyaya [17] were dedicated to the development of an off-grid passive container house. Although the study is not directly related to temporary shelters, the sustainable design ideas presented are useful for this application, such as the photovoltaic power system, rainwater harvesting, and onsite wastewater treatment. A similar study was conducted by Dumas et al. [43] who developed a model of container housing heated by circulating geothermal water inside the building's outer walls called ZETHa (zero energy temporary habitation). The main objective of the system is to minimize thermal bridges in order to improve the building's internal comfort with less energy consumption. Zhang et al. [44] took a qualitative approach to analyze the societal factors that affect the suitability of containers as temporary post-disaster shelters with a focus on case studies following Hurricane Katrina in the US, the Christchurch Earthquake in New Zealand, and the 2009 Black Saturday bushfire-affected communities in Australia.

Finally, it is important to emphasize that there are other available technologies to be used as temporary post-disaster shelters. Caia et al. [45], for example, investigated the psychological effects suffered by victims of the earthquake in Italy in 1997. The study investigated people's satisfaction with temporary shelters, comparing a control group with people housed in shipping containers and dachas. The results indicated that the victims occupying the containers were very dissatisfied, while those allocated to the dachas felt only a little uncomfortable with the shelter. The authors attributed this feeling to the constructive typology of dachas, more similar to a "real home", that is, built-in wood, with large windows, and traditional sloped roofs. Unlike containers, which are made up of metal boxes with small windows and flat roofs. However, it must be considered that almost two decades have passed and container-building technology has evolved a lot. Currently, containers, despite still being made of metal, can receive thermoacoustic insulation, ensuring the comfort of their occupants. In addition, shipping containers can be customized to resemble traditional construction with large openings, different types of roofing, and even special elements, such as balconies and terraces. However, other technologies must be continuously explored, allowing the constant development of comfortable, cheap, and quick-to-execute solutions.

The analysis of the literature presented in this section shows that there are few studies dedicated to developing sustainability and energy efficiency concepts for temporary container shelters. Therefore, this study constitutes a starting point for a series of research focused on obtaining a standard design for a sustainable, energy-efficient, and autonomous container house to be used as a temporary shelter in tropical countries. In this context, it was decided to first analyze the two most used insulation materials in the cities of Macaé and Uberlândia where the Brazilian authors of the text reside since the next steps of the research will involve field measurements to assess the actual performance conditions of the containers. In future research, other types of containers, changes in architectural design, use of rainwater, photovoltaic and wind energy microgeneration, and the perception of users concerning this type of structure will also be considered. Indeed, when compared with the need to allocate people as quickly as possible, given their emergency use, these issues are usually considered of low importance [2,5]. However, they are fundamental aspects when related to user comfort and the life cycle of this type of building [5].

3. Materials and Methods

This research aims to evaluate the energy efficiency of a temporary shelter project built in a container through computer simulation. Therefore, a standard design was developed in a 40-ft HC container where two types of thermal insulating materials were simulated for two different climatic conditions. Figure 2 illustrates the research methodological framework.

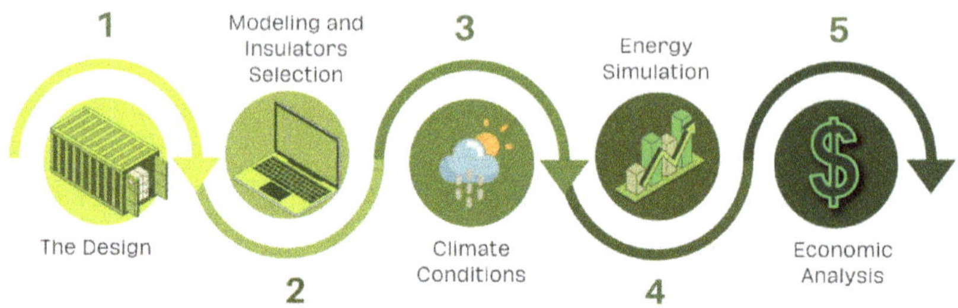

Figure 2. Research methodological framework.

3.1. The Design

The proposed temporary shelter consists of a retrofitted 40-foot dry high cube (HC) shipping container module with external dimensions of 12.192 m in length, 2.438 m wide, and 2.590 m in height with an approximate area of 28.28 m^2 and an approximate volume of 76.99 m^3. The container is made of corten steel, has hinged doors at one end, and its walls are corrugated. This model was chosen because its area allows for greater layout flexibility and better spatial room arrangement in accordance with the recommended minimum dimensions for housing. Furthermore, this is the model that the researchers have at their disposal for carrying out future studies which will involve field measurements. Considering the container's dimensions, the sectorization of the layout began. A 6.98 m^2 kitchen integrated into the 7.36 m^2 living room, a 3.01 m^2 bathroom, and a 7.74 m^2 bedroom were defined. Figure 3 shows the final layout configuration.

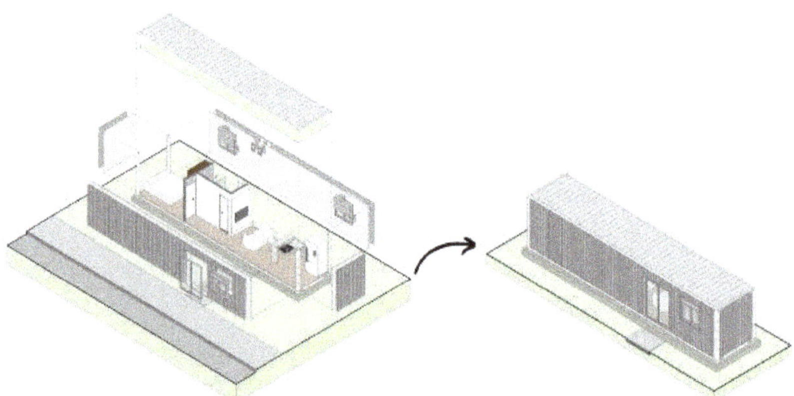

Figure 3. Final layout configuration.

3.2. Modeling and Insulation Material Selection

Once the layout was defined, the project was modeled in Autodesk's Revit® 2023 software [46]. Building Information Modeling (BIM) is an information modeling technology that is integrated into a single file, and with that, it is possible to create associations between design, analysis, and documentation as well as communication between the systems that constitute the building. According to Eastman et al. [47], it is a digital model that contains the exact geometry and information needed to support the design, construction, manufacture, and supply of resources necessary to produce a building. Parametric modeling allows the simulation and evaluation of different design solutions even during the design phase [48]. In addition, professionals from different areas can effectively participate in the design process, generating solutions that contribute to the definition of assertive and efficient choices, a situation that can also benefit the building's energy efficiency. Thus, the model has developed in such a way that each type of wall had its thermal characteristic configured, encompassing the properties of thermal conductivity, specific heat, and density.

Three scenarios for simulating thermal insulation were defined (Table 2). The first was considered a reference model in which thermal insulation was not used. In the other two scenarios, PET wool (polyethylene terephthalate) and mineral wool were used. All other project characteristics were maintained, including the type of connection between the container and the floor. Considering that this is the only uninsulated part of the structure, the option of positioning it on a large concrete base guarantees the reduction of thermal exchanges.

Table 2. Summary of the analyzed scenarios.

Analyzed Scenario	Insulator Thickness (mm)	Insulator Type
1	0	No insulator
2	50	PET wool
3	50	Mineral wool

PET wool is manufactured from plastic bottles and its development is specifically aimed at thermal and acoustic insulation in dry construction. It is a substance that does not absorb water or humidity, therefore, it does not mold and maintains its original characteristics for a long time with a lifespan of up to 100 years. Hence, PET wool is an excellent option for thermal insulation and can be used in several civil construction environments. Mineral wool is made from a volcanic rock called diabase. The manufacturing process starts with the production of fibers that are superheated to transform them into filaments that are agglomerated with resin solutions and result in products that can be light and flexible or very rigid, depending on the degree of compaction. The material is versatile and can be produced in different densities. In the three scenarios, the other materials were considered equal with the external coating painted directly on the container and the internal one in drywall (Figure 4). These two types of insulators are the most used in the regions where the model was simulated. Thus, considering that the future of this research will involve the purchase of this material and the construction of a prototype to perform field measurements, we chose to simulate only products that are easily found for sale in the local market.

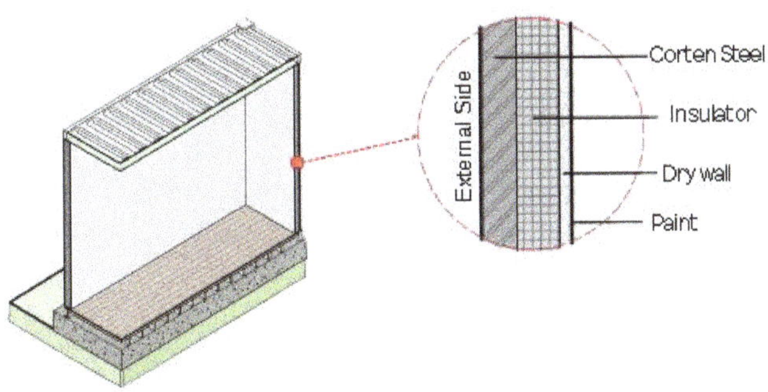

Figure 4. Assembly of the analyzed container walls.

The used PET wool was Wall 50 mm polyester blanket, from Ecofiber. This insulator is fully recyclable and being self-extinguishing, that is, it does not propagate flames and it is a lightweight material compared to other insulators. For mineral wool, the material from ISAR was selected. The product is THERMAX® PSL—32 of 50 mm with good thermal resistance, good acoustic insulation, chemical inertia, water resistance, incombustibility, and resilience. A thickness of 50 mm was chosen because it is the standard dimension for different suppliers of these two materials in our region.

Table 3 summarizes the properties of the materials proposed as thermal insulators.

Table 3. Insulating materials properties.

Material	Insulator Thickness (mm)	Thermal Conductivity (W/m.K)	Thermal Resistance (m^2.K/W)	Density (Kg/m^3)	Thermal Transmittance (W/m^2.°C)
Mineral wool	50	0.031	1.61	32	0.62
PET wool	50	0.041	1.20	30	0.83
Wood floor	30	0.12	0.25	450	4.00
Drywall	18	0.35	0.05	720	19.44
Corten Steel	2.6	55	4.72×10^{-5}	7800	21,153.85

Density, thermal conductivity, thermal resistance, and thermal transmittance are fundamental parameters in determining the efficiency of a thermal insulator, a low transmittance on the external walls of a residence is desirable, thus preventing the large amplitudes of the external environment from reaching the internal environment. In this context, the table above indicates that, despite the two materials having similar densities, the thermal characteristics of mineral wool are slightly superior to those presented by PET wool.

3.3. Climate Conditions

Climate files from two Brazilian cities were selected to be loaded into the eQuest software (version 3.65) [49]. The equipment used to run the simulation was a notebook with 8 GB of memory, 1 TB HDD, Intel Core i7 processor, and 2 GB NVIDIA GEFORCE graphic card. The objective of this engine is to present results in a fast and objective way. Therefore, simulation results were obtained in a few minutes. Both are located in the southeast of Brazil, 1126 km apart. Uberlândia, which belongs to the State of Minas Gerais, is located at an altitude of 863 m and has a tropical high-altitude climate, with heavy rains in summer and droughts in winter. Macaé, belonging to the State of Rio de Janeiro, is located just 2 m above sea level and has a predominantly tropical climate, whose main characteristics are the large volumes of rainfall throughout the year, with summers and winters with high temperatures. These cities were selected because they had similar climatic characteristics in spring and autumn but with greater variations in the summer and winter months (Table 4). Therefore, considering that this paper aims to evaluate the energy efficiency of a container building project, based only on the insulation materials used, that is, disregarding the effects of other architectural decisions, it was considered pertinent to choose cities that present greater variations only in the most extreme temperature months.

Table 4. Average, maximum, and minimum temperatures of the analyzed cities.

City	Medium Temperature	Jan.	Feb.	Mar.	Apr.	May	Jun.	Jul.	Aug.	Sep.	Oct.	Nov.	Dec.
Uberlândia	High	28 °C	28 °C	28 °C	28 °C	26 °C	25 °C	26 °C	27 °C	29 °C	29 °C	29 °C	28 °C
	Average	23 °C	24 °C	23 °C	23 °C	21 °C	19 °C	20 °C	21 °C	23 °C	24 °C	23 °C	23 °C
	Low	20 °C	20 °C	20 °C	18 °C	16 °C	15 °C	15 °C	16 °C	18 °C	19 °C	20 °C	20 °C
Macaé	High	31 °C	32 °C	31 °C	30 °C	28 °C	27 °C	27 °C	27 °C	28 °C	28 °C	29 °C	31 °C
	Average	27 °C	27 °C	26 °C	25 °C	23 °C	22 °C	22 °C	22 °C	23 °C	24 °C	25 °C	26 °C
	Low	23 °C	23 °C	23 °C	21 °C	19 °C	18 °C	18 °C	18 °C	19 °C	20 °C	22 °C	23 °C

3.4. Energy Simulation

The process of computational energy simulation in a building is known as Building Energy Simulation (BES). BES allows designers to conduct the necessary analyses in the various stages of building modeling in order to predict the real behavior of the building. This makes it possible to select the best design solution.

The BES software chosen for the analysis was the eQuest (Quick Energy Simulation Tool) developed by DOE.com. The software is free and allows calculating the energy consumption of a building throughout the year based on data from climate files in the region where the building is located. Interoperability between modeling (BIM) and energy simulation (BES) tools is usually achieved through data exchange protocols, such as Industry Foundation Classes (IFC) and Green Building Studio XML (gbXML). Existing BIM tools, such as Revit and ArchiCAD, as well as BES softwares, such as Green Building Studio, eQUEST, and IES-VE, support the IFC and gbXML format. As a BIM–BES exchange mechanism, Green Building Studio was applied which is used by Autodesk Revit.

After completing the project modeling in Revit, occupancy data, location, and all the parameters of the materials used were adjusted. The areas were defined as conceptual masses in a total of five (bedroom, bathroom, hallway, living room, and kitchen), and the reference floor was considered on the ground floor. Then, the necessary settings and adjustments were made, and the energy model was generated and exported in gbXML format which was created as an open-source project to facilitate data transfer between BIM files and building energy analysis (BES) software. To produce a Revit gbXML file, the Energy Analysis tool is used which builds an energy simulation model that can be loaded into Autodesk Green Building Studio (GBS), where it will be analyzed for errors, inconsistencies, or flaws in the export. After that, it is possible to export the gbXML file from the GBS cloud service and import it into a BES software that supports gbXML. The benefits of this energy analysis method are the accurate extraction of non-geometric data, such as occupancy, equipment, lighting, thermostat, daily weather data, and outside air information. After checks in Green Building Studio, the files were converted to DOE2 format and analyzed in eQuest software.

The occupancy of two people was considered for simulation, following the minimum pattern recommended by national standards. The use of one air conditioner per room was also considered except for the bathroom. The temperature of 23 °C was considered as the thermal comfort standard to be achieved by the model in the two analyzed cities. That is, the energy consumption for cooling simulated by the model was necessary to maintain the building's internal temperature at 23 °C. This value is within the operating temperature range considered by Brazilian regulations for both summer and winter [50]. To activate the HVAC system (Heating, Ventilation, and Air Conditioning), the recommendation of the ISO 17772-1:2017 [51] was adopted which provides suggestions for the occupancy schedule of single-family residences. Table 5 presents data from the occupancy schedule.

Table 5. Occupancy schedule.

Hour	Devices Use	Lighting	Hour	Devices Use	Lighting	Hour	Devices Use	Lighting
01:00	50	0	09:00	70	15	17:00	50	20
02:00	50	0	10:00	50	15	18:00	70	20
03:00	50	0	11:00	50	5	19:00	70	20
04:00	50	0	12:00	60	5	20:00	80	20
05:00	50	0	13:00	60	5	21:00	80	20
06:00	50	0	14:00	60	5	22:00	80	20
07:00	50	15	15:00	60	5	23:00	60	15
08:00	70	15	16:00	50	5	24:00	60	15

Source: ISO 17772-1:2017—Energy performance of buildings [51].

Table 6 shows the technical parameters adopted in the eQuest software to perform the simulations:

Table 6. Other parameters adopted in the eQuest.

Parameter	Adopted Option
Heat transmission through opaque exterior surfaces	Delayed method via conduction transform functions.
Heat transmission through transparent surfaces	84% glass solar factor
Weather data	Based on the Revit database for stations located in the analyzed cities.
Occupancy schedules	Based on ISO 17772-1:2017 [51]
HVAC System	Residential split/compact system gas residencial 14 SEER/0.9 AFUE < 5.5 ton.

3.5. Economic Analysis

The economic analysis aimed to identify the cost of each of the insulators used, and the energy savings resulting from its implementation. Initially, in the modeling software itself, the wall, and ceiling areas where the insulators will be applied were determined, totaling 114.60 m^2.

Then, the energy consumption (EC) of each simulated model (reference, mineral wool, and PET wool) was verified and multiplied by the cost of kWh in each analyzed city. In Uberlândia, R$0.65313/kWh and in Macaé, R$0.75411/kWh. Based on these data, the lag between the reference model and the models with insulators was calculated and multiplied by the tariff of the energy operators in each city. Monthly values were obtained and then summed to obtain annual consumption. Finally, the initial investment of each system was divided by the proportional annual savings of each one to discover the payback of each system. After determining the payback time, it was verified from which year the investment system would become profitable which would have the best cost–benefit in the long term. For this, the annual savings (AE) was subtracted from the initial investment (II) for a period of 20 years according to Table 7.

Table 7. Payback calculation.

Year	PET Wool	Mineral Wool
1	II—AE1	II—AE1
2	II—(AE1 + AE2)	II—(AE1 + AE2)
(. . .)	(. . .)	(. . .)
20	II—(AE1 + (. . .) + AE20)	II—(AE1 + (. . .) + AE20)

4. Results and Discussion

4.1. Results for Uberlândia

The simulation was performed over a period of one year, a parameter that was defined in the e-Quest. The results of monthly and annual energy consumption in kWh of electrical energy for cooling were analyzed in the reference model and in the models with thermal insulation. Table 8 indicates that during the months of May, June, July, and August, the energy consumption of the systems using PET wool and mineral wool are virtually the same. This is because in these months the average temperatures are lower, therefore, the demand for cooling decreases and, consequently, the energy consumption. However, it is possible to state that both insulators produced much better results than the reference model (no insulation). In this context, mineral wool had the lowest consumption with savings of 17.13% compared to PET wool and 42.80% compared to the reference model.

Table 8. Annual electricity consumption results in each simulated model for Uberlândia.

Month	Consumption (kWh)		
	No Insulation	PET Wool	Mineral Wool
January	178.20	119.60	98.50
February	175.50	119.90	95.30
March	239.60	172.90	127.20
April	216.40	145.80	113.40
May	126.60	79.60	79.60
June	107.90	70.10	75.60
July	94.30	62.50	68.50
August	133.60	88.20	90.30
September	200.00	142.20	117.10
October	201.40	146.70	110.10
November	178.40	127.30	98.60
December	229.70	161.90	116.40
Annual consumption	2081.60	1436.70	1190.60

The next step is to calculate the percentage reduction in primary energy consumption (RedCEP) of the housing unit in the real condition compared to the same housing unit in its reference condition. Applying Equation (1) in the scenarios with PET wool and mineral wool, the following results are obtained:

$$RedCEP = \frac{(CEP, ref \times Fce) - (CEP, real \times Fce)}{(CEP, ref \times Fce)} \times 100 \quad (1)$$

- *RedCEP* is the percentage reduction in primary energy consumption of the housing unit in the real model compared to the housing unit in the reference model;
- *CEP, ref* is the annual primary energy consumption of the housing unit in the reference model (kWh/year);
- *CEP, real* is the annual consumption of primary energy of the housing unit in the real model (kWh/year).
- *Fce* is the energy conversion factor.

After applying the conversion factor, there was a 30.98% reduction in energy consumption with PET wool and a 42.80% reduction with mineral wool (Table 9) when compared to the reference model without insulators. For the economic viability analysis, first, the annual expenditure on electricity was calculated, and, then, the cost of implementing each system. From these data, the payback time of each system was obtained. The cost per kWh used for the city of Uberlândia was R$0.65313. Thus, it was found that the annual cost of the system without insulation was R$1359.56; the PET wool system was R$938.35, and the mineral wool system was R$777.62. The average monthly cost of the system without isolation was R$113.30; in the system with PET wool, it was R$78.20, and in the system with mineral wool, it was R$64.80. Figure 5 shows the cumulative cost of consumption of each system over a year.

Table 9. Reduction in primary energy consumption (RedCEP).

Material	CEP, Ref	CEP, Real	Fce	RedCEP
Mineral wool	2081.60	1190.60	1.6	42.80%
PET wool	2081.60	1436.70	1.6	30.98%

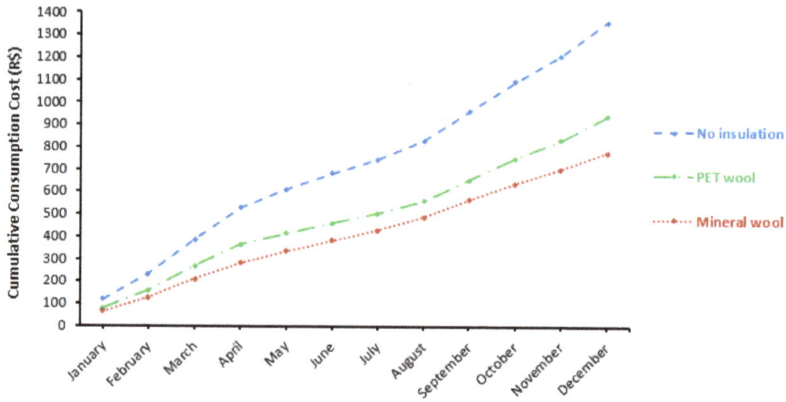

Figure 5. Cumulative consumption cost of each scenario for Uberlândia.

For the analyzed period of one year, the system with PET wool generated savings of R$421.20 compared to the system without insulation, and the mineral wool system generated savings of R$581.94, resulting in a difference between the two insulators of R$160.74. Table 10 shows the percentage of savings that each scenario achieved in relation to the scenario without insulation through the monthly comparison between the scenarios that used PET wool and mineral wool. PET wool achieved better results in June, July, and August which are the months with the lowest average temperatures. This is because the thermal resistance of PET wool is lower, so it "retains" less external heat.

Table 10. Percentage reduction in electricity costs between the simulated systems for Uberlândia.

Month	PET Wool X No Insulator	Mineral Wool X No Insulator	Mineral Wool X PET Wool
January	32.88%	44.73%	11.84% [1]
February	31.68%	45.70%	14.02% [1]
March	27.84%	46.91%	19.07% [1]
April	32.62%	47.60%	14.97% [1]
May	37.12%	37.12%	0.00% [2]
June	35.03%	29.94%	5.10% [3]
July	33.72%	27.36%	6.36% [3]
August	33.98%	32.41%	1.57% [3]
September	28.90%	41.45%	12.55% [1]
October	27.16%	45.33%	18.17% [1]
November	28.64%	44.73%	16.09% [1]
December	29.52%	49.33%	19.81% [1]

[1] Percentage of reduction that mineral wool achieved in energy costs compared to PET wool. [2] No difference between scenarios. [3] Percentage of reduction that PET wool achieved in energy costs compared to mineral wool.

In order to calculate the payback time, a market survey was carried out to obtain the m^2 value of each thermal insulator. The unitary value of each insulator was then multiplied by the areas of the walls and ceilings where the insulators were applied, obtaining the total value of the investment for each material (Table 11).

Table 11. Insulators cost.

Material	Unit Cost (R$/m^2)	Area (m^2)	Total Cost (R$)
Mineral wool	25.83	114.60	2960.12
PET wool	14.59	114.60	1672.01

It is, then, concluded that for a PET wool system, an investment of R$1671.98 in materials will be necessary, and for a mineral wool system, an investment of R$2960.12 will be required. Subsequently, these values were divided by the annual savings generated by each system, according to Equation (2).

$$Payback\ time = \frac{Initial\ Investment}{Annual\ Savings} \tag{2}$$

The payback time calculated for the scenario that used PET wool will be approximately four years (3.97 years), while in the mineral wool scenario, it will be about one year longer (5.09 years). Thus, Table 12 indicates that up to the eighth year of installation, PET wool is more economical, but from the eighth year onwards the situation is reversed. That is, mineral wool provides greater long-term savings for the city of Uberlândia.

Table 12. Annual savings and payback time for the city of Uberlândia over the next 20 years after installation.

Year	PET Wool (R$)	Mineral Wool (R$)
1	−1250.78	−2378.12
2	−829.57	−1796.18
3	−408.37	−1214.24
4	12.84	−632.30
5	434.04	−50.36
6	855.24	531.58
7	1276.45	1113.52
8	1697.65	1695.45
9	2118.85	2277.39
10	2540.06	2859.33
11	2961.26	3441.27
12	3382.46	4023.21
13	3803.67	4605.15
14	4224.87	5187.09
15	4646.07	5769.03
16	5067.28	6350.97
17	5488.48	6932.90
18	5909.68	7514.84
19	6330.89	8096.78
20	6752.09	8678.72

4.2. Results for Macaé

For the city of Macaé, the format in which the analysis was carried out in the city of Uberlândia was repeated. Table 13 indicates that during the months of June, July, and August, the energy consumption of the systems using PET wool and mineral wool are virtually the same. This is because in these months the average temperatures are lower, therefore, the demand for cooling decreases and, consequently, the energy consumption. As in the analysis performed for the city of Uberlândia, it is possible to state that the two insulators produced better results than the reference scenario. Considering that the generated consumption was 2539.3 kWh for the reference model, 1810.7 kWh for the PET wool model, and 1396.8 kWh for the mineral wool model, the mineral wool had the lowest consumption among those analyzed with savings of 22.86% compared to PET wool.

Table 13. Annual electricity consumption results in each simulated model for Macaé.

Month	Consumption (kWh)		
	No Insulation	PET Wool	Mineral Wool
January	269.60	202.50	142.00
February	298.00	221.10	149.60
March	307.20	223.90	153.30
April	209.10	141.30	112.00
May	163.80	107.70	96.50
June	103.90	63.30	68.40
July	110.70	71.80	74.10
August	125.30	85.40	85.70
September	170.60	120.60	103.60
October	183.60	131.30	106.00
November	300.20	222.50	154.30
December	297.30	219.30	151.30
Annual consumption	2539.30	1810.70	1396.80

The parameters for the calculation of primary energy consumption (RedCEP) for the city of Macaé according to Equation (1) are described in Table 14:

Table 14. Reduction in primary energy consumption (RedCEP).

Material	CEP, Ref	CEP, Real	Fce	RedCEP
Mineral wool	2539.30	1396.80	1.6	44.99%
PET wool	2539.30	1810.70	1.6	28.69%

After applying the conversion factor, there was a 28.69% reduction in energy consumption with PET wool and a 44.99% reduction with mineral wool when compared to the reference model without insulators. For economic viability analysis, the same process applied to the city of Uberlândia was used. The cost per kWh used for the city of Macaé was R$0.75411. Thus, it was found that the annual cost of the system without insulation was R$1941.91, the PET wool system R$1365.47, and the mineral wool system R$1053.34. The average monthly cost of the system without isolation was R$159.58, in the system with PET wool, it was R$113.79, and in the system with mineral wool, it was R$87.78. Figure 6 shows the cumulative cost of consumption of each system over a year.

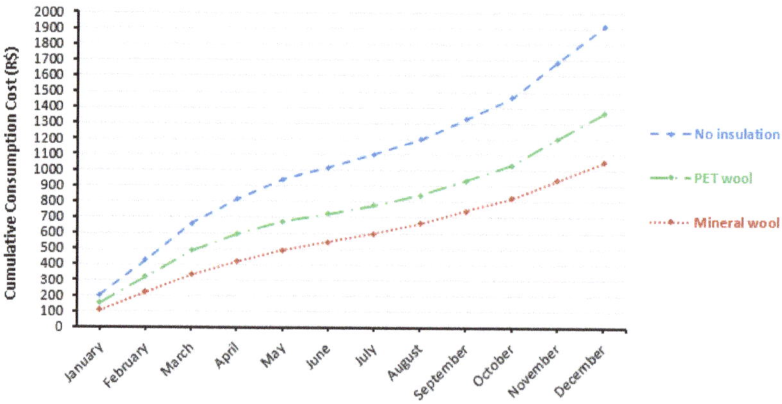

Figure 6. Cumulative consumption cost of each scenario for Macaé.

For the analyzed period of one year, the system with PET wool generated savings of R$549.44 compared to the system without insulation, and the mineral wool system generated savings of R$861.57, resulting in a difference between the two insulators of R$312.13. Table 15 shows the percentage of savings that each scenario achieved in relation to the scenario without insulation through the monthly comparison between the scenarios that used PET wool and mineral wool. PET wool achieved better results in June and July which are the months with the lowest average temperatures. This is because the thermal resistance of PET wool is lower, so it "retains" less external heat.

Table 15. Percentage reduction in electricity costs between the simulated systems for Macaé.

Month	PET Wool X No Insulator	Mineral Wool X No Insulator	Mineral Wool X PET Wool
January	24.89%	47.33%	22.44% [1]
February	25.81%	49.80%	23.99% [1]
March	27.12%	50.10%	22.98% [1]
April	32.42%	46.44%	14.01% [1]
May	34.25%	41.09%	6.84% [1]
June	39.08%	34.17%	4.91% [3]
July	35.14%	33.06%	2.08% [3]
August	31.84%	31.60%	0,24% [2]
September	29.31%	39.27%	9.96% [1]
October	28.49%	42.27%	13.78% [1]
November	25.88%	48.60%	22.72% [1]
December	26.24%	49.11%	22.87% [1]

[1] Percentage of reduction that mineral wool achieved in energy costs compared to PET wool. [2] No difference between scenarios. [3] Percentage of reduction that PET wool achieved in energy costs compared to mineral wool.

The payback time calculation was performed based on Equation (2) and on the insulator installation values indicated in Table 11. The payback time calculated for the scenario that used PET wool will be approximately three years (3.04 years) while in the mineral wool scenario, it will be about five months longer (3.44 years). Thus, Table 16 indicates that up to the fourth year of installation, PET wool is more economical but from the fourth year onwards the situation is reversed. That is, mineral wool provides greater long-term savings for the city of Macaé.

Table 16. Annual savings and payback time for the city of Macaé over the next 20 years after installation.

Year	PET Wool (R$)	Mineral Wool (R$)
1	−1122.54	−2098.49
2	−573.10	−1236.92
3	−23.66	−375.35
4	525.78	486.22
5	1075.22	1347.79
6	1624.66	2209.36
7	2174.10	3070.93
8	2723.54	3932.50
9	3272.98	4794.07
10	3822.42	5655.64
11	4371.86	6517.21
12	4921.30	7378.78
13	5470.74	8240.35
14	6020.18	9101.92
15	6569.62	9963.49
16	7119.06	10,825.06
17	7668.50	11,686.63
18	8217.94	12,548.20
19	8767.38	13,409.77
20	9316.82	14,271.34

4.3. Comparison between Scenarios

The comparison of energy consumption results between the analyzed scenarios for the two cities indicates that in Macaé where average monthly temperatures are higher, the use of thermal insulation becomes more advantageous, providing greater savings, especially in the medium and long term. Table 17 compiles the obtained results and highlights the difference between the analyzed scenarios.

Table 17. Difference in electricity consumption in kWh between the cities of Uberlândia and Macaé for the three analyzed scenarios.

	Uberlândia (kWh)	Macaé (kWh)	Consumption Difference (kWh)	Consumption Difference (%)
No insulator	2081.60	2539.30	457.70	18.02
PET wool	1436.70	1810.70	374.00	20.65
Mineral wool	1190.60	1396.80	206.20	14.76

Table 18 shows the comparison of the payback time. It is noticed that in Macaé, the payback time is shorter than in the city of Uberlândia, since the annual savings in electricity is greater and, consequently, the financial savings. In other words, in the city of Uberlândia, the investment with the installation is paid in four years for PET wool and five years and one month for mineral wool. In Macaé, both insulators pay for themselves in about three years. From the payback onwards, the application of the two insulators starts to generate passive savings for the system.

Table 18. Payback time for each system in the two cities.

	Uberlândia	Macaé
PET wool	4 years	3 years
Mineral wool	5 years and 1 month	3 years and 5 months

Regarding the cost-effectiveness of each system, it is noticed that in Uberlândia the PET wool system has a greater advantage over the mineral wool system up to the eighth year. After the eighth year, the relationship is inverted. In Macaé, very similar results were obtained, but this inversion occurs in the fifth year of installation. It is important to point out that the greater savings generated in Macaé are also due to the fact that the cost of kWh is higher in this city than in Uberlândia, that is, a difference of approximately 13.40%.

5. Conclusions

The main objective of this paper was to analyze alternatives for thermal insulation applied to the inner part of the envelope of a shipping container with the purpose of using it for construction of emergency housing construction. It is noticeable that in recent years the frequency and intensity of natural disasters have been increasing, influenced by the lack of urban planning, disorderly growth of cities, and climate change. Therefore, it is important that more studies are conducted both to mitigate the damage caused by disasters and to prevent them from recurring frequently over the years.

The results indicated that mineral wool had an advantage in reducing electricity consumption for cooling the housing unit. For the city of Uberlândia, the system with PET wool reduced the consumption of electricity by 30.98%, while the system with mineral wool achieved an annual reduction in consumption of 42.80%. In the city of Macaé, the reduction in electricity consumption with PET wool was 28.69%, and the system with mineral wool saved 44.99% within a year. Furthermore, the analysis of the scenarios when the average monthly temperatures are lower indicates that PET wool obtains better results when compared to mineral wool.

Regarding the life cycle of the materials, both PET wool and mineral wool do not suffer deformations and deterioration over the years, therefore, the insulators do not have an estimated lifespan. Therefore, this criterion does not exempt the use of any material following the building life cycle. Concerning the installation cost, it was evident that mineral wool is the most expensive insulation option, being 45.52% more expensive than PET wool. However, in the long term, the mineral wool system is more cost-effective in both cities compared to PET wool.

The results indicate that in the city of Uberlândia, PET wool has a payback time of approximately four years while for mineral wool it is five years and one month. In turn, in the city of Macaé, the PET wool system had a payback time of three years while for mineral wool it is three years and five months. In this context, in Uberlândia, the mineral wool system obtained a better cost–benefit ratio from the eighth year onwards while in Macaé, from the 5th year onwards. Thus, in terms of energy efficiency, mineral wool stands out in relation to PET wool in both cities, however, in financial terms, the system with PET wool proved to be more efficient during the first eight years of installation in the city in Uberlândia and the first five years in the city of Macaé.

Although computer simulation is being increasingly used to analyze different alternative techniques and construction materials, thus guiding design decisions, this research is subjected to some limitations that should be considered, and some may serve as a stimulus for future work. First, this paper considers only two types of insulating material, PET wool and mineral wool which are the most common in the studied cities. Future research should consider a wider variety of insulators, thus expanding the validity of the results. Second, computer simulation has evolved a lot in recent years, but the reliability of the results directly depends on the parameters adopted during the modeling and simulation steps. In this context, the results of similar studies may vary according to the adopted material parameters, climatic data, and software used. It is suggested that future research analyze the effect of each of these characteristics on this model. Third, this research only considered the energy consumption related to the cooling of the housing unit due to the tropical climate of the analyzed cities which rarely requires heating of buildings aiming at thermal comfort. Future research should analyze the insulator's behavior in regions with different climatic conditions.

Finally, this research focused on analyzing the effect of insulating the inner surfaces of the container. Possible treatments related to the structure's external surface were not considered. Corten steel, the material that makes up the container's sides, has high thermal conductivity and reflectance. However, these two properties can be improved using special paints or coatings. Since the thermal conductivity of this material has a direct impact on the building's internal environment, its high reflectance can impact the entire surroundings. The re-emission of long-wave radiation from the sun reflected in containers can result in an unwanted increase in the temperature of the local microclimate, mainly in regions with a high concentration of buildings in containers. Future research should analyze the impact of changes in the outer covering of container houses and their influence on the surroundings. Therefore, current research can be extended in several directions, and one of them is the comparison of simulation results with field data. Research on the subject will follow this direction for this group.

Author Contributions: Conceptualization, C.F.P.S. and A.C.F.M.; methodology, C.F.P.S., A.C.F.M. and B.B.F.d.C.; software, C.F.P.S.; validation, H.D.P.C., G.M. and A.N.H.; formal analysis, H.D.P.C., G.M. and A.N.H.; investigation, B.B.F.d.C., C.F.P.S. and A.C.F.M.; resources, C.F.P.S., A.C.F.M. and B.B.F.d.C.; data curation, C.F.P.S.; writing—original draft preparation, B.B.F.d.C. and C.F.P.S.; writing—review and editing, B.B.F.d.C., A.C.F.M., H.D.P.C., G.M. and A.N.H.; visualization, B.B.F.d.C. and C.F.P.S.; supervision, B.B.F.d.C. and A.C.F.M.; project administration, B.B.F.d.C. and A.C.F.M. All authors have read and agreed to the published version of the manuscript.

Funding: This research received no external funding.

Institutional Review Board Statement: Not applicable.

Informed Consent Statement: Not applicable.

Data Availability Statement: Not applicable.

Acknowledgments: Assed Haddad would like to acknowledge Conselho Nacional de Desenvolvimento Científico e Tecnológico (CNPq), and Fundação Carlos Chagas Filho de Amparo à Pesquisa do Estado do Rio de Janeiro (FAPERJ), which helped in the development of this work. Bruno B. F. da Costa would like to acknowledge Prefeitura Municipal de Macaé which helped in the development of this work.

Conflicts of Interest: The authors declare no conflict of interest.

References

1. Norwegian Refugee Council. *Children and Youth in Internal Displacement*, 1st ed.; The Internal Displacement Monitoring Centre: Geneva, Switzerland, 2022; p. 12.
2. Ling, P.C.H.; Tan, C.S.; Saggaff, A. Feasibility of ISO shipping container as transitional shelter—A review. *IOP Conf. Ser. Mater. Sci. Eng.* **2019**, *620*, 012056. [CrossRef]
3. Hong, Y. A study on the condition of temporary housing following disasters: Focus on container housing. *Front. Archit. Res.* **2017**, *6*, 374–383. [CrossRef]
4. Lin, H.; Cheng, J. A study of the simulation and analysis of the flow field of natural convection for a container house. *Sustainability* **2020**, *12*, 9845. [CrossRef]
5. Tanyer, A.M.; Tavukcuoglu, A.; Bekboliev, M. Assessing the airtightness performance of container houses in relation to its effect on energy efficiency. *Build. Environ.* **2018**, *134*, 59–73. [CrossRef]
6. Dumas, A.; Trancossi, M.; Madonia, M.; Bonnici, M. A novel concept of container house with zero energetic consumption. *SAE Int.* **2012**, *1*, 1507. [CrossRef]
7. Bertolini, M.; Guardigli, L. Upcycling shipping containers as building components: An environmental impact assessment. *Int. J. Life Cycle Assess.* **2020**, *25*, 947–963. [CrossRef]
8. Shi, M. Assessment of the impact of windows and building orientation on the energy intensity of container houses using BIM. *IOP Conf. Ser. Earth Environ. Sci.* **2021**, *769*, 022023. [CrossRef]
9. Pereira-de-Oliveira, L.A.; Bernardo, L.F.A.; Marques, A.R.A. Architectural building design with refurbished shipping containers: A typological and modular approach. *J. Eng. Res.* **2022**, *10*, 1–20. [CrossRef]
10. Elrayies, G.M. Thermal performance assessment of shipping container architecture in hot and humid climates. *Int. J. Adv. Sci. Eng. Inf. Technol.* **2017**, *7*, 1114–1126. [CrossRef]
11. Shen, J.; Copertaro, B.; Zhang, X.; Koke, J.; Kaufmann, P.; Krause, S. Exploring the potential of climate-adaptative container building design under future climates scenarios in three different climate zones. *Sustainability* **2020**, *12*, 108. [CrossRef]
12. Koke, J.; Schippmann, A.; Shen, J.; Zhang, X.; Kaufmann, P.; Krause, S. Strategies of design concepts and energy systems for nearly zero-energy container buildings (NZECBs) in different climates. *Buildings* **2021**, *11*, 364. [CrossRef]
13. Dara, C.; Hachem-Vermette, C.; Assefa, G. Life cycle assessment and life cycle costing of container-based single-family housing in Canada: A case study. *Build. Environ.* **2019**, *163*, 106332. [CrossRef]
14. Tan, C.S.; Ling, P.C.H. Shipping container as shelter provision solution for post-disaster reconstruction. *E3S Web Conf.* **2018**, *65*, 08007. [CrossRef]
15. Islam, H.; Zhang, G.; Setunge, S.; Bhuiyan, M.A. Life cycle assessment of shipping container home: A sustainable construction. *Energy Build.* **2016**, *128*, 673–685. [CrossRef]
16. Giiriunas, K.; Sezen, H.; Dupaix, R.B. Evaluation, modeling, and analysis of shipping container building structures. *Eng. Struct.* **2012**, *43*, 48–57. [CrossRef]
17. Bowley, W.; Mukhopadhyaya, P. A sustainable design for an off-grid passive container house. *Int. Rev. Appl. Sci. Eng.* **2017**, *8*, 145–152. [CrossRef]
18. Musa, M.F.; Yusof, M.R.; Mohammad, M.F.; Samsudin, N.S. Towards the adoption of modular construction and prefabrication in the construction environment: A case study in Malaysia. *ARPN J. Eng. Appl. Sci.* **2016**, *11*, 8122–8131.
19. Chatzimichailidou, M.; Ma, Y. Using BIM in the safety risk management of modular construction. *Saf. Sci.* **2022**, *154*, 105852. [CrossRef]
20. Lacey, A.W.; Chen, W.; Hao, H.; Bi, K. Structural response of modular buildings—An overview. *J. Build. Eng.* **2018**, *16*, 45–56. [CrossRef]
21. Ye, Z.; Giriunas, K.; Sezen, H.; Wu, G.; Feng, D. State-of-the-art review and investigation of structural stability in multi-story modular buildings. *J. Build. Eng.* **2021**, *33*, 101844. [CrossRef]
22. Rakotonjanahary, M.; Scholzen, F.; Waldmann, D. Summertime overheating risk assessment of a flexible plug-in modular unit in Luxemburgo. *Sustainability* **2020**, *12*, 8474. [CrossRef]
23. Cao, X.; Li, X.; Zhu, Y.; Zhang, Z. A comparative study of environmental performance between prefabricated and traditional residential buildings in China. *J. Clean. Prod.* **2015**, *109*, 131–143. [CrossRef]
24. Kristiansen, A.B.; Zhao, B.Y.; Ma, T.; Wang, R.Z. The viability of solar photovoltaic powered off-grid zero energy buildings based on a container home. *J. Clean. Prod.* **2021**, *286*, 125312. [CrossRef]

25. Jeong, G.; Kim, H.; Lee, H.; Park, M.; Hyun, H. Analysis of safety risk factors of modular construction to identify accident trends. *J. Asian Archit. Build. Eng.* **2022**, *21*, 1040–1052. [CrossRef]
26. Ahn, S.; Crouch, L.; Kim, T.W.; Rameezdeen, R. Comparison of worker safety risks between onsite and offsite construction methods: A site management perspective. *J. Constr. Eng. Manag.* **2020**, *146*, 05020010. [CrossRef]
27. Manuel, B.; Arif, B.T.K.T.; Thibaut, G.; Paul-George, I.; Joseph, R.; Kein, V.B. Sea Container Building. Bachelor's Thesis, Universitat Politècnica de Catalunya, Barcelona, Spain, 14 June 2017.
28. Moore, C.M.; Yildirim, S.G.; Baur, S.W. Educational adaption of cargo container design features. In Proceedings of the ASEE Zone III Conference, Springfield, IL, USA, 23–25 September 2015.
29. Su, M.; Yang, B.; Wang, X. Research on integrated design of modular steel structure container buildings based on BIM. *Adv. Civ. Eng.* **2022**, *2022*, 4574676. [CrossRef]
30. International Residential Code. Available online: https://archive.org/download/gov.law.icc.irc.2009/icc.irc.2009.pdf (accessed on 28 January 2023).
31. *ISO 668:1995*; Series 1 Freight Containers-Classification, Dimensions and Ratings. International Organization for Standardization: Geneva, Switzerland, 1995.
32. *ISO 830:1999*; Freight Containers-Vocabulary. International Organization for Standardization: Geneva, Switzerland, 1999.
33. *ISO 6346:1995*; Freight Containers Coding, Identification and Marking. International Organization for Standardization: Geneva, Switzerland, 1995.
34. *ISO 1496-1:1990*; Series 1 Freight Containers-Specification and Testing—Part 1: General Cargo Containers for General Purposes. International Organization for Standardization: Geneva, Switzerland, 1990.
35. *ISO 1161:1984/Cor 1:1990*; Technical Corrigendum 1:1990 to ISO 1161:1984. International Organization for Standardization: Geneva, Switzerland, 1990.
36. *ISO 2308:1972*; Hooks for Lifting Freight Containers of up to 30 Tonnes Capacity-Basic Requirements. International Organization for Standardization: Geneva, Switzerland, 1972.
37. *ISO 3874:1997*; Series 1 Freight Containers Handling and Securing. International Organization for Standardization: Geneva, Switzerland, 1997.
38. International Maritime Organization. *International Convention for Safe Containers*; International Maritime Organization: London, UK, 1996.
39. Zafra, R.G.; Mayo, J.R.M.; Villareal, P.J.M.; De Padua, V.M.N.; Castillo, M.a.H.T.; Sundo, M.B.; Madlangbayan, M.S. Structural and thermal performance assessment of shipping container as post-disaster housing in tropical climates. *Civ. Eng. J.* **2021**, *7*, 1437–1458. [CrossRef]
40. Obia, A.E. Architectural adaptation of the shipping container for housing the internally displaced persons in South-South Nigeria. *Int. J. Archit. Eng. Constr.* **2020**, *9*, 1–9. [CrossRef]
41. Wong, E.K.H.; Tan, C.S.; Ling, P.C.H. Feasibility of using ISO shipping container to build low cost house in Malaysia. *Int. J. Eng. Technol.* **2018**, *7*, 933–939. [CrossRef]
42. Sun, Z.; Mei, H.; Ni, R. Overview of modular design strategy of the shipping container architecture in cold regions. *IOP Conf. Ser. Earth Environ. Sci.* **2017**, *63*, 012035. [CrossRef]
43. Dumas, A.; Trancossi, M.; Madonia, M.; Coppola, M. Zero emission temporary habitation: A passive container house acclimatized by geothermal water. *J. Sol. Energy Eng.* **2014**, *136*, 044505. [CrossRef]
44. Zhang, G.; Setunge, S.; Elmpt, S. Using shipping containers to provide temporary housing in post-disaster recovery: Social case studies. *Procedia Econ. Financ.* **2014**, *18*, 618–625. [CrossRef]
45. Caia, G.; Ventimiglia, F.; Maass, A. Container vs. dacha: The psychological effects of temporary housing characteristics on earthquake survivors. *J. Environ. Psychol.* **2010**, *30*, 60–66. [CrossRef]
46. Revit: BIM Software for Designers, Builders, and Doers. Available online: www.autodesk.com (accessed on 15 January 2023).
47. Eastman, C.; Teicholz, P.; Sacks, R.; Liston, K. *BIM Handbook: A Guide to Building Information Modeling for Owners, Managers, Designers, Engineers and Contractors*, 1st ed.; Bookman: Porto Alegre, Brazil, 2014.
48. Filho, M.V.A.P.M.; da Costa, B.B.F.; Najjar, M.; Figueiredo, K.V.; Mendonça, M.B.; Haddad, A.N. Sustainability assessment of a low-income building: A BIM-LCSA-FAHP-based analysis. *Buildings* **2022**, *12*, 181. [CrossRef]
49. eQUEST: The Quick Energy Simulation Tool. Available online: www.doe2.com/equest (accessed on 15 January 2023).
50. *ABNT 16401-2:2008*; Central and Unitary Air Conditioning Systems. Part 2: Thermal comfort. Brazilian Association of Technical Standards: São Paulo, Brazil, 2008.
51. *ISO 17772-1:2017*; Energy Performance of Buildings—Indoor Environmental Quality—Part 1: Indoor Environmental Input Parameters for the Design and Assessment of Energy Performance of Buildings. International Organization for Standardization: Geneva, Switzerland, 2017.

MDPI
St. Alban-Anlage 66
4052 Basel
Switzerland
www.mdpi.com

Designs Editorial Office
E-mail: designs@mdpi.com
www.mdpi.com/journal/designs

www.ingramcontent.com/pod-product-compliance
Lightning Source LLC
Chambersburg PA
CBHW040812250926
51850CB00007BA/1146
* 9 7 8 3 0 3 6 5 8 5 6 6 6 *